The Impact of Emerging Economies on Global Energy and the Environment

The Impact of Emerging Economies on Global Energy and the Environment

Challenges Ahead

Edited by
Bruce M. Bagley, Dina Moulioukova,
and Hanna Samir Kassab

LEXINGTON BOOKS
Lanham • Boulder • New York • London

Published by Lexington Books
An imprint of The Rowman & Littlefield Publishing Group, Inc.
4501 Forbes Boulevard, Suite 200, Lanham, Maryland 20706
www.rowman.com

Unit A, Whitacre Mews, 26-34 Stannary Street, London SE11 4AB

British Library Cataloguing in Publication Information Available

Library of Congress Control Number: 2015949336
ISBN: 978-1-4985-1911-3 (cloth : alk. paper)
eISBN: 978-1-4985-1912-0

♾™ The paper used in this publication meets the minimum requirements of American National Standard for Information Sciences—Permanence of Paper for Printed Library Materials, ANSI/NISO Z39.48-1992.

Printed in the United States of America

Contents

Preface

No doubt, one of today's most pressing international issues is energy security. Defined by the International Energy Agency as the "uninterrupted availability of energy sources at an affordable price," energy security is not only politically important but has become a critical driver of economic growth. Unfortunately, current events pose serious threats to energy security, including political instability, the manipulation and deliberate interruption of energy supplies, and attacks on oil infrastructure.

However, the outlook is not at all bleak. North America is a secure, responsible, and reliable producer and supplier of energy. The United States has never before produced as much energy as it is producing today. Canada, with oil reserves of 173 billion barrels, has the third largest proven crude oil reserves in the world. Furthermore, the United States and Canada enjoy the world's largest, most important and integrated energy market—at $140 billion a year. As well, Mexico's energy reforms offer tremendous opportunities to enhance North America's energy security. Going forward, there is potential for greater energy trade and infrastructure integration, as well, cooperation on renewable energy and clean technologies to move closer to low-carbon economies.

As Consul General of Canada in Florida, it has been an honor to participate in this informative conversation on the future of energy security in North America and on the global stage. I want to acknowledge and thank Dr. Bruce Bagley, professor of International Studies at the University of Miami's College of Arts & Sciences, for leading this initiative. I also want to congratulate the expert panelists, including the two Canadian experts: Oskar Sigvaldason, Director of the Energy Council of Canada, and Harrie Vredenburg, professor and Suncor Energy Chair in Strategy and Sustainability at the University of Calgary's Haskayne School of Business. They carefully examined the effects of the ongoing North American energy revolution, the rapidly changing dynamics of energy production and consumption in the BRICS countries (Brazil, Russia, India, China, and South Africa), and energy security in the Western Hemisphere as a whole.

The North American continent is well positioned to contribute to global energy security, and to advance responsible resource development globally through innovation and energy efficiency. We have integrated economies, abundant reserves, shared energy infrastructure, and common values, in other words, all the required elements for a bright future.

Louise L. Léger
Consul General of Canada

Introduction

Bruce M. Bagley, Dina Moulioukova, and Hanna Samir Kassab

Changing patterns of energy production and consumption are transforming twenty-first-century geopolitics. Traditional Western powers continue to influence these changing patterns, but with less relative significance, in a period when a handful of the world's largest developing countries have become major energy consumers. Among these largest developing countries are Brazil, Russia, India, China, and South Africa,[1] a loose conglomeration of emerging powers also known as the BRICS. This edited volume captures our analysis with particular emphasis placed on the rapidly changing role of the BRICS in the world energy system. In this volume, energy experts, international relations analysts, and political economists examine the role of state and nonstate actors in energy production and consumption, in energy trade. Of equal concern to the contributors is an apparent increase alternative energy strategies and to what extent these strategies guide continuity or change in the international system.

Energy production in the Americas, especially the United States and Canada, is altering global energy frameworks. These transformations are defying twentieth-century orders and constraints. One cornerstone of the former order, the OPEC (Organization of the Petroleum Exporting Countries) cartel with its practice of price fixing may now be considered irrelevant. The United States and Canada are increasing energy production and are projected to become energy independent in the very near future. The rapid decline in the price of oil (at $52.64 from a high of $102.53 per barrel as of December 18, 2014)[2] is proving to be very difficult for many countries economically dependent on oil and gas exports.

Despite the temporary drop in the price of oil, the BRICS themselves wield much energy power as well. Specifically, Russia's oil policy and China's coal policy are indeed creating new systemic energy policy. These rising powers are creating for the world a new infrastructure within which middle and weaker countries may consider as the future. These phenomena are further contributing to the reorganization of power in the international system. Scholars must take time to analyze these fundamental and drastic changes to relative economic and political power.

In the Western Hemisphere, the oil boom is having a noticeable geopolitical impact. The most significant dimension of this impact is on the Middle East and on the US relationship with Saudi Arabia. For the United States, being a global energy producer means enjoying secured access to oil, reducing the need to protect oil sources in foreign lands. Other non-geostrategic benefits include a positive balance of trade, the encouragement of technological innovation and domestic investment, and stronger relationships with neighbors, especially Canada. In this light, this book is one of the first to take on the subject of the energy revolution and its political impacts, energy production, supply chains (pipelines and export), and consumption practices. Contributors to this volume also take on discussions of sustainable environmental policies in key countries of the Americas like Mexico, Colombia, and Venezuela.

This work answers more pertinent questions involving shifts in the distribution of capabilities across states. Such matters have a direct influence on the structure of the international system. What are the emerging patterns in energy given the rise of the BRICS? How are these states transforming energy policy for other countries? Can the BRICS truly be considered a club of equals or is this grouping simply a loose amalgam of emerging actors seeking to further similar interests such as securing access to energy? The authors suggest the impacts of the BRICS to be quite differential, depending on the size and resilience of each players. There also seems to be serious friction between members of the BRICS alliance. There are still unresolved issues between China and Russia and between China and India. This fact may impact the future of the alliance. Further, Brazil, India, and South Africa cannot be considered equal participants of the alliances; they are peripheral members that require help from the much larger China. The contributors address these questions in the chapters that follow, and each says something different about these actors and their behavior concerning energy. Indeed, the BRICS are relevant actors within the changing international system and must be studied.

Also considered are the cases of massive and significant environmental deterioration due to the preferred use of carbon-based energy, specifically oil and coal. With regard to coal, the emissions from use of cheap energy lowers standards of living. Life-threatening carbon emissions and pollution, may spur social unrest, migration, and other realities detrimental to the country's survival.

The main thrust behind this book is to make sense of energy security in a time of international systemic uncertainty. The contributors to this volume agree that the twenty-first century is very different from the past. The rise of new actors, changes in energy and energy use, and the need for environmental protections against these major alterations are all part and parcel of this new international order. This work is therefore dedicated to analyzing these changes, explaining the patterns of behavior of important states and predicting the future if certain patterns of behavior persist over alternatives. By taking these systemic transformations seriously, we put forth a unified framework for the development of pragmatic and practical foreign policy energy strategies. These strategies look to the creation of a more peaceful, stable, and secure world.

ENERGY POLICIES OF STATE ACTORS

This work starts from the basic observation that currently the international system is in a period of systemic transition. Within a system of transformation, we will inevitably have competition, a push and pull between status quo powers and revisionist powers to create the international system. Arguably, this is a struggle between the Western powers and the important, yet imperfect, BRICS alliance. This volume understands the international system to be of the Gilpinian logic: the international system like any other social system: ". . . actors enter social relations and create social structures in order to advance particular sets of political, economic or other types of interests . . . the particular interests that are most favored by these social arrangements tend to reflect the relative power of the actors involved" (Gilpin 1981, 90). Competition between actors revolve around the establishment of norms and behaviors which serve the interests of those actors: ". . . the efforts of individuals or groups to transform institutions and systems in order to advance their interests . . . the political system will be changed in ways that will reflect these underlying shifts in interests and power" (Ibid., 10). It becomes essential to understand the behaviors and expectations of competing states to explain not only the goals of such behavior, but the means to such goals. Energy policies are an integral part of these social arrangements as the practices more suited to the interests of the more powerful actor will most likely be enforced on other, less powerful actors. Such mechanisms in place will help to reinforce certain systems at the expense of others. This scenario is illustrated by the mechanisms advanced by Western powers that challenge the rise of the BRICS system. Hence, this project sees this period of systemic transformation as having a direct impact on world energy policies, of production and consumption, energy trade, and even the proliferation of alternative strategies for the sake of either maintaining the current hegemonic order or advancing a counterhegemonic order.

Conceptualizing BRICS and Systemic Change

While many see the BRICS as a firm alliance, our research has come to view the BRICS as a loose collection of emerging powers with similar geostrategic interests; it is certainly not an alliance. Scholars point to the creation of the BRICS development bank as proof of a union. Despite the bank, our analysis shows that the BRICS are not as unified as described. The BRICS states are locked in asymmetrical relationships: Russia and China are the more powerful members while the other BRICS members are peripheral. They tend to cooperate in a coordinated fashion, to avoid negative outcomes rather than a collaborative grouping of states. It must also be pointed out that the BRICS states have still some outstanding conflicts that must first be resolved before we can consider them a unified alliance. China and Russia and China and India have problematic traditional border conflicts. These conflicts stem from the politics of the twentieth century. What role will these border conflicts play in the politics of the twenty-first century?

Further, it must be appreciated by serious students of international relations, that Brazil, India and South Africa are indeed fringe members of the BRICS alliance. These do not wield major systemic force compared to other members, specifically China and Russia; their behavior is influenced by these larger, more powerful members. Hence, this volume answers the question: which BRICS actors form the major producers, consumers and trend setters? This work breaks down the grouping for analytical purposes and sees BRICS nations as separate entities with differing and conflicting interests: we can see this in their energy policies and sectors.

Russia and China are the major actors in terms of energy production; they are shapers of international systemic energy culture within an environment of international systemic transformation. Russia's role in energy security is integral as the only major supplier of energy. Russia can create closer ties with other states with its abundant sources of oil and gas. The country can also use its resources as a weapon against those countries dependent on it. Countries of Eastern Europe as well as members of the European Union are vulnerable to the whims of Russia. Currently, the issue with Ukraine creates for the European continent an untenable situation. Ukraine is a major supply route for Europe's as 25% of their energy come from Russia and a large quantity come through Ukraine. If pipelines are turned off as they have been in the past, the world could potentially witness a humanitarian crisis. Russia is thus important for EU member states especially in South and Western parts of Europe. Therefore, we must begin to understand how Russian decision-making procedures, especially with regard to its grand strategy, influence the energy insecurity of neighboring countries. Further, we must see how such a policy cannot last given international economic and reputational costs. The falling ruble was a result of sanctions placed on Russia because of its actions in Ukraine. The fall in the price of oil and gas is worsening Russia's power position. Indeed, the Russian economy remains vulnerable and undiversified. Aggressive moves in Eastern Europe only aggravates the situation for Russia and the Russian people.

Looking to China, we try to understand that energy concerns are pursued via an overly tight relationship between industry and the state. We examine their coal and oil industries and their particular historical evolution of cooperation with the world's major energy producers. Chinese energy concerns are headed by a state corporation which create energy policy for itself and, as a result, for the world. Increasing engagement with producer countries have shaped perceptions in state pursues its interests and energy sources. In other words, China politicizes oil and coal investment through its state-owned oil companies. These are not socially and environmentally responsible and it presents a challenge, not just for China, but also for the world. State-owned enterprises introduce twin pricing systems to the world economy and managing of incentives ultimately created winners and losers economically.

Regarding oil, China is increasing exchange and investment in oil producing countries like Mexico and Venezuela. Venezuela holds the largest amount of oil reserves in the world. Chinese oil companies are mainly invested through direct acquisitions especially in areas where they can spread their expertise in technological production methods. China is doing the same in coal producing countries. As such, China is shaping the structure of energy systemic behavior across states for its own developmental purposes: it is now at the core of a counter energy hegemonic movement.

India, Brazil, and South Africa, while three parts of the BRICS amalgam, are not trendsetters. These cannot be considered in the same power bracket as Russia and China for the main reason that their economies are not as developed and powerful in size, scope and systemic power; these BRICS powers are not challengers to the international system as Russia and China. The mentioned three are secondary and can even be considered "bandwagoners" to their major power counterparts. Brazil, India, and South Africa are ultimately benefactors of Chinese expansionism. For instance, they all receive assistance to produce coal for Chinese consumption; especially South Africa, the most peripheral member of the BRICS. The endemic corruption and political scandals coming out of Brazil continues to mar confidence in the economy and is a continual source of instability. Brazil needs to further explore its deep waters for energy resources but lacks the resources to do so because of the mentioned issues. India simply has no strategy and is currently reacting to the changing energy environment. The systemic power of coal is a force of Chinese systemic influence and the three minor BRICS partners are followers in this wake.

Like Western powers, there is a need for China and Russia to realize that their dependence on hydrocarbons is dangerous and short-term. The governments must come to this realization sooner or later. They must somehow put aside some investment to develop cleaner and more sustainable energies as a long-term strategic decision. The addiction to cheap energy, especially coal, is currently choking the residents of major Chinese industrial centers. It is only a matter of time until the citizens begin to riot for their basic human right to clean air. Dirty energy, therefore, has the potential to become a source of domestic political and social turmoil and instability in the years to come.

In the larger picture, states, whether or not they are competing over systemic dominance, must consider a broader position of sustainability beyond economic and political demands. There must be a realization that sustainable energy policy is not only attainable, but altogether feasible in the drive for economic stability and political security. All states share this common concern and by establishing common interests, they can begin to work together toward a more prosperous global economy that fosters sustainable growth, and a more peaceful world. By cooperating in cleaner energy matters, spillover may allow for a better understanding of common ground in other interest areas, especially since energy and environment is at the forefront of human security. Ultimately, it is up to states to encourage the growth of the necessary technological innovation in more feasible and affordable energy sources that would compete with the least costly fuel sources. Undercutting the seduction of fossil fuels, the cheapness of coal for instance, is at the core of this clean energy policy. Once this is accomplished, this volume sees the chance for progressive change in terms of cleaner and sustainable energy.

Changing Energy Polices of Major Western Powers

After a period of economic instability in the wake of the 2008 financial crisis, the United States remains mired in political uncertainty. The explosion of energy production in oil and gas in the country has helped mitigate some of the economic costs. Technological revolutions in harvesting natural gas aided rapid and unprecedented windfall from the scarce resource that it has become a possibility for the United States to become not only energy independent but a net exporter of energy in the very near future. This "energy revolution," as it has been called, is an ongoing and fragile process, which depends on continual investment and confidence. This forms a second focus of the volume in which contributors examine energy policy in the Western Hemisphere and its international political implications.

Given the boost in production of oil and gas, other cleaner, more sustainable energy technologies are seemingly put on the backburner. This is a cause for concern in the realm of environmental insecurity, more specifically, the destruction of natural landscapes and, of course, increases in carbon and other greenhouse gases in the atmosphere. In Canada, the controversy over the Alberta oil sands is palpable. Concerns over health and environmental degradation over the project continue to dominate the agenda of Canadians. However, the perception remains that while the oil sands project is an inherently good source of wealth for the country, Canada has to prioritize certain environmental goals such as lowering of emissions and encouraging development of the latest technologies. These may correct negative externalities of oil production.

Energy projections for Canada, like the United States, are positive. Canada will increasingly play a role in providing new energy to other states on a technical and expert level. Canadians acknowledge that this cannot be a unilateral effort: climate change and lower the carbon footprint is pointless if the endeavor is done alone. The international system must also be part of the effort. For this, there must be increasing collaboration between industry and government.

There are two important implications that result from this energy revolution. One is political, the other environmental. First, politically, the sudden increase in the production of oil has led to a stark decrease in the price of this commodity. This has created a declining relevance of OPEC member countries. OPEC, an oil cartel, has attempted to cooperate in fixing the price of oil since the 1970s. However, repeated dissention and struggle within the cartel and between member states over matters of production, and due to the large periodic drops in price, have marred effective coordination. Faced with a rapid decline in price, OPEC is torn between two opposing strategies. Iran and Venezuela are significantly and dangerously dependent on the resource for wealth and are exceedingly vulnerable to price shocks such as the one we are seeing today. These countries prefer to cut back dramatically on production while Saudi Arabia, a relatively more developed and diversified economy, hopes to win market share in the long-run, rather than increase the price of oil over the short-run. This discord has profound implications for the cartel's survival and the survival of these oil producing states. Economics disparities arising from this range from balance of payments imbalances, spiraling government debt, financial crisis, and ultimately default.

Geopolitically, the decline in the importance of oil may cause a shift in United States' foreign policy away from the Middle East and toward other parts of the world such as Asia. This pivot would leave Arab countries to fend for themselves against threats such as Iran and international terrorist networks. All of this because production has shifted to North America. The energy revolution is certainly ongoing and the implications are a source of excitement for researchers of international politics.

The danger, however, is that continued use of oil and gas and its decline in price help exacerbate dependence. The fact remains that the revolution has made the use of fossil fuels more attractive compared to more expensive and underdeveloped cleaner energies. Climate change just became even more of a challenging affair. Sustainable energy is needed more than ever as estimates show that there is about fifty years of oil of proven reserves left in the world. This fact makes this work an even more pressing project.

CONCLUDING REMARKS

The contributors are currently planning new projects which hope to open discussion to other parts of the world neglected by this study. While this volume ultimately, but not purposely, ignores Latin America, Caribbean, Middle East, and Central Asia, we hope to address them in a following volume. The energy policies of states within these regions must also be analyzed, especially in terms of the dramatic changes taking place currently. By studying these countries, we will begin to further understand the how the international system is transforming, whether in the service of Western powers or the emerging powers. We hope that this volume's contributors would be able to take part in this project. It is our desire that energy economics, politics, and policy will continue to be a source of interest given the importance of energy and environment in our world today in the midst of systemic change.

The editors would like to thank all participants involved for their time, energy, and especially their intellectual prowess in what proved to be a very exciting conference. The level of productive exchange in an environment of mutual respect and admiration resulted in the organization of this volume as well as plans for future collaboration. We welcome contributors to return to Miami, and we hope this volume attracts more scholars of international relations and security in broader terms to participate in future gatherings.

NOTES

1. South Africa not discussed in this volume.
2. http://www.marketwatch.com/investing/future/crude%20oil%20-%20electronic.

Part I

ENERGY AND THE GLOBAL SYSTEM

Chapter 1

The Globalization of the Western Hemisphere Energy Revolution

Edward Glab

The Western Hemisphere is the driving force behind a tectonic upheaval in global energy. Driven largely by new technology, over the past several years the Americas, from Canada to Argentina, have been undergoing a dramatic increase in the discovery and production of oil and gas. This has profound geopolitical, economic, and national security implications for the entire world that could ultimately prove to be as significant as any development in the fossil fuels area since the dawn of the oil age 155 years ago.

Led by Canada and the United States, the Western Hemisphere has the potential to become both energy self-sufficient and a world-class exporter of oil and gas. The key factor in this energy revolution has been the advance in technology. New technologies, coupled with ever more powerful computers, have unlocked billions of barrels of oil and trillions of cubic feet of gas trapped in shale formations in North and South America, as well as the oil sands of Canada and Venezuela. The heavy and extra-heavy oil in these latter two countries constitute the largest oil deposits in the world.

Ten years ago few thought it possible to profitably produce shale oil. Today, billions of dollars are being invested in its production. Since 2010, North Dakota's crude production has surged 177%, and as shown in the following chart, in 2014 the state's Bakken shale region produced its 3 billionth barrel of oil, mostly from shale.

North Dakota is not the only state to see a dramatic surge in oil and gas production. In Texas, oil production alone has grown 119% since 2010. According to the US Energy Information Agency (EIA), Texas produced an average of over 2.9 million barrels per day of oil in the first quarter of 2014, making it the eighth largest producer of oil in the world, just ahead of the United Arab Emirates (UAE) and Kuwait.

Figure 1.1 North Dakota Oil Production. *Source*: ExxonMobil "Perspectives," May 2014.

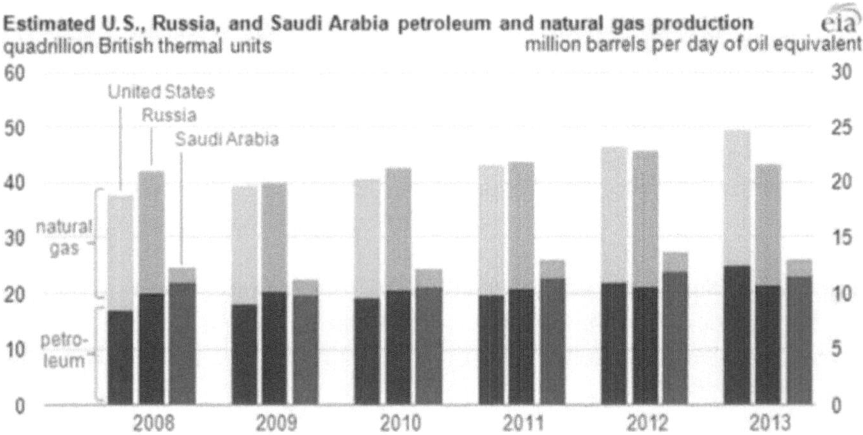

Figure 1.2 Estimated US, Russia, and Saudi Arabia Petroleum and Natural gas Production. *Source:* US Energy Information Administration.

In its June 2014 "Short-term Energy Outlook," the EIA predicted US crude oil production would average 8.5 million barrels per day this year and 9.3 million barrels per day in 2015, the highest level since 1972.[1] This would be almost 2 million barrels more than the 2013 average of 7.4 million barrels per day. Perhaps most startling of all, in 2013, the United States produced more oil and gas combined than any other nation on earth, ahead of both Russia and Saudi Arabia (see Figure 1.2). All of the above makes it abundantly clear that the United States is in the midst of an energy revolution. However, this revolution is taking place elsewhere in the Americas as well.

CANADA

Canada's proven oil reserves of 173 billion barrels are the third largest in the world after Venezuela's and Saudi Arabia's reserves (Figure 1.3). Oil production started to grow rapidly in Canada beginning in 2007, as the chart below from the Canadian National Energy Board indicates. Oil sands production from Alberta province began adding significantly to production in 2011. The latest data show that in 2014, oil production will exceed 4 million barrels per day. According to the EIA's most recent "International Energy Outlook," Canadian production is projected to grow to 6.6 million barrels per day by 2035, due largely to increased production from the oil sands.[2]

The EIA also reports that Canada ranks third in natural-gas production and is the fourth largest exporter of natural gas, behind Russia, Norway, and Qatar. Overall, Canada is the world's sixth largest crude oil producer and also the largest crude oil and petroleum products supplier to the United States.

MEXICO

Mexico has over 10 billion barrels of proven oil reserves, the most in Latin America after Venezuela and Brazil. However, Mexico's total resource base of oil and gas is far greater. According to the Mexican national oil company, Pemex, there are an estimated 27 billion barrels of oil in the Gulf of Mexico alone, while the EIA estimates the country's recoverable shale oil resources at 13 billion barrels and shale gas at 545 trillion cubic feet. Those are the eighth largest shale oil and sixth largest shale gas resources in the world, according to the EIA. Mexico's shale gas is the energy equivalent of 97 billion barrels of oil. In sum, Mexico possesses the equivalent of approximately 147 billion barrels of oil.[3]

Although rich in resources, Mexico's oil production has been declining for the past ten years, per the chart below, due to a lack of technology and proper investment in the oil sector. Recognizing the need to reverse this trend, Mexico has approved historic legislation to end the monopoly of Pemex in the oil business and to open the Mexican energy sector to exploration and production by foreign companies.

Before the reforms can take effect, however, Mexico's legislature must finalize the secondary laws detailing the fiscal regime, including contract terms for the exploration and production models and local content requirements. The reforms will include model contracts for exploration and development, service contracts, production-sharing agreements and licensing. Pemex will be allowed to reserve some of the best acreage for itself, but it will also have to compete with other firms during future bid processes. Although

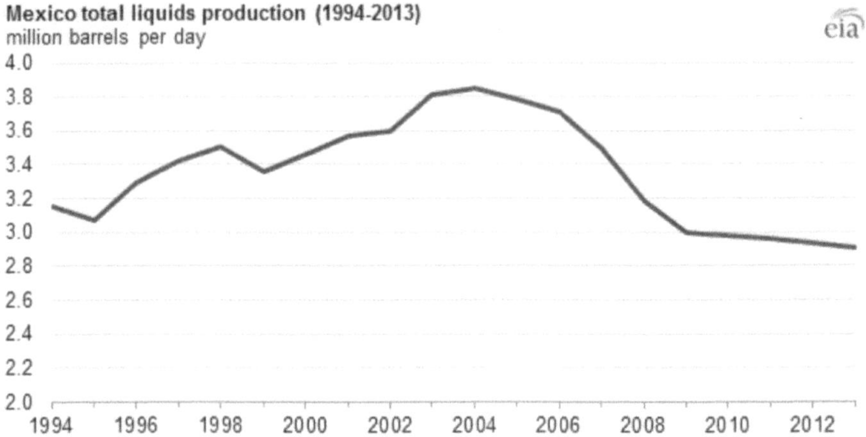

Figure 1.3 Mexico Total Liquids Production (1994–2013). *Source*: EIA "Mexico" September 2014, http://www.eia.gov/countries/cab.cfm?fips=MX.

Mexico will retain ownership of its oil, companies will be allowed to book reserves so the values can be registered on a company's balance sheets.

If the reforms are successful, Pemex CEO Emilio Lozoya and Pemex Board Member Hector Moreira estimated they could generate 50 billion dollars in new investment over the next decade, and by 2025, oil production would grow from the current level of approximately 2.5 million barrels per day to over 4 million barrels per day, boosting annual GDP growth by 1% annually by 2018 and up to 2% annually by 2025. Table 1.4 demonstrates current past/present production.

BRAZIL

According to the EIA, Brazil has 13 billion barrels of proven reserves, second after Venezuela in Latin America. Most of Brazil's oil is very heavy and 90% is located offshore, where it is estimated that there are another 50–60 billion barrels in the very deep-water, pre-salt areas discovered by the Brazilian national oil company (Petrobras) several years ago. Brazil's proven gas reserves are 13.7 trillion cubic feet, or the equivalent of approximately 2.4 billion barrels of oil. In addition, the pre-salt area is estimated to have significant gas resources that will undoubtedly add the energy equivalent of billions of barrels more oil to Brazil's reserves as the pre-salt area goes into production.[4] In sum, between its oil and gas, Brazil has the equivalent of between 65–75 billion barrels of oil. According to the Renewable Fuels Association, Brazil also produces on average 411,000 barrels of ethanol per day and, although seldom mentioned, Brazil has coal reserves equal to the energy equivalent of 110 billion barrels of oil. In sum, Brazil is a supremely energy-rich nation.[5]

Drilling began in Brazil's offshore Santos basin in 2004, with the first discovery in 2006. Others followed, and in 2011, nineteen additional discoveries were made including seventeen in the pre-salt area. The country was euphoric with its newly found oil bonanza and then-president Inâcio "Lula" de Silva declared Brazil had "won the lottery." The country's ten-year energy plan predicted production of 6 million barrels per day by 2021, but this was later reduced to 5 million and more recently to 4.2 million by 2020 due to delays in development. For several reasons, it remains open to question whether or not Petrobras can hit its 2020 production target.[6]

First, in 2010, when Petrobras made one of the largest share issues in history, the company predicted its oil production would rise 50% to 2.98 million barrels per day by 2014.[7] In the first quarter of this year it was only 1.92 million barrels per day, with the pre-salt area contributing about 500,000 of those barrels. In fact, instead of increasing production for export, Brazil has been headed in the opposite direction over the past few years. For example, according to the Agencia Nacional de Petróleo (Brazilian Petroleum Authority), Brazil imported nearly 470,000 barrels of refined products per day in 2012 and by the first quarter of 2014, was importing 793,000 thousand barrels per day. Brazil has not built a new refinery in three decades and will likely have to keep importing refined products.

Second, investor interest in the pre-salt has slowed to a crawl due to many challenges, from technical and financial to political and regulatory. First, pre-salt areas are excluded from foreign concessions, with Petrobras having exclusive access. As a result, Brazil's 2013 bid round for new exploration acreage was a total failure and there is no real market competition in the Brazilian upstream oil sector. In June 2014, the Brazilian government handed Petrobras a no-bid production agreement, awarding the company an additional 10–15 billion barrels of offshore reserves in exchange for a payment to the government of $6.8 billion dollars.[8]

Third, there are disagreements among Brazil's states about how oil revenue should be divided and many political groups are against increased private-sector involvement in oil and gas development under any circumstances, reflecting a widespread sense of resource nationalism. Brazilian environmental regulations are also very strict and the approval process is slow.

Fourth, Brazil needs an estimated $175 billion in capital over the coming decades to develop the pre-salt area, but the country's most recent bond issue raised only 8 billion dollars. Unattractive investment terms, such as domestic content rules as high as 60% according to the Petroleum Economist, have also slowed development of the Brazilian oil sector along with traditional concerns such as corruption and lack of transparency, among others. Petrobras's profits were down 30% in 2014 and the company is worth less than half of what it was when it went public in 2010. Investor confidence is also not helped by Brazil's subsidized gasoline prices that cost the country $2.5 billion per year.[9]

Finally, according to the *International Resource Journal*, there are also serious technical and economic challenges related to commercial recovery of the pre-salt oil many thousands of feet below the ocean floor. With pre-salt oil around twice as expensive to extract compared to conventional oil recovery today, these great depths demand technological investments and a strong focus on exploration.

In sum, investor interest in Brazil has shifted substantially since the big pre-salt discoveries due to the country's investment climate. At the same time, this shift is due to the dramatic change in the global energy outlook and competition for investment dollars and technology from many other countries in the Americas and elsewhere in the world. Hydraulic fracturing has also made oil from pre-salt less critical to global supply. While the resources are in place for Brazil to be a world-class major oil producer, it remains to be seen if the country can actually hit its production target of 4.2 million barrels per day by 2020.

ARGENTINA

Argentina has 2.7 billion barrels of proven oil reserves, according to the EIA, as well as an estimated 27 billion barrels of recoverable shale oil (fourth in the world) and approximately 802 trillion cubic feet of shale gas (second only to China). Argentina's shale gas is the energy equivalent of 143 billion barrels of oil. This easily makes Argentina's shale oil and gas resources the largest in Latin America.[10]

Argentina has historically been blessed with abundant energy resources, but not the political skills to properly develop them for maximum benefit to the country. Argentina's oil and gas sectors have been nationalized and privatized on a number of occasions over the past several decades, something that has discouraged investors, along with a lack of transparency, political instability, poor financial management, an inadequate regulatory system, and corruption. Nonetheless despite these challenges, Argentina has gotten off to an excellent start in attracting investors to its shale oil and gas resources.

A number of things have encouraged investors. One was the five billion-dollar settlement with Spanish oil company Repsol over Argentina's 2012 nationalization of Repsol's 51% ownership of YPF (Yacimientos Petrolíferos Fiscales; English: "Treasury Petroleum Fields").[11] In addition, as reported in the Buenos Aires Herald last year, oil companies investing in shale oil and gas development will be allowed to export tax free up to 20% of the oil and gas they produce and those investing at least $1 billion over five years will be exempt from foreign-exchange controls and be allowed to retain export earnings outside of Argentina. The most important factor of all, however, is the size of the prize in Argentina. International gas and oil companies are willing to take enormous risks when the prize is one of the biggest shale oil and gas plays on the face of the earth.

Last year Chevron announced it would be investing $1.2 billion in a joint venture with YPF to develop shale oil and gas deposits in the "Vaca Muerta" formation in Neuquén province and this year announced it would be investing an additional $1.6 billion. ExxonMobil also acquired about 900,000 exploration acres in the same province and in May 2014 made its first discovery. Shell, Total, and other companies are also active in the shale oil and gas plays in Argentina.[12]

VENEZUELA

The EIA estimates Venezuela's proven oil reserves at around 298 billion barrels and its conventional gas reserves as the second biggest in the Americas after the United States. The US Geological Survey estimates the country's recoverable oil reserves at 513 billion barrels. In spite of having more oil resources than any other nation in the world, Venezuela's oil sector has been in steady decline for the past several years. Oil output at the Venezuelan state oil company, PDVSA, has fallen 25%, from 3 million to 2.4 million barrels per day, since 1998.[13]

Former Venezuelan president Hugo Chávez and current President Nicolás Maduro have done great damage to the oil sector. After firing over 20,000 employees a decade ago, PDVSA has been featherbedded with 110,000 employees. By comparison, ExxonMobil operates the largest and most profitable private oil company in the world with fewer than 80,000 employees and has revenues nearly four times greater than PDVSA and oil production nearly two times greater.[14]

Venezuela's policies have greatly weakened the economy and oil sector by feckless spending and statist intervention in the economy, resulting in inflation, capital flight, and shortages of everything from food and water to consumer goods and foreign exchange. The government's policies have also driven off major investors such as ConocoPhillips and ExxonMobil. Finally, oil subsidies to Cuba cost Venezuela $2 billion per year.[15] If Venezuela hopes to play a major role in the developing global energy revolution, it must undertake major political changes and dramatically improve its investment environment in order to benefit from its abundant oil and gas resources over the coming decades.

SUMMARY

Latin America's proven oil reserves are approximately 335 billion barrels, or 68 billion barrels more than Saudi Arabia's proven reserves. If we add to that total the probable offshore reserves of Brazil and Mexico, plus the estimated shale-oil reserves in the region, Latin America possesses about 465 billion barrels of recoverable oil. That total does not include the US Geological Survey's estimate of Venezuela's mean recoverable oil reserves of 513 billion barrels. If we include Canada and the United States, proven reserves of oil in the Americas are approximately 541 billion barrels. Estimated US shale oil reserves would bring the amount of recoverable oil in the Americas to 600 billion barrels.

Gas resources are equally impressive throughout the Americas. Latin America has the energy equivalent in gas reserves of an estimated 268 billion barrels of oil. The gas reserves of Canada and the United States would add the energy equivalent of another 74 billion barrels for a total of 342 billion barrels of oil. In sum, the total oil and gas reserves for the Americas is almost the equivalent of 1 trillion barrels of oil. To put that number in perspective, most authoritative sources, including the Oil Depletion Analysis Center (ODAC) in London, estimate that the world has consumed a bit more than 1 trillion barrels of oil since the dawn of the oil age over 150 years ago.

It is clear that the Americas have the necessary oil and gas resources to become a major producer and exporter of energy and, in the process, have a profound impact on the regional and global geopolitical environment. The challenges facing development of Latin America's energy resources are above ground, not below. The resources, technology and capital are there. The challenges that must be overcome revolve around issues such as political instability, resource nationalism, corruption, legal uncertainty, regulatory red tape, contract stability, environmental objections, and misguided economic policies.

The EIA has estimated that to meet future energy demand, Latin America must invest a total of between $3 and $4 trillion in energy infrastructure needs between today and 2035, with nearly $2 trillion in the oil sector alone. According to an Organization of the Petroleum Exporting Countries (OPEC) estimate,[16] Latin America needs to invest $100 billion in refining alone over the next generation. It is by no means certain that the needed investment will be forthcoming.

Today's energy revolution is global and so is the competition for investment dollars and technology to develop not only traditional fossil fuels, but also, the next generation of alternative energy sources such as solar, wind and biomass, among others. Given the massive fall in the price of oil, there may be significant political ramifications that will affect the structure of global investment. The focus of this paper dwells on the business side of the issue. In this global competition, capital and technology will flow to those countries that provide the most attractive investment environments. How many nations in the Americas will benefit from investment in the next generation's energy future remains an open question.

NOTES

1. EIA "Short Term Energy Outlook," June 2014, http://www.eia.gov/forecasts/steo/.
2. EIA "International Energy Outlook," September 2014, http://www.eia.gov/forecasts/ieo/.
3. EIA "International Energy Outlook."
4. EIA "Brazil," December 2014, http://www.eia.gov/countries/cab.cfm?fips=BR.
5. EIA "Brazil."
6. EIA "Brazil."
7. EIA "Brazil."
8. EIA "Brazil."
9. EIA "Brazil."
10. EIA "Argentina," June 2012 http://www.eia.gov/countries/cab.cfm?fips=AR.
11. EIA "Argentina."
12. EIA "Argentina."
13. EIA "Venezuela," June 2014 http://www.eia.gov/countries/cab.cfm?fips=VE.
14. EIA "Venezuela."
15. EIA "Venezuela."
16. IEA "Oil Market Report," January 2015, https://www.iea.org/oilmarketreport/omrpublic/currentreport/.

Chapter 2

Canada–US and the Global Energy Markets of the 21st Century

Harrie Vredenburg

CHANGE OF ENERGY AND ENVIRONMENT *STATUS QUO* IN CANADA AND THE UNITED STATES

In this chapter I will start out by making an argument for what we are likely to see in the coming years of the twenty-first century in terms of Canadian energy policy in contrast, I argue, to US energy policy. After making this initial statement I will develop the argument for why I expect this to unfold in the coming years of the twenty-first century.

Canada and the United States have historically been complementary energy partners; at least that has been the case for the last number of decades, probably over the last possibly half a century. That role has been characterized by Canada, particularly Western Canada, being the producer of oil and gas, and the United States, particularly the eastern seaboard and the mid-west of the United States, serving as the customer base for the Canadian energy resources.

This arrangement of a major seller of energy resources and a major buyer of energy resources in two adjacent countries is at a point of inflection and is likely to be changing over the next several years. The reason for this is the recent market dynamics that are the basis for the end of this historical half a century partnership.

I argue that what we are likely to see over the next number of years is that Canada will forge separate future in energy based on a very different energy and environment reality in Canada in contrast to that in the United States. Where Canada is at in respect to climate policy is a question that must be addressed because that is the basis of where the two countries are likely to be following different paths going forward.

The first point is that Canada was an early signatory to the Kyoto Protocol and initially, back in the 1990s when Kyoto was first under way, was taking a leadership position. Then this position was changed over time. Although Canada ratified the Kyoto protocol, it ultimately took a position that it would not act on Kyoto and the Canadian climate policy at that point then became one of following the US climate policy in a lock-step fashion. The argument put forth by the Canadian federal government was that Canada and the US economies were far too intimately linked that Canada could not afford to go its own way because the two economies were so integrally linked. Canada's only course of action was to simply do what the United States did. This has been the policy that Canada has been following for the last number of years since Kyoto started to unravel.

As we all know, the US policy on climate change was initially to be a participant at the Kyoto meetings, and making a strong rhetorical position in terms of policy and the need for global agreements such as Kyoto. Then with change of government in the United States that policy was not followed through, and the United States never became a signatory to Kyoto and has never taken much stock in being a participant in Kyoto discussions. So in the post-Kyoto era in which we all live now, Canada still has a policy of following US climate policy in a lock-step fashion, rather than in fashioning its own climate policy independent of the United States.

Now, however, Canada is coming to the realization, not from any change in high level philosophical perspective on Canada's role in the world with respect to environment or with respect to climate policy, but from energy market realities and from the realities of the power of activists or nongovernmental organizations' role in the energy markets, that the policy that it has been undertaking over the last number of years post-Kyoto or at least since the unraveling of the Kyoto protocol, that this policy is putting Canada in a rather vulnerable position with respect to its hugely important national energy industry.

So whereas it appeared that following the United States in a lock-step fashion with respect to climate policy was good policy, now it is becoming apparent to Canadian policy makers that as a policy it presents unique problems to Canada and places Canada in a position of vulnerability with respect to the energy markets.

REPOSITIONING CANADA VIS-À-VIS CLIMATE POLICY

So what I see developing now is the recognition that Canada repositions itself with respect to climate policy, and that it repositions itself as a stronger environmental and social leader more in line with what it has been historically.

At one point in time Canada used to be seen as a leader in environment and a leader in progressive social programs, and that position has been eroded over the last number of years. I see this policy now changing to one where Canada reestablishes its historic environmental position. And again the reason for this is not a high level policy premised on philosophical change, but one dictated by the realities of energy markets.

So what can we expect over the coming decade or two of the twenty-first century with respect to Canadian energy and environmental policies? Well I believe that we will see increasing technology and management leadership in low greenhouse gas hydrocarbon production from Canada. I believe that we will see Canada increasingly being a country that takes the lead in this area. Canada will not be shutting down its hydrocarbon production facilities, however, simply because the oil and gas industry—and particularly the oil sands industry—is far too important to the Canadian economy. But I do see Canada taking a position of leadership with respect to technologies and management regimes that will lower greenhouse gas emissions from the hydrocarbon industry. I think we will see less industry and government rhetoric arguing with climate change and arguing for the status quo with respect to the hydrocarbon industry because that policy has not paid any dividends for Canada.

I also see an increasing focus on renewable energy production in Canada. Canada has significant renewable energy resources; it has large wind resources and it has a large solar resources. The challenge in Canada is the proximity of these resources to its major urban markets, but that is something I see in the coming decade or so, increasing policy change to support renewable energy production. This comes at a time when renewable energy production is increasingly competitive with hydrocarbon produced grid power in many parts of Canada and the United States. So I see over the next decade or so an increasing focus on renewable energy in Canada.

I also see over the next decade or so a reevaluation of the role of nuclear energy. There was a time when Canada was a global leader in terms of nuclear energy. Canada has some of the largest and most prolific resources in the raw materials for nuclear energy, uranium. Uranium is found in large quantities of very high grade ore in the province of Saskatchewan and historically Canada, through its Atomic Energy of Canada Limited, a Crown Corporation or state-owned enterprise, was a leader in developing nuclear technology, particularly the CANDU reactor, which was at one point a global leader and is still seen by many as having many properties that are very attractive. But Canada has over the past decade or so squandered its national lead in this area. But I predict that over the next decade or so, that we will see a reexamination of nuclear energy in Canada. This may come nationally or Federally, or it may be initiated by a province such as Saskatchewan which is obviously pro-nuclear.

I also see over the next decade or more, an increase in exports not only in raw energy resources such as oil and gas, but also of technology and management services. Already Canadian companies operate in many parts of the world. A preferred strategy of Canadian oil and gas companies operating in different parts of the world is to bringing in cutting edge state-of-the-art technology and management approaches from Canada to those countries where those resources are found. That is a business model that has worked very effectively for Canadian companies working in Latin America, Africa, and Asia. Canada is already and is likely to be more of a leader in technology and in management services based on it highly competitive energy industry domestically and this will be an export in terms of management and technology skills being exported to the global energy industry. Canada's energy finance sector will increasingly follow it energy sector around the world and become more competitive with London energy financial markets.

BACK TO THE FUTURE: INTER-FIRM AND INTER-SECTORIAL COLLABORATION

I also see, over the next decade or so, what I would call a policy of going back to the future. I see what has been happening in Canada for the past 10 or 15 years as a sole reliance on markets to operate in a manner with a minimal role of government; that has been the philosophy or ideology that has underscored both the federal and many of the provincial governments in Canada. I see an increasing questioning of this approach and I see what some might call a turning of the pendulum, a going back to more of a focus on both industry collaboration as well as industry and government collaboration. We already see in the energy industry the establishment of an organization with the acronym COSIA, which stands for the Canadian Oil Sands Innovation Alliance. This is a formal collaboration between leading oil sands companies to deal with the environmental aspects of development of bitumen or developing oil sands in Alberta and Saskatchewan. This organization has had some initial successes in areas of tailings ponds and some other environmental areas. It has to date not been as successful in the area of reducing greenhouse gas emissions because of the fact that greenhouse gas emissions are very closely correlated with energy inputs and energy inputs, of course, are closely related to costs. So companies have been much more hesitant to collaborate in the greenhouse gas area because of the perceived competitive advantage of being a first mover with a technology that will reduce energy input and therefore reduce costs as it is a very cost driven industry. But nevertheless COSIA has been established and it is quite a remarkable innovation. There have been previous attempts to collaborate in a much smaller fashion which have also had similar outputs in dealing with some environmental issues but not the major climate change ones. There is a great deal of anticipation and expectation on COSIA as an organization and it is to be seen yet what it will finally result in.

On the notion of back to the future, I also see a reexamination of Canada's historical industry-government collaboration model. In the past economic and policy success in Canada has been premised on not just an unfettered free market but also on a serious role of Canadian government, both at the Federal and provincial level. What I expect over the next decade or so is a re-examination of how Canada can, as a country, re-establish those types of initiatives. In the past, government clearly played a role. In the oil sands it can be argued that the oil sands today would not exist as an industry had it not been for an organization known by its acronym of AOSTRA, or the Alberta Oil Sands Technology Research Authority, which is now a defunct organization. It has various successor organizations but the various successors have not been as well funded, nor have played as serious a role.

AOSTRA was established by the Province of Alberta in order to develop with government funding a significant role of University-based researchers to develop technologies that would allow Alberta to commercially produce its oil sands or bitumen resources and it is out of the Alberta Oil Sands Technology Research Authority, that the technology currently used in the oil sands was developed. Both the mining techniques for mining oil sands that are currently still being utilized both by Suncor and Syncrude, the largest oil sands producers, were developed as part of this government-funded authority, AOSTRA. But more importantly the so called *in situ* technologies which are being employed to access the 80% of the oil sands or bitumen resource which is buried too deep to access by open pit mining, is being developed by a technology known as Steam-Assisted-Gravity-Drainage, or SAGD by its acronym. That technology was developed by university-based engineering scientists working with AOSTRA and funded by AOSTRA to develop these new technologies. The patents developed for SAGD were owned by AOSTRA and the patents were made available free of charge to all Canadian oil sands industry firms. AOSTRA was used as a stimulus to develop the Canadian oil sands.

Similarly there was a federally funded Crown Corporation known as AECL, or Atomic Energy of Canada Limited. AECL was completely funded by the Canadian Federal government and it was owned by the government. It was the government who invested in AECL and the CANDU reactor technology was developed in Canada and was successfully marketed around the world. AECL was effectively dismantled about four or five years ago when the reigning ideology was one of government getting out of an active role in technology and business development. What exists as AECL today is but a small remnant of what its original role was as a nuclear energy organization. The current AECL organization essentially is an after-sales-service organization to service existing reactors in operation worldwide. It no longer engages in research and development nor in sales and project development worldwide.

Canada has in the past been successful with developing resources, developing economically important industries such as oil sands and nuclear energy through fairly significant government investment. Over the past ten or fifteen years that Canadian policy has been abandoned for philosophical or one could say ideological reasons and I would suggest that in the coming decade or so that we will see a reexamination of that traditional Canadian historical industry-government collaboration model and how it might work going forward. Again, the reason for governments reexamining the government-business model now is market driven, in that the market now calls for a successful link between energy development and environment, specifically greenhouse gas emissions as they relate to climate change.

ENERGY TRANSPORT: PIPELINES AND OIL-BY-RAIL

So what are we likely to see over the next decade or so in terms of energy transport? We are likely to see, in my opinion, that the lowest profile and least politicized tidewater pipeline proposals will get the nod in response to Canada's expected climate change policies.

We will also see that several of the natural gas line proposals will succeed. There are a number of proposals that are being put forward by different proponents at this point in time; they cannot all succeed because they are in fact competitive and redundant with one another. But we will see the strongest ones succeed. Natural gas pipelines prove to be much less contentious than oil pipelines because of the commodity that flows through the pipe. Oil, particularly heavy oil or bitumen or oil sands oil, is a particular target of climate change activists and it also has a higher level of concern for communities, particularly aboriginal communities in western Canada, as the spills are more serious particularly when they are in water off the coast of British Columbia. Natural gas is less of a target, although it is a hydro carbon, it is less of a target for climate activists then oil and heavy oil, and it is also less of a concern in terms of pipeline ruptures than oil is.

We will also see in my opinion over the next decade or so a continuation of oil by rail filling the transport void in the short term. Oil by rail is more expensive than oil by pipeline, and it is less safe as we have seen in the Lac Megantic accident some six months ago. But in the absence of pipeline transport availability, rail will continue to be a viable short-term option for getting oil to tidewater at the coast. Oil by rail will play a different and more long term role in the United States than in Canada, because in the United States it may have a longer term opportunity because of its unique attribute of being able to deliver oil to very specific markets and being much more flexible than oil by pipeline. That is an attribute of oil by rail that will be more valuable in the United States where individual markets will be accessed than in Canada where the transport issue is more one of bulk transport of oil and gas resources to the coast, or tidewater, so as to access international markets in emerging economies of the world.

Next having made our sort of forecast of what we are likely to see in Canadian energy policy in the next decade to decade and a half I will develop the argument for why I believe we are going to be seeing what I just said over the next decade to decade and a half.

ENERGY DEMAND IN THE TWENTY-FIRST CENTURY

The first question we need to address is where energy demand is likely to be coming from over the next number of years.

First, energy demand is going to be increasing as we go forward over the next couple of decades. There are no forecasts that predict or forecast the leveling off of total global energy demand.

This total global energy demand however will consist of an increase in demand from the developing countries or the emerging economies. In the industrialized world of Canada, the United States, and Western Europe, we have already seen a flattening of demand. In fact there are some articles in media over recently stating that some toll roads had been declaring bankruptcy in the United States because of the fact that car traffic has fallen off. In the United States, Canada, and Europe demographics and consumer tastes are rapidly changing; that is in terms of younger consumers, they are not as enamored with the automobile as older consumers were and are increasingly using other forms of transport. And the fact that the demographics are such that smaller populations and lower growth is the reason why we are seeing flat energy demand growth in the industrialized world. So all the global demand increase is coming from the developing or emerging economies, and that is where the market is. That is happening for two reasons. One of the reasons is the emerging economies are still an area where we are seeing continuing population growth. The global population has surpassed seven billion people. Depending on which forecast one subscribes too, we are likely to see an increase in population over the next several decades of the planet to approaching eight to nine possibly to ten billion people.

Virtually all of this growth will come from the developing countries, not from the industrialized countries. There is a fair bit of variance in the forecast of what we are likely to see over the next number of decades, but even the most conservative forecasts still call for a significant increase in global population.

But global population growth is not the only reason we continue to see growth in energy demand in emerging economies. The other reason that we are seeing growth in energy demand in emerging economies, and probably the more important reason, is that world GDP or Gross Domestic Product per capita is continuing to increase. It has increased dramatically since the 1980s and it continues to increase.

Despite the fact that the Gross Domestic Product per person in the emerging economies has already risen dramatically over the past several decades, is the fact that can be seen in a graph which suggests that there is still a large gap between the GDP per capita in the United States and that of the emerging economies. This is a gap which I refer to as the aspiration gap; the emerging economies aspire to having a GDP per capita comparable to that of the United States, Canada, and Western Europe. Despite the fact that the GDP per capita has been increasing rapidly it is still well short of that of the United States.

The GPD per person growth around the world differs dramatically between countries. There is a graph which shows GPD per capita growth in different parts of the world over the first decade of the twenty-first century. GDP per capita growth in the United States over the previous ten years ending in 2011 was 1.8%. The GDP per capita growth in Canada was 2.0%. The GDP per capita growth in the Euro area of Western Europe was 1.1%. In contrast, over that period the GDP per capita growth in China was 10.6%. China's growth is well recognized and understood but China is not alone. When we look at India the GDP per capita growth over the same ten year period was 8.0%. The percent for GDP per capita growth in the Asian former Soviet republics was 6.8%. Interestingly, sub-Saharan African, which has for decades been seen as the "basket case" of the world in terms of persistent poverty and persistent lack of development, the sub-Saharan region of Africa saw a growth in GDP per person of almost 6% or 5.8%.

So clearly the growth in wealth, and thus the growth in energy demand, is to be found in the emerging economies. There has always been a strong link between increasing wealth of a person and their increasing energy consumption and this growth is all occurring in the emerging economies of Asia and arguably sub-Saharan Africa and Latin America. It is not in the United States, Canada, and Western Europe. So I am fairly certain that we will continue to see growth in the emerging economies of the world and primarily in the cities of those emerging economies, the major metropolises of those areas like Beijing, Shanghai, Delhi, Mumbai, Lagos, and Johannesburg. So that is the demand side of the equation. Also on the demand side things have been fairly consistent for the past ten or fifteen years; we have all recognized that energy demand continues to rise unabated driven by the emerging economies.

REVOLUTION ON THE SUPPLY SIDE

On the supply side however, things have been undergoing a dramatic change. Here I like to refer back to a scientist-thinker of the 1950s, Dr. M King Hubbard, who presented a scientific paper at an energy conference in San Antonio, Texas, in 1956 where he effectively coined the term "peak oil." In this presentation, which is reproduced in Figure, he forecast that US oil production would peak around the year 1970.

He made this prediction based on straight forward geological forecasts. That is, that production from any given well takes the form of a normal distribution. Initially there is very little production, then it accelerates and it goes up to a peak and then it falls off in sort of a classic normal distribution. One can get more production out of the well by using various technologies to enhance the oil recovery. These technologies require heat and chemicals and sophisticated engineering processes and most importantly they require more money. Left to itself, without enhanced oil recovery methods, a well will produce in a normal distribution fashion.

He took this basic concept or model of oil production and suggested that a similar type of normal distribution production distribution curve would be found logically at a reservoir level that is found at a well level. That is a reservoir would initially produce very little and then increase up to a peak and then would fall off. This concept and logic would hold in the Permian Basin in Texas or the Western Canadian Sedimentary Basin.

Hubbard then took the same concept and the same model of the normal distribution of oil production, or gas production for that matter, and extrapolated it to the national basis. He argued that taking a straight-forward extrapolation of his well and basin model one would conclude that the lower forty-eight states of the United States would reach peak oil production around the year 1970. This was deemed to be quite preposterous in the 1950s when all that anyone could imagine was increasing oil production because that is what people were seeing at that time. But, in fact, come the 1970s he was actually pretty close to the mark; oil production in the United States did peak and the US oil production saw its highest levels in the early 1970s. That is, until very recently when it was surpassed due to the technological revolution in multistage fracturing.

Hubbard also extrapolated the model to the entire planet and he argued that if the normal distribution that holds at the well level and that holds at the resource basin level and at the nation level, it should also hold at the global level. He predicted that would see global oil production peak around the year 2000.

When US oil production peaked in the early 1970s it was not heralded or reported on in a big fashion because it went past almost unnoticed because it wasn't until much later that statisticians would look back and see that, yes indeed, oil production had peaked in the early 1970s. But Hubbard's model and his concept of peak oil was reintroduced in the late 1990s when scientists publishing in the most prestigious journals in the world, namely Nature and Science, published articles with titles such as "Oil back on the global agenda" and "The next oil crisis looms large." What these articles did was stimulate a discussion by policy makers and by the media that the world was facing a challenge of peak oil production. There were books written on the topic; there were articles in popular newspapers and magazines, and the common understanding was that we were running out of oil, although technically that was not the case. But nevertheless oil was going to be much more expensive.

What we see in fact is that the peaking phenomenon is actually interesting in that conventional oil and gas in fact probably did peak in the early 2000s. But just as one can enhance the oil recovery from a particular well one can access more oil globally by using new technologies and spending more money. So what happened was when in the first decade of the twenty-first century, there was great concern about oil shortages, the notion of peak oil was very popular and the price of oil went up dramatically and stayed higher as did the price of natural gas. What that did was the market worked and stimulated research and development in finding new processes for producing hydrocarbon energy.

In terms of classic economists' supply and demand curves, we saw demand from the emerging economies increasing and moving the demand curve to the right in the classic supply and demand curve. Consequently with static supply curve and a rightward shifting demand curve the price of oil went up and stayed up. What that high price of oil did was that it stimulated technology and financial investment which in short order then shifted the supply curve to the right increasing the quantity, and again reducing—but only reducing in a small way—the price. What was very interesting about this phenomenon in the first decade of the twenty-first century was that the rightward-shifting demand curve and the resulting higher price of oil stimulated new technology development and investment. It actually had impact on the development of the Alberta oil sands in Canada. Once the price of oil was significantly high it made economic those resources that had been worked on for fifty years and had been only been marginally economic. Very rapidly the high price of oil made oil sands a very economic resource and Canada's reserves were recognized and Canada became a major player on the global oil market with its oil sands.

Even more rapidly was what happened with multistage fracking and so-called tight oil and tight gas. Fracking, or fracturing technology, had been worked on for many years to access oil resources trapped in tight rock formations but it suddenly succeeded and almost overnight brought on-stream a very large supply of tight oil in tight rock formations which now became accessible suddenly with this new technology. So almost overnight the supply of oil in North America was increased as was the supply of natural gas. The increased supply of natural gas caused the price of gas to plummet. The price of oil did not go down as significantly and there are other reasons for that both related to supply and demand and to the price stabilization policies of the Organization of Petroleum Exporting Countries (OPEC) to maintain prices, which we will not go into more detail here.

What we saw was concern about peak oil and an actual increase in oil price driven by emerging economy energy demand and lack of conventional oil supply and then a rapid investment in and success in new technologies of developing heavy oil and bitumen, and developing tight rock formations of oil which became accessible with the multistage fracking technology. So oil sands and multistage fracking has brought a supply of oil on to the North American market in a period of about a decade in very large quantities. What this has done is change the dynamic of oil markets sand gas markets in Canada and the United States dramatically.

One might normally expect an innovation as fracking, as well as innovations in Steam Assisted Gravity Drainage for accessing heavy oil and bitumen, to diffuse worldwide rapidly because resources similar to the tight oil formations that multistage fracking allows us to access and heavy oil formations are found in many parts of the world. But interestingly that is not happening so quickly simply because of the fact that Canadian and US oil markets are the most liberal and the most open markets and it is the easiest in which to diffuse a new innovation. We are likely to see much more resistance to a classic diffusion of the fracking innovation as per researches such as in Carl Rogers classic 1950s diffusion of innovation studies in that we will continue to see real opposition to fracking technology in Europe. In many countries in Europe opposition will prevent the technology from taking hold there, meaning that oil and gas markets in Europe are unlikely to see as dramatic a revolution as we have seen in Canada and the United States.

In other parts of the world we still have other factors which will prevent the rapid adoption and diffusion of the new technologies of both heavy oil and tight oil because of much more rigid regulation regimes and foreign direct investment regimes. The country that is probably the best example of that is Mexico as both heavy oil resources and tight oil resources are believed to be found there. But Mexico is only now, in late 2014, making changes to its oil and gas regime and fiscal regime and possibly opening up the Mexican oil industry to new technology and new investment. But the rigid regulatory and foreign direct investment scenario there is also still found in other countries like Venezuela that has large heavy oil resources but is not very friendly to foreign investment. There are many other countries that have either heavy oil or tight oil resources which are not as easily accessible by foreign finance and technology due to more rigid fiscal regulatory regime.

The historic pattern of crude oil production and distribution over the last century shows that oil is sourced in western Canada and sent to markets in the Mid-West and the Eastern Seaboard of the United States, as well as to the west coast in California.

Similarly, in the case of Canadian and US crude oil pipelines, connections are seen between the source of oil in western Canada and the demand or market for oil in Eastern Seaboard and the America mid-west.

Further, it is noted that the proposed pipelines all lead from Western Canada to the coast or to what is increasingly referred to tidewater.

The pattern for natural gas production and distribution is much the same: from Western Canada to eastern US and Mid-West as well as to Eastern Canada. And that has been the pattern for the last five decades.

The links between supply and demand in different parts of the world show that most parts of the world are much more connected in terms of across the water connections between supply and demand, but North America in effect is an island on to itself and there are very few connections off the American continent.

When we then juxtapose that fact with the fact that energy markets of the twenty-first century are not in North America but rather in the emerging economies of Asia, sub-Saharan Africa and Latin America, it describes fully the Canada and the United States energy scenario.

So the United States and Canada effectively have more than enough oil and gas to supply their own needs. In fact the United States likely has enough oil and gas to satisfy its own needs and is now for the first time in fifty years to be concerned about export markets and dealing with export markets.

So these issues from the United States energy perspective are three fold. One, the United States is dealing with anti-fracking activists. These activists or nongovernmental organizations protest the impact of fracking on the water table, and also on increasing small earth quakes. These activists have legitimate concerns, but the concerns or more focused on the legitimacy of the concerns for earthquakes than on water table. Much of the research has been shown that because the oil and gas formations are at very different depths than the water table, and that a properly executed multistage fracking operation is very unlikely to have any impact on the water table.

But more likely it is possible that small earthquakes that are being experienced currently in places like Oklahoma are the consequence of multistage fracking, and that is an issue that is only starting to be addressed through concepts of refilling the spaces in which natural gas and oil have been evacuated.

Chapter 3

The European Challenge

Balancing Energy Security and Economic Growth in the 21st Century[1]

Maxime H. A. Larivé

> In the European Union, the largest importer and second largest consumer of energy in the world, energy policy is not abstract ideas; it is everyday life
>
> Benita Ferrero-Waldner, former Commissioner for External Relations and European Neighborhood Policy

> Climate action is central for the future of our planet, while a truly European energy policy is key for our competitiveness. Today's package proves that tackling the two issues simultaneously is not contradictory, but mutually reinforcing.
>
> José-Manuel Barroso, President of the European Commission

The European Union (EU) is a largest world importer of energy. 53.4% of EU-28 growth inland energy consumption in 2012, and 61% for the Euro Area-18 members, was from imported sources.[2] The EU imports 80% of its oil and 64% of its natural gas.[3] The EU is facing a series of challenges in assuring the reliance of energy imports and production in order to maintain it economic engine. The question of energy security for the EU and its 28 Member States is directly intertwined with their economic growth and stability. The EU is confronting several challenges: assuring the flows of energy imports despite the growing global competition with emerging markets; a divided internal energy market; and, the responsibility to address climate change policy.

The European energy market is structured around two dimensions: the primary production of energy by the EU-28 through nuclear energy and renewables, and the extensive energy imports of hydrocarbons. In terms of primary production of energy, the European market is composed of different sources of energy production: nuclear energy (28.7%), renewable energy sources (22.3%), solid fuels (20.9%, mostly coal), natural gas (16.8%), and crude oil (8.9%).[4] The second dimension of sources of energy is the imports composed essentially of three sources of energy: crude oil, solid fuels and natural gas. From 2002 to 2012, the energy dependence of the EU-28 on energy imports has increased by almost 6 percentage points. It went from 47.5% in 2002 to 53.4% in 2012 (see Table 3.1).

The concept of energy security has been a core component of European strategy. Since the turn of the century, energy security has taken some depth in terms of threat perceptions. For instance, in the 2003 and 2008 strategic documents,[5] the concept of energy was only mentioned five times in the 2003 versions compared to 12 times in the 2008. In the 2003 document, the concept of energy was from a dependency point of view. The 2003 document stipulated that the Council of the EU agreed on the fact that "Energy dependence is a special concern for Europe."[6] In the 2008 strategic document, the Council identifies "energy security" as an important security issue composed of three dimensions: dependence problem, lack of a unified energy domestic market, and promotion of renewable energies.[7]

When analyzing and studying the concept of energy security and the European Union, one should recognize that energy security is composed of a triple dimension. As illustrated in Figure 3.1, energy security is understood in Europe along three aspects: geopolitical realities (energy flows); economic competitiveness (shift to a greener economy); and environmental concerns (climate change).

The concept of energy security is complex and multifaceted. The European Commission defines the concept of energy security as "the ability to ensure that future essential energy needs can be met, both by means of adequate domestic resources worked under economically acceptable conditions and maintained as strategic reserves, and by calling upon accessible and stable external sources supplemented where appropriate by strategic stocks."[8] The

Table 3.1 Energy Imports of EU-28 (2002–2012)

	2002	*2003*	*2004*	*2005*	*2006*	*2007*	*2008*	*2009*	*2010*	*2011*	*2012*
All products	47.5	49 6	50.1	52.2	53.6	52.9	54.7	53 7	52.7	53.9	53 4
Solid fuels	33.3	35.0	38.2	39.4	41.7	41.5	44.9	41.1	39.4	41.7	42.2
Crude oil	76.3	78.5	80.7	82.3	83.8	83.5	84.9	84.1	85.1	85.9	88.2
Natural gas	50 9	52.0	53.6	57.1	60.3	59.5	61.7	63.4	62.1	67.1	65.8

Source: Eurostat. 2014. "Energy Dependency Rate, EU-28, 2002–2012 (percent of net imports in gross inland consumptions and bunkers, based on tonnes of oil equivalent) YB14." http://epp.eurostat.ec.europa.eu/statistics_explained/index.php/File:Energy_dependency_rate,_EU-28,_2002%E2%80%9312_(%25_of_net_imports_in_gross_inland_consumption_and_bunkers,_based_on_tonnes_of_oil_equivalent)_YB14.png [Accessed on 16 September 2014].

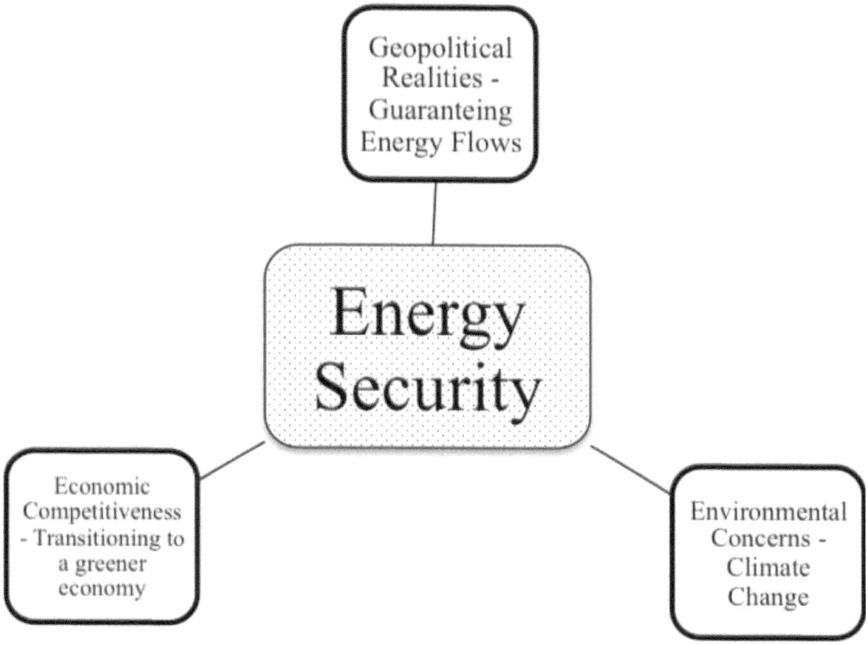

Figure 3.1 Variables in the Equation of European Energy Security. Energy Trilemma.
Source: Maxime H. A. Larivé. 2014.

imports and production of energy in order to fuel the EU common market have to remain stable; otherwise, it could have direct impacts on 28 national economies and the entire competitiveness of a continent.

In order to decide on the question of energy security, the energy policy in the EU has progressively changed since the 2009 Treaty of Lisbon, which sought to increase the shared competencies between the Commission and the Member States.[9] The fight between the Commission, defending the EU interests, and the Member States is making the decision-making process complex and messy.

Ultimately this paper seeks to understand the different dynamics shaping the overarching energy security policy in the European Union and within the 28 EU Member States (EU-28). In order to study the questions of energy security and economic growth in the twenty-first century, this paper is structured around three themes and is principally an empirical study. In a first part, the geopolitical context and the different sources of energy are analyzed in order to understand the degree of dependency of the EU-28 on foreign sources of energy. The second and third parts are contributing to the understanding of the complex relationship between economic competitiveness of the EU-28 and the commitment of the EU of leading the march toward environmental responsibilities. Additionally, the European Council had agreed on having a functioning Internal Energy Market (IEM) by the end of 2014, supposed to make the energy sector more efficient.[10] But the search of competitiveness and environmental responsibilities is taking place at one of the hardest time in European history considering the continuous economic crisis affecting the Eurozone and the rest of the Union.

INFLUENCES OF GEOPOLITICS ON ASSURING THE FLOW OF ENERGY

In the last decade, the energy dependence of the EU-28 on energy imports has increased by almost 6 percentage points. It went from 47.5% in 2002 to 53.4% in 2012 (Table 3.1). The European energy market needs to be divided into two sections: on the one hand, the primary, meaning the energy productions by EU Member States; and, the other, the most important flow of energy, the imports. The EU-28 import three major sources of energy: solid fuels (mainly coal), crude oil and natural gas. As illustrated in the table below (Table 3.1) the imports have increased in the last decade.

The EU is in a unique position as one of the major economic blocs in the world to be highly dependent on foreign energy sources. This dependency is ultimately affected by the geopolitics and regional realities. This section reflected on two dimensions: the impacts of regional geopolitics realities on the stability of energy flow; and, the composition of the energy market in the EU-28.

REGIONAL ENERGY REALITIE

In the post-9/11 global environment, incorporating the 2008 global financial crisis and the Arab Spring, the geopolitical dimension has become an important variable to account for Europeans. Since the turn of the century, geopolitics has played a role in affecting the imports and reliance of energy sources for the EU-28. Russia, the largest exporter of gas to the EU-28, has used gas as an energy weapon in order to increase its influence in its neighborhood, while underscoring European unwillingness to confront Russia's foreign policy. In terms of crude oil, the Arab Spring has complicated the reliance on crude oil produced in the Middle East despite the current decline in crude oil prices. Last, the shale gas revolution taking place in the US has changed the structure of the natural gas market.

Russia: The Militarization of Gas

Russia under the leadership of Vladimir Putin has used and transformed its production of gas into a foreign policy instrument, an energy weapon. One reason is that Putin views the 1990s as a dark period in Russian history leaving the country "weak and frustrated."[11] Ultimately, Putin has sought to rebuild Moscow's lost empire by first controlling energy resources and then seeking regional economic domination.[12] The energy shortages of the winters of 2006[13] and 2009[14] have underscored the degree of dependence of EU-28 on Russian gas.

Following the 2006 gas crisis, EU Energy commissioner, Mr. Piebalgs, declared that "it is clear that Europe needs a clearer and more collective and cohesive policy on security and energy supply."[15] After the 2009 crisis, the EU and its Member States increased their attention in developing pipelines in order to avoid transit countries like Ukraine.[16]

The EU-28 dependence on Russian hydrocarbons (solid fuel, crude oil and natural gas) is undeniable. In the last decade, as illustrated above (in Chart 3.1), Russian exportations of solid fuel have double from 13.1% in 2002 to 25.9% in 2012. In terms of crude oil, Russian exportations to the EU-28 are greater than other regions with 33.7% in 2012.[17] With natural gas, Russia tops the list of exporters with 32% to the EU-28 in 2012.

Despite the realities of the first decade of the twenty-first century and the current tensions in Ukraine, this dependency on Russian gas is not going to decrease. For instance, several EU Member States, like Germany and Italy, have signed long-term bilateral energy agreement with Russia. The signature

Chart 3.1 EU-28 Dependence on Russian Sources of Energy. *Source*: Eurostat. 2014. "Energy Dependency Rate, EU-28, 2002–2012 (percent of net imports in gross inland consumptions and bunkers, based on tons of oil equivalent) YB14." http://epp.eurostat.ec.europa.eu/statistics_explained/index.php/File:Energy_dependency_rate,_EU-28,_2002%E2%80%9312_(%25_of_net_imports_in_gross_inland_consumption_and_bunkers,_based_on_tonnes_of_oil_equivalent)_YB14.png [Accessed on 16 September 2014].

of bilateral agreements with individual EU Member States falls under the strategy laid out by Putin to divide the EU.

THE AMERICAN GAS REVOLUTION

In recent years the United States has seen an important rise in production of natural gas. The current gas revolution taking place in the US, under the name of shale gas, is progressively changing the global market of liquefied natural gas (LNG). The shale gas revolution is already affecting the global market of LNG, and especially in Europe.[18]

The reason behind the sudden increase of shale gas production in the US was the technological revolution through hydraulic fracturing and horizontal drilling. According to the US Energy Information Administration (EIA), between 2008 and 2011, the production increased by 51%. Gas production is foreseen to double from 2013 to 2040.[19] The Energy Information Administration of the US Department of Energy has estimated that by 2035 45% of US gas supply will come from shale gas.[20]

With the recent gas revolution in the US, the LNG previously exported to the US market, has now been shifted to the European markets. The increase of LNG in Europe is affecting the influence and the market shares of Gazprom, the Russian state gas company, and other gas producers in North Africa seeking to lower gas production in order to increase gas prices.[21] The US is progressively becoming one of the largest gas exporters in the world.

Additionally, the current negotiation on the Transatlantic Trade and Investment Partnership (TTIP) between the EU and the US is touching on the question of energy. The TTIP is supposed to create the largest economic bloc in the world.[22] With the new domestic energy realities of the US, which is soon foreseen to overtake Russia and Saudi Arabia in gas exports, the TTIP would enable to the US to export gas to Europe. Such flow of gas from the US would allow the EU-28 to decrease their dependency on Russian hydrocarbons. However, there are some limitations such as the need to create new infrastructures requiring large investments.[23] Aside from the requirement of new infrastructures, the main tension between the US and the EU is about the concept of energy security. In the US, it is mostly from a dependency point of view, while in the EU, the environmental dimension is central.

MIDDLE EAST AND NORTHERN AFRICA COMPLEXITIE

With the rise of instabilities due to the Arab Spring in 2011, the Middle East and North Africa (MENA) have been facing serious domestic and regional challenges. The Arab Spring brought political changes in three of the largest energy exporters to the EU, Algeria, Libya and Egypt.[24] Concerning North Africa, the main exporter of crude oil and natural gas to the EU-28 are Algeria and Libya. On the international stage, the shares of North African countries (Egypt, Libya and Algeria) represent 4% in oil and natural gas reserves. But these countries are important energy exporters to EU Member States like France, Spain, Italy and Greece.[25] Both country combined (see Table 3.2) represents 11% of EU-28 imports in crude oil and 15.4% in natural gas.

The 2011 Western-led war to Libya, under the NATO umbrella, had a direct impact on the production of gas and crude oil in Libya affecting Libyan exports to the European continents.[26] "Libya may have the greatest potential to increase natural gas exports to Europe once a new regime is established and possibly a new state oil and natural gas company in a post-Qadhafi Libya."[27] In 2011, the Libyan gas production dropped by over 90%, but it has since re-increased.

Concerning Algeria,[28] the country has neither seen Western military intervention nor a civil war like Libya, but the domestic situation has remained unstable affecting the production of energy and increasing the risks for foreign companies. For example the terrorist attack orchestrated by Islamic militants on 13 January, 2013 on the In Amenas gas plant operated by BP of Britain, Statoil of Norway, and Sonatrach the Algerian national oil company. Such attack on a gas plant in Algeria has raised some concerns about the cost of energy production in Algeria for European companies.[29] The attack has certainly launched a procedure forcing European governments and companies to rethink their policies in Northern Africa. The volatility of the region does not favor reliable energy production and exports. The terrorist attack in Algeria[30] sent the message around the world that "it is going to get more expensive to entice hydrocarbons from the ground in unstable parts of the world."[31]

Table 3.2 Principal North African Exporters of Energy to the EU-28 (in % of EU-28 imports)

	Crude Oil										
	2002	*2003*	*2004*	*2005*	*2006*	*2007*	*2008*	*2009*	*2010*	*2011*	*2012*
Libya	7.4	8.4	8.8	8.7	9.1	9.7	9.9	8.9	10.1	2.8	8.2
Algeria	2.7	3.0	3.3	3.5	2.5	1.9	2.5	1.6	1.2	2.6	2.9
	Natural Gas										
	2002	*2003*	*2004*	*2005*	*2006*	*2007*	*2008*	*2009*	*2010*	*2011*	*2012*
Algeria	21.1	19.8	18.0	17.6	16.3	15.3	14.7	14.2	14.0	13.0	13.5
Libya	0.3	0.3	0.4	1.6	2.5	3.0	2.9	2.9	2.7	0.7	1.9

Source: Eurostat. 2014. "Energy Dependency Rate, EU-28, 2002–2012 (percent of net imports in gross inland consumptions and bunkers, based on tonnes of oil equivalent) YB14." 28,_2002%E2%80%9312_(%25_of_net_imports_in_gross_inland_consumption_and_bunkers,_based_on_tonnes_of_oil_equivalent)_YB14.png [Accessed on 16 September 2014].

The political tensions and military problems in MENA are a threat to Europe's dependence on these energy producers. The continuous violence in Libya is a concern to Europe and the rise of radical Islamism in Algeria are affecting the stability of the region as whole.

COMPOSITION OF THE ENERGY MARKET IN THE EU-28

The European energy market is structured around several sources of energy: natural gas, crude oil, renewable energies, solid fuels and nuclear energy. Figure 3.2 illustrates the proportion of production of the different types of energy.

Additionally, the EU-28 are dependent on three major sources of energy that are heavily imported as illustrated below (Chart 3.2). In all major types of fuels EU-28 dependency has increased in the last 20 years.

The three main sources of energy imported and consumed in the EU-28 are solid fuels, petroleum, and natural gas. The consumptions and imports have increased in all three sectors in the last 20 years.

Gas[32]

The EU has always been highly dependent on foreign gas. As illustrated in the graph below (Graph 3.1), the dependence on foreign gas has increased in the last decade by 15.8%.

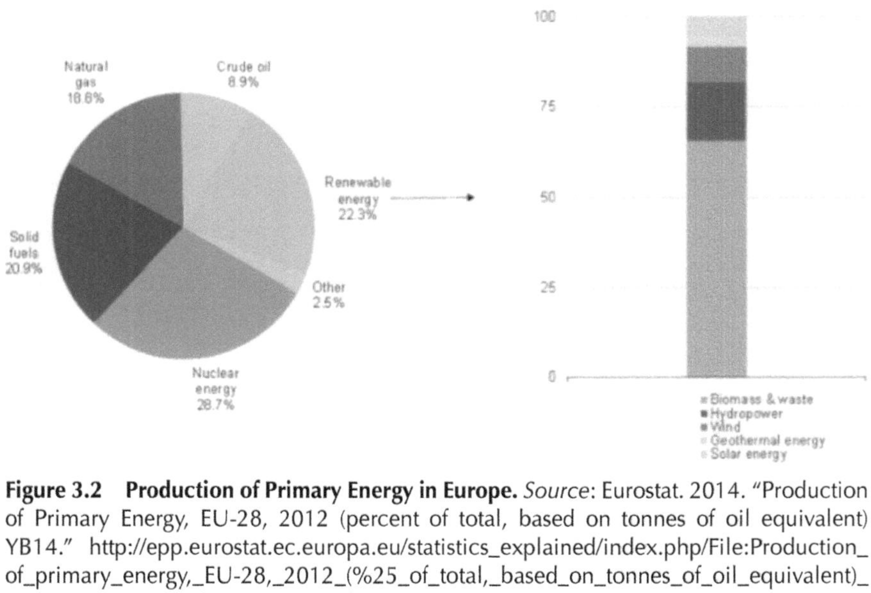

Figure 3.2 Production of Primary Energy in Europe. *Source*: Eurostat. 2014. "Production of Primary Energy, EU-28, 2012 (percent of total, based on tonnes of oil equivalent) YB14." http://epp.eurostat.ec.europa.eu/statistics_explained/index.php/File:Production_of_primary_energy,_EU-28,_2012_(%25_of_total,_based_on_tonnes_of_oil_equivalent)_YB14.png [Accessed on 16 September 2014].

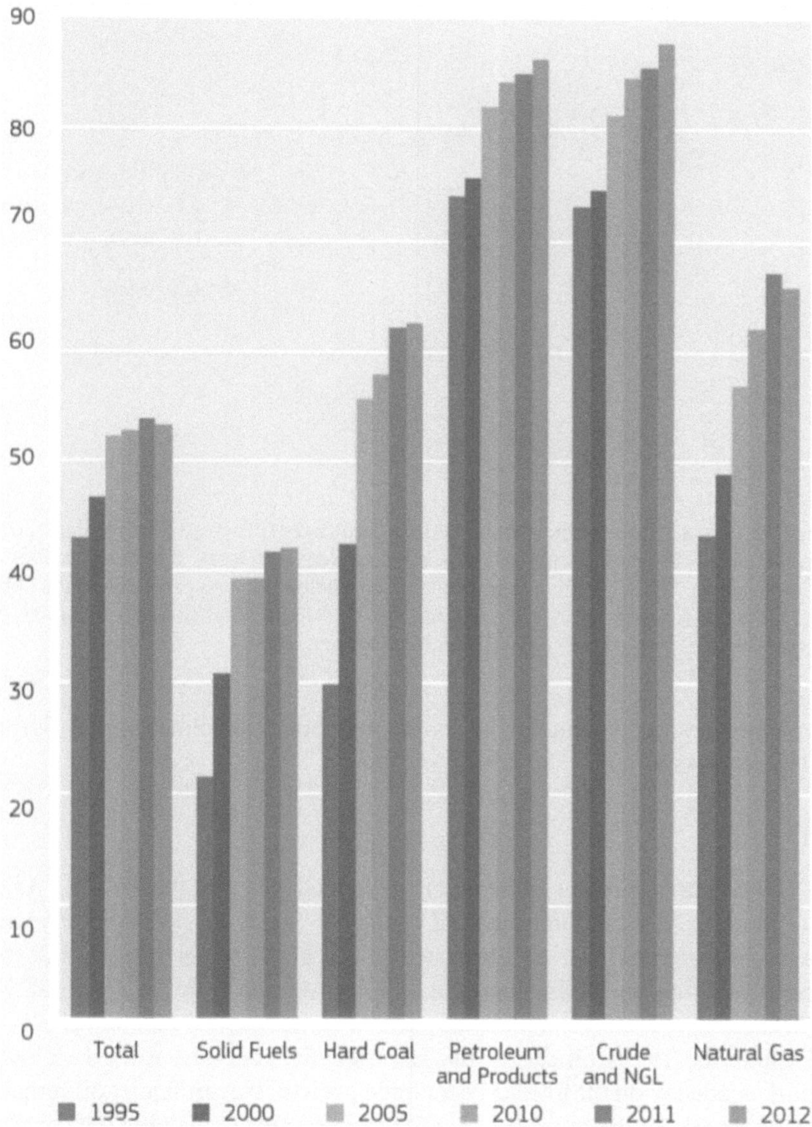

Chart 3.2 EU-28 Energy Import Dependency (1995–2012). *Source*: European Commission. 2014. "EU Energy in Figures. Pocketbook 2014." Belgium, 2013: 24. http://ec.europa.eu/energy/publications/doc/2014_pocketbook.pdf [Accessed on 22 September 2014].

Such high dependency rate illustrates the problem for the EU-28 considering the last geopolitical tensions in Eastern Europe and North Africa and Middle East (MENA) regions. The EU, and especially the powerful Member States have addressed the problem of security of imports through the

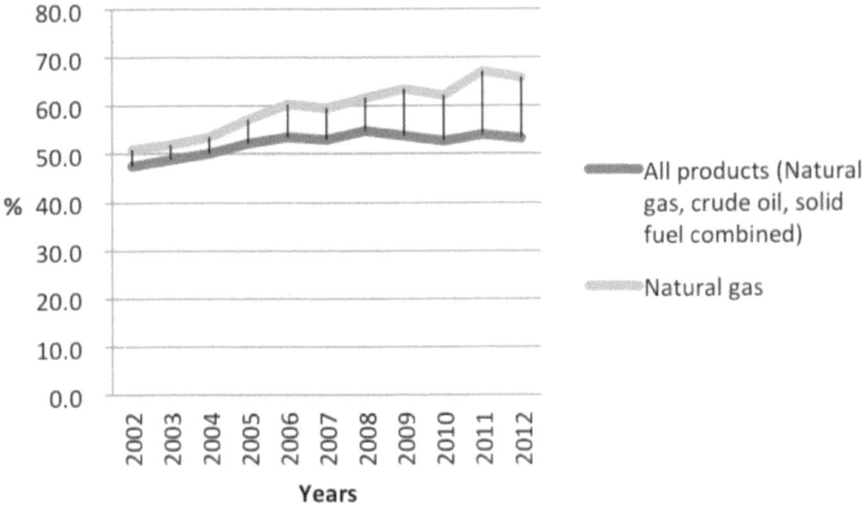

Graph 3.1 Gas Dependency Rate, EU-28 (2002–2012). *Source*: Eurostat. 2014. "Production of Primary Energy, EU-28, 2012 (percent of total, based on tonnes of oil equivalent) YB14." http://epp.eurostat.ec.europa.eu/statistics_explained/index.php/File:Production_of_primary_energy,_EU-28,_2012_(%25_of_total,_based_on_tonnes_of_oil_equivalent)_YB14.png [Accessed on 22 September 2014].

development and construction of a wide network of pipelines across Europe and Central Asia.[33]

Oil

Oil is another important dimension of dependence for the EU-28. Aside from the United Kingdom, most EU Member States do not have any serious reserve of crude oil. As illustrated in the graph below (Graph 3.2), the dependency rate on oil has increased the last decade from 76.3% to 88.2% in 2012.

Despite the 1970s oil crises, the EU Member States have tried to lower their dependency on oil for their economic growth. Nevertheless, oil remains the benchmark for the transportation sector in Europe explaining the continuous growth in oil imports since 1995.

Nuclear Energy

Nuclear energy is an important component of sources of energy in Europe. At the distinction of gas and oil, nuclear energy is a domestic production. Europeans remain the global leaders in civil nuclear energy. The deep relationship between European integration and nuclear power goes back to the early years

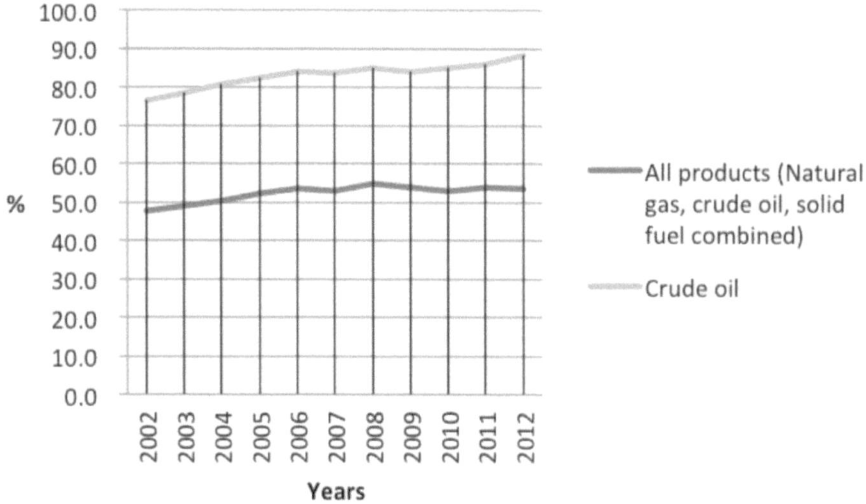

Graph 3.2 Crude Oil Dependency, EU-28 (2002–2012). *Source*: Eurostat. 2014. "Production of Primary Energy, EU-28, 2012 (percent of total, based on tonnes of oil equivalent) YB14." http://epp.eurostat.ec.europa.eu/statistics_explained/index.php/File:Production_of_primary_energy,_EU-28,_2012_(%25_of_total,_based_on_tonnes_of_oil_equivalent)_YB14.png [Accessed on 22 September 2014].

of the European integration process, with the creation of the European Atomic Energy Community Treaty (EURATOM) in 1957. Several crises, such as the 1979 Three Mile Island and 1986 Chernobyl accidents, have shaken up the popularity of nuclear energy in Europe. But the 2011 Fukushima accident in Japan changed some Member States' perceptions of the cost-benefit ratio of nuclear energy, especially Germany.[34]

According to the European Commission's data, nuclear energy represents 28.7% of the overall production of primary sources in the EU-28 (see Figure 3.2). Historically, the main producers of nuclear energy have been France, Germany, the United Kingdom and Sweden.

Following the Fukushima accident, Germany decided to phase out its nuclear production to ultimately become a nuclear-free country by 2022.[35] However, the German strategy requires an increase in the imports of energy produced by French nuclear power plants. Germany may become nuclear free in the coming decade, but will continue to buy neighboring nuclear energy?

European Shale Gas

Even Europe could advance in the production of shale gas, a major distinction ought to be addressed between the US experience and the European realities. In the US, the debate about environmental consequences of shale

Table 3.3 World Leaders in the Production of Nuclear Electricity

Net Installed Capacity	*GW*	*Country (Top-ten Producers)*	*% of Nuclear in Total Domestic Electricity Generation*
United States	102	France	76.1
France	63	Ukraine	45.4
Japan	44	Sweden	38.5
Russian Federation	24	Korea	28.3
Korea	21	United Kingdom	19.5
Canada	14	United States	18.8
Ukraine	13	Russian Federation Germany	16.6
People's Republic of China	13	Germany	16.0
Germany	12	Canada	15.0
Sweden	9	People's Republic of China	2.0
Rest of the world	58	Rest of the world*	8.1
World	373	World	10.9

Source: International Energy Agency. 2012. "Key World Energy Statistics." Paris: 17.

gas production is directly linked to the purity of water. In Europe, European citizens identify shale gas as a dirty source of energy. The popular opposition to shale gas has been greater in Europe, especially in the UK, France, Romania, Poland and the Netherlands.[36] There is no common position on the production of shale gas in Europe. EU Member States can be categorized into two groups: the pros, Poland, and to some extent the UK; and the cons, France, Bulgaria, Germany, Romania, and Czech Republic. Most European governments have implemented a wait-and-see approach.

Renewable Energy Sources (RES)

Renewable energies are perceived as the future for the production of energy inside Europe and a way to finally decrease European dependency on foreign energy producers. Renewable energies are composed of several types of energy: hydropower, wind energy, solar thermal, solar photovoltaic energy, biofuels and waste among others.

Renewables have progressively become a larger part of the energy mix of the EU-28 Gross inland consumption from 5% in 1995 to 11% in 2012.[37] In power generation, RES saw the "highest rate of growth (13%), with solar power generation registering particular high growth (63%)."[38]

The production and development of RES fall under the broader picture of energy diversification. It allows Member States to increase the production of energy through these different sources. Progressively, the RES start to become a larger part of the overall EU-28 total production of primary energy with 22.3%. However, one of the main problems with the production of RES

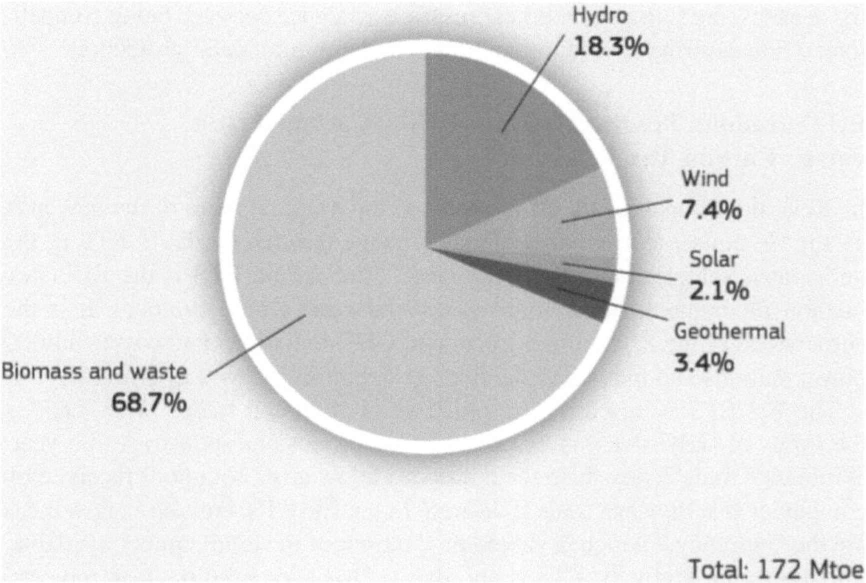

Chart 3.3 RES: Gross Inland Consumption by source (as of Mtoe 2010). *Source*: European Commission. 2012. "Energy Markets in the European Union in 2011." Belgium: 17. http://ec.europa.eu/energy/gas_electricity/doc/20121217_energy_market_2011_lr_en.pdf [Accessed on 22 September 2014].

is the high cost of energy for companies. Additionally, RES requires state subsidies considering the high cost of these technologies.

BALANCING ECONOMIC COMPETITIVENESS WITH . . .

The question energy is directly intertwined with the economic wealth of Member States and the EU as well as the competitiveness of European companies and industries. With the recent gas revolution occurring the US, the cost of natural gas and electricity has decreased compared to the costs in Europe. European companies are increasing becoming worried about the rising imbalance in energy costs, due to the shale revolution occurring in the US, between the US and Europe.[39] "With shale oil and gas, North America is rapidly reaching energy independence and the price gap between the US and European on gas is now so enormous as to undermine Europe's competitiveness and begin a process of re-industrialization in the US."[40] The cost of energy is affecting the global industrial competitiveness of European companies.

When studying the competitiveness of European companies, the question of green policies is closely intertwined. As underscored by the Commission for decades, the European market must find a balance between being competitive, while assuring its economic growth and environmental protection.

EU Emissions Trading System (ETS)—Carbon Trade versus Carbon Tax

In 2008, the Council of the EU warned on the eventual harm of the new plan by the Commission to combat climate change through the EU's ETS as the undisputed vehicle for cutting pollution.[41] The carbon ETS is the EU's key method for reducing greenhouse gases under the Kyoto Protocol. It is the cornerstone of the EU in order to combat GHG emissions and covers 11,000 power stations and industrial plants in 31 countries, as well as airlines.[42]

The EU ETS is based on the principle of "cap and trade." The "cap" is the limits of GHG that can be emitted by factories and industries in a year. While the "trade" consists in the emissions allowances bought or received by companies that they can trade if desired. In the EU ETS, emission allowances are the "currency," which is valued on the limit of the total number available. The money raised by the ETS is not required to be invested in green projects, but most EU Member States have done so.

However, a debate has emerged concerning the debate between the ETS and Carbon tax. The European Commission has based its strategy to competitiveness and green economy on the market-based system of the ETS, while

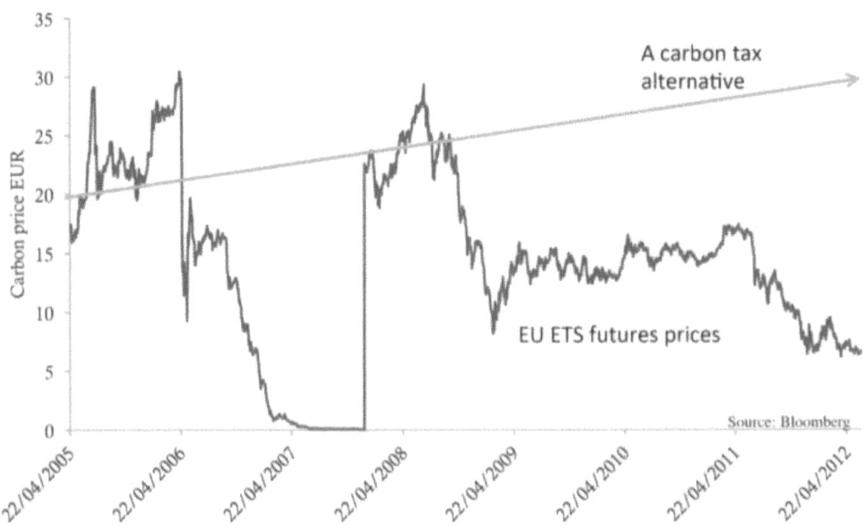

Graph 3.3 Price of Carbon (2005–2012). *Source*: Helm, Dieter. 2012. "Climate Change and EU policy—why has so little been achieved?" Bruegel Energy and Climate Exchange. October 4: 13. Available online at: http://www.dieterhelm.co.uk/sites/default/files/Bruegel041012.pdf [Accessed on 24 September 2014].

others have talked of the implementation of a carbon tax system. In recent years, the problem with the ETS has been a fluctuation of the prices of carbon and the 2013 carbon-market crash. As illustrated in the graph below (Graph 3.3), the prices of carbon emissions fluctuate and remain lower than would a carbon tax.

But experts have argued that a tax would in fact offer less flexibility to companies and countries. Additionally, the adoption and then the implementation of a carbon-tax would have to be approved by the EU-28, which is contentious domestic and political issues. In the long run, the ETS is foreseen to remain the benchmark of the Union.

The ETS is organized around a four stage process: 2005–2007, first trading period the learning by doing period; 2008–2012, second trading period, decrease of allowances by 6.5%; 2013–2020, third trading period, introduction of EU-wide cap on emissions with a progressive reduction by 1.74% per year and shift to auctioning of allowances; 2021–2028, fourth trading period.

Internal Energy Market (IEM)

One of the most considerable weaknesses of the EU in dealing with energy security is the fragmentation of its internal energy market. As underscored in a 2013 Congressional Research Service report, "in early February 2011, European heads of state pledged to complete the integration and liberalization of the internal European energy market by 2014; ensure all European member states are connected to a Europe-wide energy supply grid by 2015; boost energy efficiency throughout Europe; and better coordinate external energy policies."[43] Addressing the fragmentation of the EU internal market would contribute to on the one hand, lowering foreign dependence on energy and on the other, increasing the cohesion of EU voice on dealing with foreign energy exporters.

The integration of the internal market goes through a furthering of market liberalization, while increasing the interdependence on energy infrastructures such as electric grids and pipelines. The European Commission has estimated that it would cost approximately $1.4 trillion to realize such goals.[44] The Commission in its 2011 report underscored that "the internal energy market is not an end in itself. It is a key instrument in delivering what EU citizens aspire to most: economic growth, jobs, secure coverage of their basic needs at an affordable and competitive price, and sustainable use of limited resources."[45] The Commission argues that the nonexistence of such internal energy market, for electricity and natural gas, is affecting the competitiveness and economic wealth of the Union and its Member States.

This desire of an IEM goes back to the 1990s with two aims: first, to liberalize the electricity and natural gas markets for more competitive prices for the consumers; second, to foster the deepening process of EU integration.[46] A completed and functioning IEM will require a harmonization of national policies for the free flow of electricity. The current realities of the IEM are far from where it should be. For instance, the European energy markets are "poorly interconnected [and] the compatibility of national grids remains low, and that creates challenges like energy islands and loop flows."[47]

RENEWABLE ENERGIES—THE SILVER BULLET?

Another dimension of the IEM is the development of an European economy based on renewable energies. For instance, the new 2013 Framework for Climate Change and Energy Security emphasizes that a shift to a greener economy does not translate into a pain for economic competitiveness. "The internal energy market provides the basis to achieve this goal and I [Energy Commissioner Günther Oettinger] will continue to work on its completion in order to use its full potential. This includes the 'Europeanisation' of renewable energy policies."[48] Renewable energies are perceived as an important dimension of the European strategy to become greener while remaining competitive. The Commission bases its strategy on renewable energies for the transition of the European economies. Through renewable energies, the Commission believes that it can push Member States to develop new technologies and ultimately create more jobs and attract investments.[49]

Germany is the most obvious example of the Commission's plan. Germany has been the most committed country in Europe and the world in the promotion and development of an economy grounded on renewable energy. The nuclear accident of Fukushima strengthened this commitment. However, in reality Germany has struggled in balancing this strategy. In a recent article in the *New York Times*, Eduardo Porter underscored that at first Germany exempted its export-oriented, energy-intensive industries from the surcharges in the cost of energy produced by renewable energies. In recent years, the cost of these renewable energies has increased and Germany has lowered its subsidies, which has led to a drop in the rollout of solar power.[50] Following the 2011 Fukushima nuclear accident, Germany called for a phasing-out of its nuclear power by 2022 and targeting the use of 80% of energy produced by renewable energy by 2050.[51]

Additionally, there has been a certain lack of funding in the energy sector since the beginning of the Eurozone crisis. "The crisis in Europe's financial sector has led to most banks and many insurers pulling out of project finance in energy and other infrastructure sectors" argues David Buchan "to focus instead on re-building their capital ratios."[52] Despite the ongoing reforms in order to create an Internal Energy Market and increasing the use of renewable energies the latest economic data indicates a slowdown of the EU economic engine. For instance, "the EU's primary energy consumption declined by 0.3% in 2013 [. . .], the lowest level since 1995, with energy intensity (the amount of energy required per unit of GDP) falling by 0.4% to the lowest level on record."[53] Such record may indicate that the economic crisis of the Eurozone and the EU has had a certain impact on the productivity and growth of the economic situation of the EU-28.

. . . ENVIRONMENTAL RESPONSIBILITIES

The promotion of green policies is a contentious topic in Europe as it illustrates the perpetual tensions between the Member States and the EU. For instance, the EU ministers underscored in 2008 after the promotion by the Commission's plans to combat climate change. The 2008 plan known as the 20-20-20 energy package sought three precise targets: 20% overall reduction in CO_2 emissions from a 2005 baseline; 20% increase in the share of renewable energies in energy consumption; and 20% increase in energy efficiency, all to be achieved by 2020.[54]

The Council of the EU's claims was that the plans could not overshadow the competitiveness.[55] Six years later, the European Commission published its 2030 climate and energy blueprint. The 2030 framework seeks to balance out "climate and energy goals for a competitive, security and low-carbon EU economy."[56] The 2030 framework is the baseline for addressing climate change dimension of the energy security. The pillar of the 2030 framework are as follow: a reduction in greenhouse gas (GHG) emissions by 40% below the 1990 level; an EU-wide binding target for renewable energy of at least 27%; renewed ambitions for energy efficiency policies; a new governance system; and a set of new indicators to ensure a competitive and security energy system.[57] The main argument emanating for the European Commission is the fact that dealing climate change is not antonymous with economic competitiveness.

GHG Emissions Reduction

In 2013, the European Commission published a Green Paper, "A 2030 framework for climate change and energy policy,"[58] was adopted by the European Council on October 23, 2014.[59] The European Council agreed on the 2030 guidelines for climate and energy targets as well as adding at the last minute a "flexibility clause" to the text in order to adjust the targets after the upcoming UN talks in December 2015 in Paris. Part of the 2030 Framework deals with the reduction GHG emissions as one of its main targets.

World CO_2 Emissions* by Region

Mio ton CO_2	1995	2000	2005	2010	2011	2011 (%)
EU-28	4 062	4 099	4 279	3 968	3 846	11.9%
China	3 085	3 382	5 503	7 393	8 094	24.9%
United States	5 275	5 844	5 922	5 577	5 435	16.7%
Asia**	1 796	2 247	2 761	3 608	3 699	11.4%
Middle East	843	980	1 269	1 649	1 711	5.3%
Russia	1 573	1 510	1 527	1 600	1 682	5.2%
Africa	635	721	867	1 007	1 007	3.1%
Rest of the World	5 291	5 812	6 347	6 805	6 983	21.5%
World	22 560	24 595	28 475	31 606	32 456	100.0%

World CO_2 Emissions by Region (%)

Total 2011 = 32 456 Mio ton CO_2

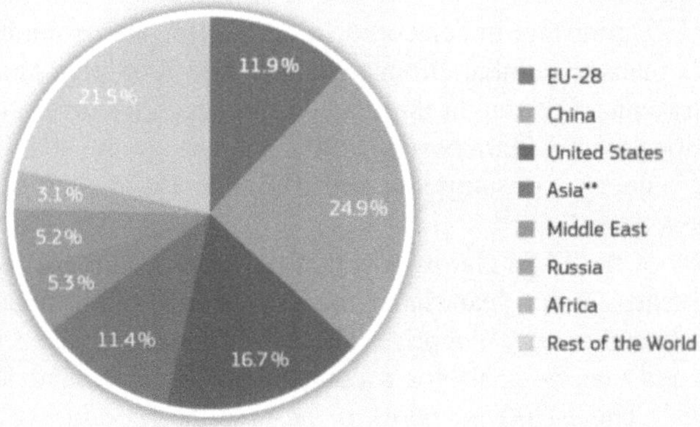

World CO_2 Emissions by Region (Mio ton CO_2)

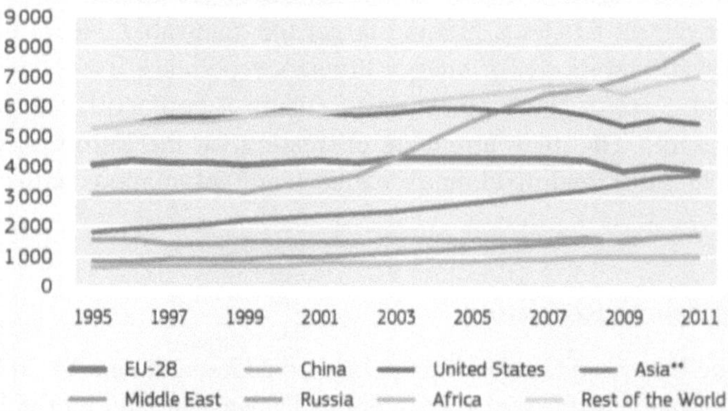

* Sectoral Approach, including Bunkers ** Excluding China – Source: IEA, May 2014
Methodology and Notes: See Appendix 13 – No 1

Figure 3.3 World CO_2 Emissions by Regions (1995–2010). *Source*: European Commission, "EU Energy in Figures. Pocketbook 2014," Belgium, 2014: 19. http://ec.europa.eu/energy/publications/doc/2014_pocketbook.pdf [Accessed on 22 September 2014].

Table 3.4 GHG[61] Emissions (1990–2012)

Index 100 = 1990	*1990*	*1995*	*2000*	*2005*	*2010*	*2011*	*2012*
EU-28	100	93	92	93	86	83	82
BE	100	105	103	100	93	85	83
BG	100	70	54	58	55	60	56
CZ	100	77	75	74	70	68	67
DK	100	111	100	94	90	83	77
DE	100	90	84	81	77	74	77
EE	100	49	42	46	49	52	47
IE	100	107	124	128	113	106	107
EL	100	105	120	128	112	110	106
E5	100	111	135	154	125	126	122
FR	100	99	101	102	94	89	89
HR	100	73	83	96	90	89	83
IT	100	102	107	112	97	95	90
CY	100	121	138	150	151	147	148
LV	100	48	38	42	47	45	43
LT	100	45	40	48	43	44	44
LU	100	81	81	108	102	100	97
HU	100	81	80	81	69	67	64
MT	100	123	130	147	150	151	157
NL	100	107	103	102	101	95	93
AT	100	103	104	120	110	108	104
PL	100	95	84	85	88	88	86
PT	100	117	138	145	119	116	115
RO	100	71	55	58	48	50	48
SI	100	101	103	110	106	106	103
SK	100	74	69	71	64	63	58
FI	100	100	99	98	107	97	88
SE	100	102	96	93	91	86	81
UK	100	93	90	89	80	75	78

Source: European Commission. 2014. "EU Energy in Figures. Pocketbook 2014," Belgium, 2014: 30. http://ec.europa.eu/energy/publications/doc/2014_pocketbook.pdf [Accessed on 22 September 2014].

GHG contribute to the natural greenhouse effect. Based on the Kyoto Protocol, there are six GHGs related to human activity: carbon dioxide, methane, nitrous oxide, hydrofluorocarbons, perfluorocarbons, and sulfur hexafluoride. EU Member States have tried to lower the GHG emissions in the last two decades. Based on the 2030 Framework, the EU ought to have reduced its GHG emissions by 40% below the 1990 benchmark in 2030 in order to be on track of the overall reduction by 85–90% reduction by 2050.[60] In order to deal with the reduction of GHG emissions, the main instrument of the EU is the ETS. The Commission is pushing the Parliament and Council to accept such number and defend such baseline in the coming global conferences and especially in the future one in Paris in 2015.

As illustrated in Table 3.4, the overall GHG emissions for the EU-28 represent an 18 percentage point decrease over 20 years. Among the most developed and active EU-28 economy, Germany has experienced a 23 percentage point decrease in 20 years. The example of the German economy demonstrates that a decrease in GHG emissions does not equate with a decrease in productivity and economic growth; after all, the German economy has weathered through the financial crisis. However, its Western neighbors, specifically France, has only managed an 11 percentage point decrease, whereas the UK emissions have decreased by 22 percentage point. Spain and Cyprus have been the bad students of Europe with an increase of 48 percentage point for Cyprus and 22 percentage point for the earlier.

As illustrated in the Figure 3.4 the GHG emission and fuel combustion in the EU-28 has decreased since the 1990 to 2012 by almost 1000 million ton CO_2 for the GHGs emissions and by around 700 million ton CO_2 for fuel combustion. In terms of the different sectors of EU-28 economy, the ones with the higher decrease in GHG emissions are the energy industries and manufacturing. The ones with lesser decrease and too some degree an increase are transport and commercial/industrial.

President of the Commission, José-Manuel Barroso underscored in the release of the 2013 Energy Framework that "An ambitious 40% greenhouse reduction target for 2030 is the most cost-effective milestone in our path towards a low-carbon economy. And the renewables target of at least 27% is an important signal: to give stability to investors, boost green jobs and support our security of supply."[62]

Part of the 2013 Framework, the Commission seeks in increasing the target of reduction of GHG of 40% below the 1990 benchmark. The Commission is pushing the Parliament and Council to accept such number and defend such baseline in the coming global conferences and especially in the future one in Paris in 2015.

The diminution of carbon dioxide emissions has been a challenge for many European countries, especially for Germany. After the Fukushima accident and the radical shift in German nuclear power, the level of carbon dioxide emissions in Germany have increased the last two years in a row with an increase by 1.2% in 2013 compared to 2012.[63] One direct reason behind the increase of carbon dioxide emissions has been the reduction of nuclear production in Germany from 22.2% in 2010 to 15.4% in 2013 by energy produced by coal.[64]

EUROPEAN PUBLIC OPINION ON CLIMATE CHANGE

In terms of public opinion on climate change, it appears that the concerns toward the threat of climate change should be seen under two spectrums: before the pain of the Eurozone crisis and after. According the results from

Million ton CO_2 equiv.	GHGs Emissions	Fuel Combustion	Energy Industries	Manufacturing Industries and Construction	Transport	Commercial/ Institutional	Residential	Agriculture/ Forestry/ Fisheries
1990	5 806	4 325	1 676	861	783	201	523	97
1995	5 451	4 058	1 519	767	838	183	508	92
1996	5 570	4 166	1 549	756	864	200	556	95
1997	5 484	4 066	1 502	746	877	185	522	92
1998	5 457	4 057	1 520	717	905	184	511	89
1999	5 349	3 994	1 477	697	924	183	498	89
2000	5 372	4 004	1 508	707	918	177	483	87
2001	5 427	4 080	1 548	683	932	194	518	87
2002	5 388	4 050	1 566	665	943	183	491	85
2003	5 481	4 136	1 617	675	953	184	505	86
2004	5 500	4 130	1 603	673	973	188	497	86
2005	5 477	4 103	1 595	663	971	185	495	86
2006	5 487	4 107	1 609	658	979	191	483	83
2007	5 440	4 044	1 617	659	989	170	429	80
2008	5 327	3 961	1 538	633	967	182	462	81
2009	4 936	3 678	1 414	528	941	178	445	79
2010	5 039	3 783	1 436	568	936	189	483	81
2011	4 900	3 642	1 413	553	926	171	408	80
2012	4 824	3 604	1 409	533	893	176	425	78

GHGs Emissions – EU-28 – Total and Fuel Combustion (Million ton CO_2 equiv.)

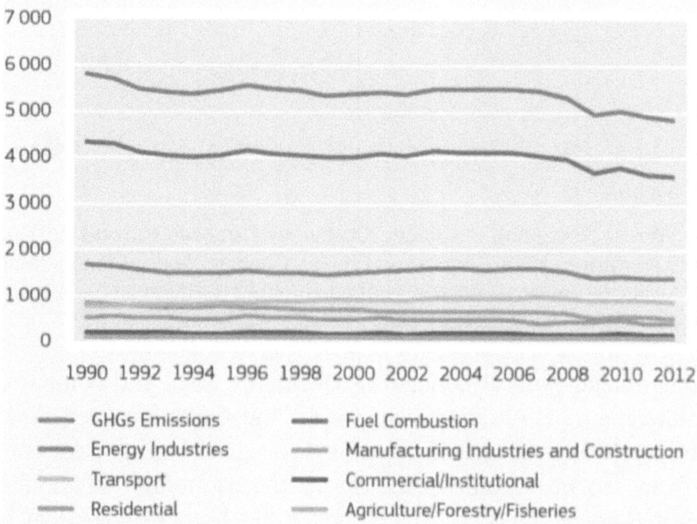

Source: EEA, UNFCCC v_15, May 2014
Methodology and Notes: See Appendix 13 – No 4

Figure 3.4 GHG Emissions in EU-28 Based on Total and Fuel Combustion in the Different Sectors (in Million Ton CO_2 Equivalent). *Source*: European Commission. 2014. "EU Energy in Figures. Pocketbook 2014," Belgium, 2014: 161. http://ec.europa.eu/energy/publications/doc/2014_pocketbook.pdf [Accessed on 22 September 2014].

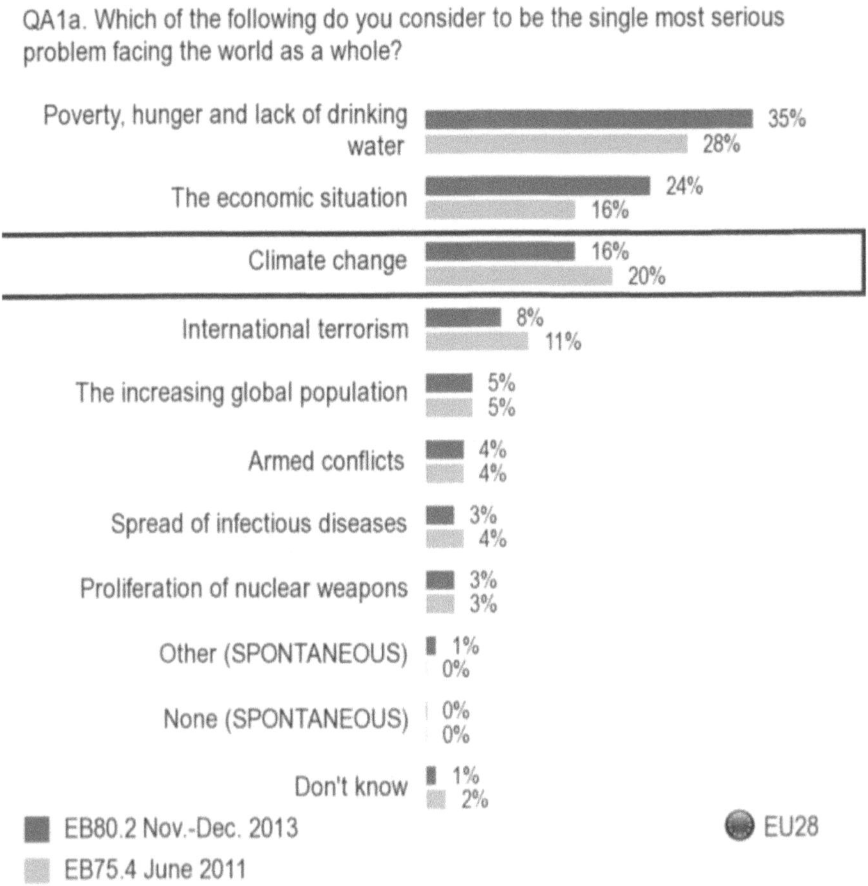

Chart 3.4 Threat Perception of Climate Change by European Citizens (2013). *Source*: European Commission. 2014. "Report on Climate Change." Special Eurobarometer 409. March: 8. Available online at: http://ec.europa.eu/public_opinion/archives/ebs/ebs_409_en.pdf [Accessed on 5 September 2014].

the Eurobarometer polls conducted in December 2013 and compared to the 2011 results, the topic of climate change as "the single most serious problem facing the world as a whole," Europeans downgraded the problem of climate change from second to third place above the economic situation.[65] In the 2014 Eurobarometer report, climate change has been surpassed by the economic situation by over 8 points.[66] In 2011, 20% of Europeans (out of the 27 Member States) identified climate change as the second single most serious problem, while three years later, only 16% of Europeans (out of the 28 Member States) do. One of the main reasons behind the shift of perceptions is the Eurozone crisis and the consequences of the austerity measures implemented domestically in the majority of European countries.

Based on the 2014 Eurobarometer, the results at the EU level approximately figures at the middle by the two "Europes": the Northern EU Member States, above the 16% is principally composed of northern EU Member States;[67] and the Southern EU Member States which is below the 16%.[68] It is undeniable that European public opinion plays a role in supporting a transition to greener policies.

CONCLUDING REMARKS

This paper has underscored the European perceptions and understanding of energy security. In the EU, energy security ought to be addressed from a triple approach: security of supply, transition to a greener economic engine, and fight against climate change. Each component is directly intertwined between one another and is inextricable from the others.

In the June 2014 European Council meeting, heads of state and government agreed on five core priorities for the EU in the coming five years, one of them was "a secure energy and climate future."[69] The commitment of the 28 heads of state and government underscore the peculiar European interpretation of energy security, which is connected with "scarce natural resources, the cost of energy and impact of climate change are major challenges; Europe's current energy dependency is a vulnerability."[70] The EU and its Member States have a series of options in order to address the issue of energy security. The commitment to greener energies, building infrastructures, diversification of energy sources and internal energy market are at the heart of the EU strategy to energy security to the twenty-first century.

Even if the EU has its own strategy in order to guarantee a supply of energy, at a relatively cheap price and a greener approach to production, the overall challenge of tackling climate change cannot be done solely by the EU.[71] The EU needs to advance and strengthen the ETS domestically and internationally. Through the commitment of an international carbon-trade market, the GHG emissions can be to some degree regulated. It can at least have more teeth in regulating international GHG emissions. For instance, in the last decade the global GHG emissions have increased by 47%. On the EU commitment to protection of the environment and curbing global warming, the attention and preparation will be in order to prepare the December 2015 United Nations COP 2015 meeting[72] in Paris and beyond, including by setting ambitious 2030 targets.

According to the French Ministry of Foreign Affairs, France will be seeking to develop a program of actions permitting to address the consumption of natural resources, namely water, waste and energy as well as the issue of greenhouse gas emissions. France, as well, wants to contribute to the creation of a post-2020 regime leading to the "creation of a universal agreement on climate" for the largest emitters of GHG. The COP 2015 meeting will be the opportunity for the EU to push and advance its 2030 Framework.

NOTES

1. Maxime H. A. Larivé, Ph.D., is a Research Associate at the EU Center of Excellence at the University of Miami. He received his Ph.D. in International Studies at the University of Miami. Ensuing his Ph.D. he completed a Post-Doctoral Fellowship as a Jean Monnet Fellow (2012–2013). He has published in several peer-reviewed journals like *European Security* and his book, "Debating European Security and Defense Policy," was published in October 2014 with Ashgate. Dr. Larivé also writes weekly commentaries on the EU and transatlantic relations for *Politipond*.

2. European Commission. 2014. "Energy Production and Imports." *Eurostat*. May. http://epp.eurostat.ec.europa.eu/statistics_explained/index.php/Energy_production_and_imports#Imports [accessed on 16 September 2014].

3. European Commission, Market Observatory for Energy, Key Figures, June 2011. http://ec.europa.eu/energy/observatory/eu_27_info/doc/key_figures.pdf

4. European Commission. 2014. "Energy Production and Imports." *Eurostat*. May. http://epp.eurostat.ec.europa.eu/statistics_explained/index.php/Energy_production_and_imports#Imports [accessed on 16 September 2014].

5. Council of the European Union. 2003. *A Secure Europe in a Better World. European security strategy,* No. 15859/03, December; Council of the European Union. 2008. *Report on the Implementation of the European Security Strategy—Providing Security in a Changing World,* No. S407/08. Brussels.

6. Council of the European Union. 2003. *A Secure Europe in a Better World. European security strategy,* No. 15859/03, December: 3.

7. Council of the European Union. 2008. *Report on the Implementation of the European Security Strategy—Providing Security in a Changing World,* No. S407/08, Brussels: 5.

8. Bahgat, Gawdat. 2006. Europe's Energy Security: Challenges and Opportunities. *International Affairs, 82*(5), 965.

9. Aalto, Pami and Korkmaz Temel, Dicle. 2014. "European Energy Security: Natural Gas and the Integration Process." *Journal of Common Market Studies 52*(4): 758; Braun, Jan Frederik. 2011. "EU Energy Policy under the Treaty of Lisbon Rules: Between a New Policy and Business as Usual." *CEPS/EPIN* Working Paper 31. Available at: http://www.ceps.eu/system/files/book/2011/02/EPIN%20WP31%20Braun%20on%20EU%20Energy%20Policy%20under%20Lisbon.pdf

10. It has not happened yet and has now been postponed. On February 6th, 2015, EU ministers and representatives were negotiating the plans for a European Energy Union allowing to lay out the foundations of an internal energy market, discuss questions of security of supply, reducing energy demand, lowering greenhouse gas emissions, and research and innovation.

11. Larrabee, Stephen. 2010. Russia, Ukraine, and Central Europe: The Return of Geopolitics. *Journal of International Affairs, 63*(2), 34–52; Fernandes, S., & Simão, L.

(2010). Competing for Eurasia: Russian and European Union Perspectives. In R. Kanet & M. Raquel Freire. (Eds.). *Key Players and Regional Dynamics in Eurasia. The Return of the "Great Game."* (pp. 103–125). New York: Palgrave Macmillan.

12. Nygren, Bertil. 2008. "Putin's Use of Natural Gas to Reintegrate the CIS Region." *Problems of Post-Communism, 55*(4), 3–15; Nygren, Bertil. 2008. *The Rebuilding of Greater Russia: Putin's Foreign Policy towards the CIS Countries*. New York: Routledge.

13. Youngs, Richard. 2009. *Energy Security. Europe's new Foreign policy Challenge*. New York: Routledge; Larrabee, Stephen. 2010. "Russia, Ukraine, and Central Europe: The Return of Geopolitics." *Journal of International Affairs, 63*(2), 34–52.

14. EurActiv. 2011. *Pipeline Politics? Russia and the EU's Battle for Energy*. 9 September.

15. Bahgat, Gawdat. 2006. "Europe's Energy Security: Challenges and Opportunities." *International Affairs, 82*(5), 962.

16. Paillard, Chirstophe-Alexandre. 2010. "Russia and Europe's Mutual Energy Dependence." *Journal of International Affairs, 63*(2), 66.

17. In 2012, the EU-28 imported 11.1% from Norway, 12.9% from the Middle East, 19.3% from Africa, and most of the rest from Central Asia.

18. Kefferpütz, Roderick. 2010. "Shale Fever: Replicating the US Gas Revolution in the EU?" *Center for European Policy Studies* 210, 1.

19. EY's Global Oil & Gas. 2013. "Shale Gas in Europe: Revolution or Evolution?" 1. Ernst & Young Global Limited, London.

20. Howarth, Robert W., Ingraffea, Anthony. 2011. "Should Fracking Stop? Yes, It's Too High Risk," *Nature*, 477, September 15: 273.

21. Kefferputz, Roderick. 2010. "Shale Fever: Replicating the US Gas Revolution in the EU?" *Center for European Policy Studies*. 210, 1.

22. See Barker, Tyson. 2013. "For Transatlantic Trade, This Time Is Different. Why the Latest U.S.-EU Trade Talks Are Likely to Succeed," *Foreign Affairs*. February 26; Hamilton, Daniel S. 2014. "TTIP: What are the Implications for Emerging Powers and the International Order?" *Bridges Africa,* 3(10), December 1; Sapiro, Miriam. 2014. "TTIP: A Fresh Start for 2015," *Brookings. Up-Front,* December 12.

23. Hamilton, Daniel. 2014. "Transatlantic Challenges: Ukraine, TTIP and the Struggle to be Strategic." *Journal of Common Market Studies 52*(Annual Review): 35.

24. Ratner, Michael, Belkin, Paul, Nichol, Jim, and Woehrel, Steven. 2013. "Europe's Energy Security: Options and Challenges to Natural Gas Supply Diversification." *Congressional Research Service*. R42405. 20 August: 23.

25. See Ghilès, Francis. 2013. "Algerian Gas' Challenges After the Attack on in Amenas." CIDOB. Notes 70. 2–3 February.

26. Anderson, Richard. 2011. "Libya Oil: The Race to Turn the Taps Back On." *BBC News*. 8 September; Cristiani, Dario. 2014. "The Libyan Oil Crisis: Social Fragmentation in an Unstable State." *Terrorism Monitor 12*(7); EIA. 2013. "Libyan crude oil production levels influence international crude oil markets." November 26. http://www.eia.gov/todayinenergy/detail.cfm?id=13951 [accessed on 17 September 2014].

27. Ratner, Michael, Belkin, Paul, Nichol, Jim, and Woehrel, Steven. 2013. "Europe's Energy Security: Options and Challenges to Natural Gas Supply Diversification." *Congressional Research Service*. R42405. 20 August: 25.

28. Idem: 24.

29. Ghilès, Francis. 2013. "Algerian Gas' Challenges After the Attack on in Amenas." CIDOB. Notes 70. February;

30. Ensuing the latest western military intervention against the Islamic State of Iraq and Syria (ISIS) and Khorasan terrorist groups, radical Islamists have called on the killing of Western citizens. The killing of a French citizen in Algeria on September 23, 2014 may have an impact on the holding of western drilling companies in Algeria and Libya.

31. Daugherty, Deon. 2013. "Algeria Terrorist Attack Example of Geopolitical Risks Increasing Energy Industry Costs." *Houston Business Journal*, 25 January.

32. Aalto, Pami and Korkmaz Temel, Dicle. 2014. "European Energy Security: Natural Gas and the Integration Process." *Journal of Common Market Studies 52*(4): 758–774.

33. The construction of pipelines throughout Europe for the supply of European markets has led to the return of the "Great Game" (Freire, Maria Raquel and Kanet, Roger E. 2010. *Key Players and Regional Dynamics in Eurasia. The Return of the "Great Game,"* New York: Palgrave Macmillan. This Great Game consists in increasing the influence of the EU in Eastern Europe and Central Asia, while Russia tries to maintain its supremacy in the region (Nygren, Bertil. 2008. "Putin's Use of Natural Gas to Reintegrate the CIS Regio." *Problems of Post-Communism 55*(4): 3–15; Nygren, Bertil. 2008. *The Rebuilding of Greater Russia: Putin's Foreign Policy towards the CIS Countries*. New York: Routledge). Pipelines have become the objects of all attention as they offer alternative of sources of energy as well as bypassing unstable transit countries such as Ukraine. Currently three routes have been prioritized: Nord Stream or "North European Gas Pipeline"; South Stream; Nabucco. The development of pipelines in order to diversify the gas imports has been one of the EU strategies. The EU has been careful to underline that the pipelines must be operated distinctly from energy production and should allow competitors to use them. However, in the reality the EU has been helpless in the area of construction of pipelines facing a powerful competition. MS have pursued their own interests at the expense of seeking a common energy policy.

34. Larivé, Maxime H. A. 2014. "Seeking for an Energy Renaissance in the Post-Fukushima Period. The European Debate on Nuclear Energy and Shale Gas." in Bagley, Bruce, Piet, Rémi, and Marcelo Zorovich (eds.) *Energy Security in the Western Hemisphere* (in press).

35. Dempsey, Judy, and Ewing, Jack. 2011. "Germany, in Reversal, Will Close Nuclear Plants by 2022." *The New York* Times. 30 May.

36. EY's Global Oil & Gas. 2013. "Shale Gas in Europe: Revolution or Evolution?" 10. Ernst & Young Global Limited, London.

37. European Commission. 2014. "EU Energy in Figures. Pocketbook 2014." Belgium, 2013: 24. http://ec.europa.eu/energy/publications/doc/2014_pocketbook.pdf [accessed on 22 September 2014].

38. European Commission. 2012. "Energy Markets in the European Union in 2011." Belgium. http://ec.europa.eu/energy/gas_electricity/doc/20121217_energy_market_2011_lr_en.pdf [accessed on 22 September 2014].

39. Marsh, Peter. 2012. "EU Groups Fear Growing US Advantage from Cheaper Energy." *Financial Times*. 29 November.

40. Helm, Dieter. 2012. "Britain's Energy Plan Is Dangerous and Out of Date." *Financial Times*. November 30.

41. Irish Independent. 2008. "Climate Change Costs' Must Not Harm Competitiveness." 1 February.

42. European Commission. 2014. "The EU ETS." Accessible online at: http://ec.europa.eu/clima/policies/ets/index_en.htm [accessed on 24 September 2014].

43. Ratner, Michael, Belkin, Paul, Nichol, Jim, and Woehrel, Steven. 2013. "Europe's Energy Security: Options and Challenges to Natural Gas Supply Diversification." *Congressional Research Service*. R42405. 20 August: 2.

44. Ratner, Michael, Belkin, Paul, Nichol, Jim, and Woehrel, Steven. 2013. "Europe's Energy Security: Options and Challenges to Natural Gas Supply Diversification." *Congressional Research Service*. R42405. August 20: 2; European Commission. 2012. "Making the Internal Energy Market Work." COM(2012) 663 final. Brussels. 15 November: 2.

45. European Commission. 2012. "Making the Internal Energy Market Work." COM(2012) 663 final. Brussels. 15 November: 2.

46. Friends of Europe. 2013. *EU's Internal Energy Market: Though Decisions and a Daunting Agenda*. Discussion Paper. Summer: 4.

47. Buzek, Jerry. 20 in Friends of Europe. *EU's Internal Energy Market: Though Decisions and a Daunting Agenda*. Discussion Paper. Summer: 15.

48. European Commission. 2014. "2030 Climate and Energy Goals for a Competitive, Secure and Low-carbon EU Economy." Press Release. IP/14/54. January 22.

49. With the sudden and current crash in oil prices since December 2014, the attraction in renewable energies has decreased sending a signal to green industries and manufacturers around the world. Some argues that the prices of oil are threatening renewables (see Reuters. 2015. "Low Oil Price Threat and Opportunity for Renewables: U.N. Climate Chief." January 22), while others claim that falling oil prices won't hurt in the long run clean energy (see Saha, Devashree and Muro, Mark. 2015. "Falling Oil Prices won't hurt Clean Energy." *Brookings Institution*. The Avenue. 23 January).

50. Porter, Eduardo. 2014. "The Benefits of Easing Climate Change." *The New York Times*. 23 September.

51. Reck, Hans-Joachim. 2014. "Europe is Coming to Terms With Energy, Climate Policies." *EurActiv*. 25 September.

52. Buchan, David. 2013. "Europe's Unresolved Energy versus Climate Policy Dilemma." In Friends of Europe. *EU's Internal Energy Market: Though Decisions and a Daunting Agenda*. Discussion Paper. Summer: 9.

53. BP. 2014. "BP Statistical Review 2014. The EU in 2013." http://www.bp.com/content/dam/bp/pdf/Energy-economics/statistical-review-2014/BP-Statistical-Review-of-World-Energy-2014-European-Union-insights.pdf [accessed on 16 September 2014].

54. Howarth, David. 2009. "Greening the Internal Market in a Difficult Economic Climate." *Journal of Common Market Studies, 47* (Annual Review): 138.

55. Irish Independent. 2008. "Climate Change Costs 'Must not Harm Competitiveness." 1 February.

56. European Commission. 2014. "2030 Climate and Energy Goals for a Competitive, Secure and Low-carbon EU Economy." Press Release. IP/14/54. 22 January.

57. Idem.

58. European Commission. 2013. *Green Paper. A 2013 Framework for Climate and Energy Policies*. COM(2013) 169 final. 27 March. Brussel

59. Euractiv. 2014. "EU Leaders Adopt 'Flexible' Energy and Climate Targets for 2030." 28 October.

60. European Commission. 2013. *Green Paper. A 2013 Framework for Climate and Energy Policies*. COM(2013) 169 final. 27 March. Brussels: 2.

61. The GHG are the six gases identified by the Kyoto Protocol.

62. European Commission. 2014. "2030 Climate and Energy Goals for a Competitive, Secure and Low-Carbon EU Economy." Press Release. IP/14/54. January 22.

63. World Nuclear News. 2014. "Coal Taints Germany's Energy Mix." 12 March.

64. Idem.

65. European Commission. 2011. *Climate Change. Report*. Special Eurobarometer 372. October; European Commission. 2014. *Climate Change. Report*. Special Eurobarometer 409. March.

66. European Commission. 2014. *Climate Change. Report*. Special Eurobarometer 409. March: 8.

67. Part of the Northern EU Member States, Sweden figures at the highest level with 39% along with Denmark (30%), Germany (27%), Belgium (24%), Finland (23%), the Netherlands (20%).

68. Part of the Southern EU Member States, Italy and France share the same number (14%), followed by the United Kingdom (12%) and ending with Spain (8%) and Portugal as last (6%).

69. European Council. 2014. "Conclusions." EUCO 79/14. Brussels. 27 June: 14.

70. Idem.

71. The US-China climate deal agreed in November 2014 is an important step in the reduction of GHGs and for an eventual global agreement and global baselines in December 2015 in Paris for the Climate Change Conference, known as COP21 (see Landler, Mark. 2014. "US and China Reach Climate Accord after Months of Talks." *The New York Times*. 11 November).

72. These are part of the global negotiation process on climate change revolving around the sessions of the Conference of the Parties (COP) to the United Nations Framework Convention on Climate Change (UNFCCC). The UNFCCC was adopted in 1992, and ratified by 195 parties. In 1997, the Kyoto Protocol was adopted and entered into force in 2005. The baselines of the Kyoto Protocol consist in agreeing on specific targets for the reductions and/or reductions of greenhouse gas emissions in developed countries and transition economies. The COP meets on yearly basis in order to review the implementation process of the Convention. The COP consists as the meeting of the Parties to the Kyoto Protocol (CMP). The CMP contributes to the adoption of decisions and resolutions.

Chapter 4

China and Latin America's Energy Future

The Case of Oil

Wenyuan Wu

On July 23, 2014, the Chinese President Xi Jinping concluded his one-week visit to four Latin American countries—Brazil, Argentina, Venezuela and Cuba, which resulted in a series of multilateral and bilateral agreements. At the China-Latin America and the Caribbean Summit, Xi proposed a new "1+3+6" cooperation framework.[1] Under the new framework, the energy field is emphasized as one of the six cooperation priorities.[2] The president promises to promote China-Latin American trade up to $500 billion and provide the region an investment stock of up to $250 billion to key areas in ten years. During his trip, Xi signed a number of oil deals with Venezuela, including an agreement of a $4 billion credit line in return for Venezuelan oil products, and a memorandum of understanding for China's EximBank to lend $1 billion to PDVSA (Petroleum of Venezuela).[3]

While domestic production stagnates, China has been accelerating its overseas quest for oil. In 2013, with a domestic supply of 4.46 million barrels per day, China satisfies its tenacious demand for petroleum and other fuel liquids by net imports of 5.05 million barrels per day.[4] In September 2013, China became the world's largest net importer of petroleum and other liquids, surpassing the United States on a monthly basis.[5] Chinese oil and gas upstream acquisitions in 2013 were valued at $38.04 billion.[6] In Latin America, in 2013, PetroChina (CNPC's listed arm) acquired 100% shares of Petrobras' Peruvian subsidiary, *Petrobras Energia Peru*. In the same year, CNOOC and CNPC jointly won a 35-year production sharing contract to develop Brazil's Libra oil field, a pre-salt oil discovery.[7] What are the push and pull factors behind China's adventure into Latin America's oil sector? Who are the main actors? What are the implications of China's oil bid for Latin America's energy future?

As China increasingly engages in global oil production, doubtful voices have emerged. Some assert that China's insatiable appetite for oil has the danger of diminishing the global supply available to other nations and people, thereby posing threats to other oil importers' energy security. There is also suspicion that Chinese National Oil Companies (NOCs), backed by various Chinese government agencies "unlevel" the playing ground and showcase undesirable social and environmental behaviors. Even worse allegations refer to the influence of Chinese oil interests in fostering signs of Dutch Disease in Latin America. This article strives to tackle these "myths" by examining China's role in Latin America's energy future. The first part provides an overview of China's evolving oil demand, the Chinese oil industry, and its relationship with the Chinese government. The second part briefly reviews oil developments in Latin America and China's engagement in the twenty-first century. Subsequently, the third part offers a systematic analysis of main actors and key drivers of China's oil pursuit in Latin America. The last part evaluates China's impact on Latin America's energy industry and related areas.

AN OVERVIEW OF THE CHINESE OIL INDUSTRY AND ITS GLOBAL ENDEAVOR

History and Evolution

The dawn of China's modern oil industry was closely linked with developments in the world oil industry. As oil became a strategic commodity fueling modern transportation in the mid nineteenth century, it was also utilized as the third major tools for the Western powers to forcefully open the Chinese market, with the first two being opium and cotton. The United States started to export crude oil and other petroleum products to China in 1867. From then until the establishment of the PRC, "*yang you*" (foreign oil) had sustained a monopolist position on the Chinese market. From 1904 to 1948, total domestic crude oil production was only estimated to be 2.78 million tons. Meanwhile, China imported about 28 million tons of crude oil from advanced countries.[8] With insufficient fund and fierce foreign competition, the domestic oil industry developed sluggishly, relying mainly on manual operation and animal power. Before 1949, China's reliance on oil imports to fuel its sprouting industrialization was more than 90%.

Since the establishment of the PRC in 1949, the country's oil sector has gone through major transformations in many aspects. Domestic production picked up at a rapid pace. In 1959, national crude oil production reached 3.73 million tones and the oil sector was systematically established based on four major oil fields: Yumen, Xinjiang, Qinghai and Sichuan.[9] An oil boom in the late 1960s and early 1970s helped China achieve its oil self-sufficiency with a 20% annual growth in domestic production and transformed the country into the world's fourth largest oil producer outside the Middle East.[10]

Over-drilling of oil wells and lack of discovery in the 1970s shadowed the newly established industry. In 1978–1979, prospectors drilled over fifteen million meters of new wells and discovered only one new field, resulting in a critical juncture in which oil output nationwide dwindled to less than 3% per year over the next decade and the short-lived oil boom was called to an end.[11] In 1993, China became a net oil importer.[12] Ever since, the gap between China's huge demand for oil and its limited domestic production capacity has been widening. Table 4.1 and Figure 4.1 illustrate the large discrepancy. In the last two decades (1990–2010), the global consumption of crude oil has been growing at 1.3% per year while China's annual consumption growth rate hit an astonishing 5.7%.[13] To solve the energy shortage and secure steady flows, the Chinese government began to restructure its oil industry by gradually introducing market principles and later encouraging its oil companies to seek overseas opportunities.

Table 4.1 China: Supply, Demand and Import of Crude Oil and Liquid Fuels (1992–2013) (Unit: million barrels per day)

Year	*Consumption/Demand*	*Production/Supply*	*Net Import*
2013	10.61	4.46	6.05
2012	10.28	4.37	5.91
2011	9.85	4.29	5.56
2010	9.19	4.27	4.92
2009	8.32	3.99	4.33
2008	7.83	3.97	3.86
2007	7.58	3.9	3.68
2006	7.27	3.84	3.43
2005	6.9	3.8	3.1
2004	6.5	3.6	2.9
2003	5.6	3.5	2.1
2002	5.3	3.4	1.9
2001	4.9	3.3	1.6
2000	4.6	3.2	1.4
1999	4.3	3.2	1.1
1998	3.9	3.2	0.7
1997	3.9	3.2	0.7
1996	3.6	3.1	0.5
1995	3.4	3	0.4
1994	3.2	2.9	0.3
1993	3	2.9	0.1
1992	2.7	2.8	-0.1

Source: US EIA, *STEO Archives*, last modified September 9th, 2014, http://www.eia.gov/forecasts/steo/outlook.cfm#issues1996.

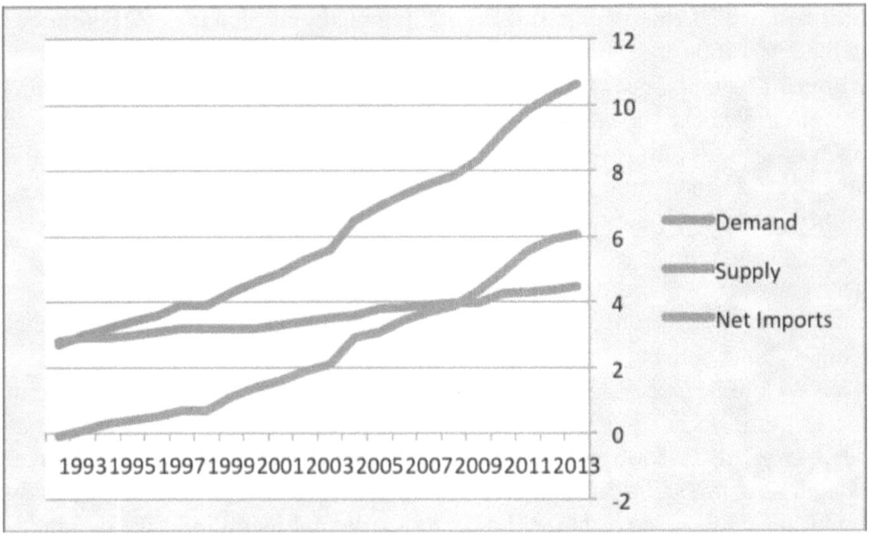

Figure 4.1 China: Supply, Demand and Import of Crude Oil and Liquid Fuels (1992–2013) (Unit: million barrels per day). *Source*: Author Self-Construction Based on Table 4.1.

Before reform, under the central-planning economy, oil exploitation, processing, and distribution in China were tightly supervised by the Ministry of Petroleum Industry (MPI). The MPI was created in 1955 to replace the Fuel Industry Ministry and to oversee both upstream production and downstream retail activities. In 1970, the MPI was merged with the Ministry of Chemical Industry and the Ministry of Coal Industry into the Ministry of Fuel and Chemical Industries (MFCI). Five years later, the MFCI was again divided into the Ministry of Petroleum and Chemical Industry and the Ministry of Coal Industry.[14] In order to promote scientific exploitation, augment production, and stabilize distribution, these succeeding government agencies controlled both market quantities and prices of petroleum products, in accordance with the State Planning Commission's economic blueprints. Nevertheless, political mandates regardless of market principles and institutional inconsistency led to inflexibility, inefficiency and cumbersome bureaucracy, worsening problems in the oil industry. Structural reforms to rejuvenate the oil industry became an imminent task for the Chinese government in the early period of open and reform. Against this backdrop, China National Petroleum Corporation (CNPC), China Petroleum & Chemical Corporation (Sinopec), Sinochem and China National Offshore Oil Corporation (CNOOC) were founded to effectively divide functions of petroleum ministries. Establishment of large NOCs constitutes an early attempt of structural reforms. However, at the beginning, these companies' operations still carried many characteristics of the command economy.

Throughout the 1990s, Chinese NOCs were organized in a fashion of "three divisions and four oligopolists." The oil industry undertook three divisions of labor among the four oligopolistic companies. The divisions were: (1) "between onshore exploration and production and offshore exploration and production," (2) "between upstream business and downstream business," (3) "between domestic trading and international trading."[15] CNPC was responsible for onshore E&P as well as upstream business. Sinopec was involved in domestic and downstream activities. CNOOC was in charge of offshore and international businesses while Sinochem was mandated for international trade of crude oil and chemical products.[16] All operations of these companies were directly financed by the Chinese government.

During the crisis of Chinese SOEs in the late 1990s, these divisions gradually disappeared and NOCs started to vertically integrate. Industry reform and restructuring dismantled the financial relationship between the state and NOCs, and introduced market mechanisms, including a twin pricing system and the manager incentive system.[17] As a result, Chinese NOCs are today, to some degree, exposed to market competition and investor oversight. In addition, although downstream prices for refined oil products still remain tightly controlled by the government, upstream exploration and production activities, especially those from abroad, as well as crude oil prices, have been "almost entirely liberalized" to follow movements on the global market.[18] These changes have created winners and losers. Sinopec, which was originally assigned to take charge of domestic downstream business has been hit the hardest. CNOOC, the most internationalized company of the four, has enjoyed the greatest returns on their overseas investment. To improve financial performances, these companies often cooperate and sometimes compete with each other to acquire upstream assets overseas. They have also found innovative ways to enhance their global competitive edge.

There is a natural connection between China's state-owned oil companies and the Chinese government. Overall, these companies at the group level are owned by the people and authorized by the state. The State Council holds the ultimate authority over all state-owned enterprises and a number of government agencies are involved in strategic planning, financing, and evaluation processes of the NOCs. The State-Owned Assets Supervision and Administration Commission (SASAC) has been these companies' dominant shareholder since 2003 and monitor their transnational ventures. The Ministry of Commerce provides ex ante review of each company and the State Administration of Foreign Exchange (SAFE) organizes NOCs' outward direct investment (ODI). On the other hand, the National Development and Reform Commission (NDRC) offers NOCs policy recommendations for overseas investment locations and its price bureau sets retail prices for refined petroleum.[19] At last, these companies need to obtain their upstream oil and gas extraction licenses from the Ministry of Land and Resources before they start every new operation.[20] Whether the complex bureaucratic structure showcases political mandates or gives NOCs room to prioritize commercial considerations is under heated situation. The issue will be further analyzed in the third part when evaluating the key drivers of China's endeavors in Latin America's oil sector.

AN OVERVIEW OF LATIN AMERICA'S OIL SECTOR AND CHINA'S INVOLVEMENT

Major oil producing countries in Latin America hold huge, untapped oil reserves and the region's relatively sound political climate helps attract foreign oil investors. According to experts, "Latin America is estimated to hold 13.5% of the world's proven oil reserves but accounts for only 6% of total output."[21] Recent discoveries in Brazil's deep-water pre-salt oil fields, as well as reforms in Mexican and Colombian oil sectors have renewed passion among resource-hungry countries, especially China. The oil sector is also a major source of government finance and a significant recipient of foreign direct investment (FDI) in countries such as Mexico, Venezuela, Colombia, Ecuador, and Brazil. For example, Mexico's oil sector accounted for 13% of the country's total export earnings and 32% of total government revenues in 2013.[22] In Colombia, the oil industry received $5.39 billion FDI in 2012, making up 34% of total FDI the country obtained.[23]

According to the US Energy Information Administration (EIA), Mexico is Latin America's top oil producer and the world's 10th largest producer. Producing 2.49 million barrels per day of total oil liquids in 2013, Venezuela is the third largest oil producer in the region and the largest exporter of crude oil.[24] The country also contains some of the world's largest oil reserves. Table 4.2 offers a comprehensive picture of oil industry development in Latin America's top oil producers.

China first began to invest in Latin America's oil industry in the early 1990s. Besides conventional oil trade, China's burgeoning demand for petroleum is also satisfied through overseas investment in various forms, including upstream mergers and acquisitions (M&As), engineering service (ES), loan-for-oil deals, and oil pipeline constructions. In many cases, Chinese oil companies, often backed by the government, are the primary facilitators of these overseas investments. The largest Chinese NOC, *CNPC*, signed a service contract on Block 7 in Peru's Talara Oilfield on October 23, 1993, marking its beginning of international operations.[25] In June 1997, CNPC won a bid for Venezuela's Intercampo and East Caracoles Oilfields.[26]

Table 4.2 An Overview of Latin America's Oil Sector Development (Data from 2013/2)

Country	*Proven Oil Reserves (billion barrels)*	*Oil Liquids Production (million barrels/day)*	*Crude Oil Production (million barrels/day)*	*Crude Oil Exports (million barrels/day)*
Mexico	10	2.9	2.5	1.19
Brazil	13	2.7	2.1	0.55
Venezuela	298	2.49	2.2	1.7
Colombia	2.4	1	1	0.68
Ecuador	8	0.505	0.505	0.36

Source: US EIA's *Country Analysis* web pages.

Initially, Chinese overseas oil investment proceeded in the fashion of direct acquisition or inter-NOCs cooperation. As late comers, Chinese NOCs often avoided competition with established international oil companies (IOCs) and chose to devote their assets to oil fields left by or uninterested to IOCs. Because of exploration and production experience from domestic oil fields, Chinese oil companies developed expertise in sweet (low sulfur) and light (high API gravity) crude oil and were constrained to invest in similar oil fields overseas. As a result, they were often "forced to operate in areas of greater political risk . . . and settle for smaller-scale onshore operations."[27]

Since 2000, as the Chinese government began to institute its implementation of the "going out" policy, Chinese NOCs have expanded their oil cooperation in Latin America and have progressively stepped out of their comfort zone to engage in innovative ways of overseas investment. Even though the majority of their foreign assets are located in conflict-burdened Middle East and Africa, Chinese oil companies are intentionally moving their operations to more politically stable climate such as OECD countries and several South American nations.[28] Furthermore, they are now more focused on unconventional oil assets and actively seek opportunities of cooperation with IOCs. For instance, CNPC and CNOOC each bought 10% of stakes in a 35-year Production Sharing Contract (PSC) in Brazil's Libra Oilfield, making themselves partners with Petrobras, Shell, and Total in exploring Brazil's pre-salt oil discovery.[29] By the end of 2013, China has generated 2.5 million barrels of oil equivalent per day in combined oil and gas overseas production.[30] Latin America accounts roughly for 11% of Chinese NOCs' overseas production. During 2011 and 2013, China invested a total of $73 billion in global upstream M&A deals and $29 billion in long-term loan-for-oil projects.[31] Of the $73 billion M&A deals, about $10 billion went to oil cooperation projects in Latin America. Table 4.3 summarizes Chinese upstream acquisitions in Latin America from 2002 to 2013.

Table 4.3 Chinese Upstream Acquisitions in Latin America 2002–2013

Company	*Assets*	*Stake(%)*	*Deal Size (bi.)*
CNPC	Oct 2013, won a 35-yr PSC to develop Brazil's Libra Oilfield with CNOOC;	10	1.4
	2006, purchased Encana for oil and pipeline interests in Ecuador with Sinopec.	100	1.47
PetroChina	Nov 2013, purchased Petrobras's Peruvian subsidiary;	100	2.6
	Aug 2012, bought Respol's Ecuadorian subsidiary;	20	0.0137
	Nov 2011, acquired 33 oil blocks in Galp Energy's Brazilian unit.	30	5.16
Sinopec	Oct 2010, purchased Brazilian subsidiary of Respol;	40	7.1
	2006, purchased Ominex de Colombia for oil sand.	25	0.4
CNOOC	Feb 2012, agreed to buy stakes in Brazil's Espirito Santo oil blocks from Perenco;	10	n/a
	Jan 2012, purchased Total SA's Colombia subsidiary;	n/a	1
	Jan 2012, acquired shares in 5 deep-water exploration blocks in Espirito Santo;	10	n/a
	Mar 2009, acquired assets of Argentina's Bridas Corp.	50	3.1
Sinochem	Mar 2010, purchased Brazil's Peregrino Oilfield from Statol;	40	3.07
	Aug 2009, acquired Emeralds' assets in Colombia, Syria;	100	0.878
	2003, purchased Block 16 in Ecuador from Conocophilips.	14	0.1

Source: Jiang and Chen, "Updates," 35–39.

Acquisition of overseas upstream assets is a main goal for Chinese NOCs. Seizing foreign oil assets through production entitlement is a long-term alternative to regular oil imports, which are vulnerable to changes in the international economic system, such as oil price fluctuations and exchange rate movements. In 2013, 59% of China's oil demand was satisfied by imports.[32] According to the 12th Five-Year Plan for Energy Industry Development, the Chinese government plans to cap the country's oil import dependency at 61% by 2015.[33] This means that China has to utilize more sustainable methods to obtain oil. While mergers and acquisitions is the most direct way to give Chinese oil companies long-term entitlements to overseas production, other modes of cooperation are also gaining significance in China's oil quest in Latin America.

After the global financial crisis, China has targeted long-term cooperation with a few oil-rich countries through the loan-for-oil supply project.

The financial meltdown caused rampant capital flight and deprived many oil-exporting countries of foreign investment necessary to develop their domestic oil industry. Under this background, these countries "were more eager to find money but reluctant to sell assets"[34] and some sought long-term supply arrangements with China. Although each loan-for-oil deal is unique, the project generally involves three parties—a Chinese bank as the financer, a Chinese NOC as the buyer, and an oil company from the resource-rich country as the supplier. In other words, the Chinese bank offers long-term financing to a foreign government or a foreign oil company from China's foreign reserve pool. In parallel, the foreign entity makes a long-term contract with a Chinese NOC to supply a guaranteed volume of oil at an agreed-upon price.

So far, China has signed loan-for-oil deals with nine countries, four of which are from Latin America. Chinese loans to Bolivia, Brazil, Ecuador, and Venezuela account for approximately 45% of the total amount of loan-for-oil deals and the majority goes to Brazil and Venezuela.[35] On the supply side, longstanding financing from China provides these countries with desired investment capital to develop their oil infrastructure and enhance their production capacity. Strategically, this alternative channel of fund helps these countries to diversify their oil trading partners.[36] For the four Latin American countries whose oil traditionally goes to North America, Chinese financing consequently lessens their reliance on the US market. Brazil, for example, is in desperate need for foreign capital to delve into its rich offshore oil reserve on its southern coast. Chinese loans greatly compliment the country's five-year investment plan of $174 billion for the new oil discovery.[37] In Venezuela, China has injected more than $40 billion into the Joint China-Venezuela Fund since 2008 to help the country build infrastructure and combat economic problems. In return, in addition to 250,000 barrels per day Venezuelan oil from regular trade and Chinese NOCs' entitlements, another 250,000 barrels are shipped to China every day to pay off the loans.[38] On the demand side, the long-term loan-for-oil arrangement serves to stabilize overseas oil supply, counteract depreciation risks of China's huge foreign reserves, evade oil trade barriers, and aid Chinese NOCs' "going global" venture. Table 4.4 summarizes China's long-term oil supply loans to Latin America since 2009.

In addition to upstream investments, China is also deepening its engagement in Latin America's oil industry in middle stream and downstream areas. NOCs "recognize that building refineries or pipelines can help them to quickly respond to local markets . . . and show their commitment to host countries, thus gaining credibility and strengthening relationships."[39] CNPC's subsidiary, BGP Inc. has been providing engineering services to Latin America since the early 1990s and so far has 179 service teams working in Venezuela, Ecuador, Cuba, Peru, Mexico, etc.[40] In 2012, CNPC won contracts on 3D seismic, drilling, and offshore mud logging for the Block Junin-4 project in

Table 4.4 China's Loan-for-Oil Deals in Latin America

Country, Date	*Lender*	*Borrower*	*Amount($)*	*Beneficiary*	*Details*
Bolivia Apr 2009	Ex-Im Bank	Government/ YPFB	1 billion 60 million		Build oil infrastructure (home gas lines, oil drilling rigs . . .).
Brazil Feb 2009	CDB	Petrobras	10 billion (6% interest)	Sinopec and its trading subsidiary Unipec	150 kb/d oil in 2009; 200–250 kb/d from 2010–2019 at market price.
Brazil Apr 2010	CDB	Petrobras	10 billion	Sinopec	Cooperation in deep-water exploration, production, refining and transport.
Ecuador Jul 2009	CDB	PetroEcuador	1 billion (6% interest)	CNPC/ PetroChina	96 kb/d for two years.
Venezuela Feb 2009	CDB	Bandes (PDVSA)	4 billion	CNPC/ PetroChina	200 kb/d to CNPC at market price, term contract, 1–2 billion discount, monthly invoice.
Venezuela Apr 2010	CDB	Bandes (PDVSA)/ Government	10 billion + RMB 70 billion	CNPC	Joint venture to develop Junin 4 block for production of 2.9 mb/d of heavy oil.
Venezuela 2012	CDB	PDVSA/ Government	10 billion		Develop the Orinco Oilfield; agricultural, industry, and housing projects, to be paid with 230 kb/d for 3 years.
Venezuela 2013	CDB	PDVSA/ Government	4 billion		Increase production level of Orinco fields, to be paid with 310 kb/d for 4 years.
Venezuela Jul 2014		PDVSA/ Government	4 billion		Further support to Orinco.

Abbreviations: CDB: China Development Bank; PDVSA: Petroleum of Venezuela; YPFB: Yacimientos Petrolíferos Fiscales Bolivianos; Sinopec: China Petroleum & Chemical Corporation; CNPC: China National Petroleum Corporation.

Sources: Jiang and Chen (2014), "Update"; Jiang and Sinton (2011), "Overseas"; MOFCOM (2009), "Bolivia Approves Contract with China for $60 million Loan," November 21; SeeNews (2010), "Petrobras negotiates USD 10 billion loan with China Development Bank"; Kevin P. Gallagher, Amos Irwin, and Katherine Koleski (2012), *The New Banks in Town: Chinese Finance in Latin America,* Inter-American Dialogue Report.

Venezuela.[41] Earlier this year, CNPC has expressed interest in a $4 billion gas pipeline project in Peru.[42] The Sinopec-Petrobras joint venture, financed by CDB, aims for cooperation in transportation and refinery in Brazil's oil fields. In Nicaragua, after the government granted a 50-year concession for a Chinese company to build and manage a cross-national canal, plans were revealed that an oil pipeline will be constructed linking the Caribbean and Pacific coasts.[43]

AN ANALYSIS OF MAIN ACTORS AND KEY DRIVERS IN CHINA'S OIL PURSUIT IN LATIN AMERICA

Undeniably, multiple actors and interests are at play in China's inroad in Latin America's oil development. Chinese oil companies spearhead China's oil investments in the region, as field operators or joint venture partners. Behind these giant NOCs are a multitude of government agencies that hold major shares, authorize production, certify overseas practices, and review performances of these companies. The Chinese Central government, in general, provides important political and strategic guidelines to NOCs. Moreover, Chinese NOCs need to comply with rules and laws of the host countries and cooperate with their local counterparts through joint ventures, mergers and acquisitions, etc. Questions still remain: who is in the driver's seat in China-Latin America oil cooperation? What factors motivate them to actively seek partnerships with Latin America?

As demonstrated in Part I, Chinese oil companies are state-owned and answer to the government as their biggest stakeholder. In spite of the state ownership, each of these major Chinese NOCs has a subsidiary (PetroChina, Sinopec Corp., CNOOC Ltd. and Sinochem Ltd.) listed on the New York and Hong Kong Stock Exchanges and all aim to raise their global influences. The parent companies (CNPC, Sinopec, CNOOC, and Sinochem) collect regular dividends from the listed arms and sometimes use them to offset losses elsewhere. The Chinese government's "going out" policy facilitates and supports NOCs' global ambitions by offering easy financing options, diplomatic assistance and tax rebates. Today, PetroChina is the world's 5th largest petroleum company in terms of oil production level[44] and Sinopec ranks 5th among 500 multinational corporations on the Fortune Magazine.

Many argue that Chinese NOCs, albeit state-owned, are not state-run. Erica S. Downs pinpoints that "China's oil majors—with their subsidiaries listed on foreign stock exchanges, global business portfolios, and vast profits earned from the high oil prices of recent years—are powerful and relatively autonomous actors with their own domestic and international interests that do not always coincide with those of the party-state."[45] Layers of state regulatory institutions often with overlapping functions created fragmented bureaucracy that leaves substantial room for NOCs to maneuver. While NOCs' general managers assume both commercial and political roles, they are given party ranks of ministers, the same position given to their majority shareholder—SASAC. In other words, managerial personnel at NOCs have more direct and concentrated administrative capacity than the bundle of government agencies designed to oversee NOCs activities separately.

CNPC's decision to internationalize its operations by investing in Peru in 1992 preceded the "go out" policy. Back then, Chinese government officials paid little attention to overseas upstream investments and CNPC's decision did not obtain government approval.[46] Instead, the company took the initiative to enter Latin America. On the grand level, oil companies have the same objective as the government, to expand oil reserves, diversify supplies and ensure energy security of the country. However, their interests are also deeply rooted in commercial grounds. NOCs want to reap the benefits of high energy prices and make profits on the international market to compensate their losses at home due to retail price control.

More importantly, they desire to improve their international profile, develop a matured supply chain, and obtain technical know-how and managerial capabilities.[47] At times when industry interests are at odds with government interests, NOCs managers are increasingly prioritizing "their economic identities over their political ones, representing corporate interests."[48] In 2006, as world crude oil prices rose, Sinopec and CNPC successfully persuaded the NDRC to impose a 15% increase on retail gasoline and diesel prices. Emphasizing economic interests at the cost of political ones is more obvious in their overseas ventures where their market shares are not under the influence of the command economy. Moreover, NOCs often compete with each other to acquire overseas assets. For instance, Sinopec collaborated with a Malaysian company MMC and outbid CNPC to build a Sudanese pipeline in 2003, while the latter had been the Chinese government's designated company to develop oil fields in Sudan since 1995.[49]

It is valid to say that Chinese NOCs have a mind of their own and have a certain degree of autonomy from the central government. Reforms of the 1990s and 2000s corporatized these companies and two decades of international experience shaped their behavior more toward standard practice of IOCs. Nevertheless, it will be unrealistic to assume that they have gained complete independence from the party-state. Being state-owned and given political functions, Chinese oil enterprises are intricately interconnected with the Chinese government through multiple channels.

To start, their transformation into corporate actors has deep institutional backgrounds in the central government's reforms to dissolve the companies' financial dependence and to create oligopolistic competition among them.

Although NOCs respond to market mechanisms better nowadays, their managers are still appointed by the Department of Organization and the SASAC. Their foreign investments in excess of $30 million need to be signed off by the NDRC and those over $200 million need approval by the State Council.[50] More recently, NOCs are "working with the Chinese government under the framework of the Shanghai Cooperation Organization to engage in dialogue and enhance co-ordination with host countries."[51] In Latin America, major loan-for-oil deals were often reached during official visits through *energy diplomacy*.

In cases of conflicted interests, although NOCs have increasingly chosen to focus on their commercial concerns, the central government has also taken political measures to counteract. In the Sudanese case, the Ministry of Commerce authorized CHINCA to negotiate an agreement between Sinopec and CNPC to moderate competition. Also, under the pressure to balance angry NOCs wanting to charge more and angry citizens over higher prices, Beijing collected "a windfall tax on (NOCs') profits to compensate farmers and cab drivers."[52] Rather than a passive owner, the state keeps its hands off NOCs' day-to-day operations but still retains control over essential matters, such as personnel appointment, long-term blueprint, and comprehensive plans of cooperation.

The pairing between the oil-hungry China and the oil-rich Latin America responds to the basic economics of demand and supply. But beyond economics 101, more profound strategic concerns are in action. After all, oil is a commodity crucial to national planning and sustaining of economic development. As the largest net oil importer, China is eager to secure overseas energy supply via stable partnerships. Security concerns and ethnic tensions in its traditional partners—Central Asia, Middle East and North Africa, pose great challenges to China and force it to search for new land of opportunities. Resource-abundant countries in Latin America, with tremendous untapped oil resources and relatively stable sociopolitical environment stand out in China's global oil pursuit.

On the other hand, in spite of great natural endowments, these countries are short of refining capacity and long for influxes of additional foreign investment as fund from the US and Europe dries out. China, with its economic might, suits their request to upgrade their infrastructure and production level. In this sense, the need to diversify becomes a two-way street. For China, Latin America is an attractive alternative in terms of geography and unconventional assets (Brazil deep-water oil). For Latin America, China represents a strong source of financing to develop its oil sector and dilute US influence. Under the current situation of current downward trend in global oil prices, coupled with the Mexican government's effort to liberalize the country's oil sector, Chinese oil companies could embrace an opportunity to lock in new oil reserves with generous contract terms as the global oil market shifts its leverage toward consumers.

AN ANALYSIS OF THE IMPLICATIONS OF CHINA FOR LATIN AMERICA'S ENERGY FUTURE

Undoubtedly, China's enormous demand for oil and primary commodities in general led to a "commodity boom" in Latin America during the first decade of the twenty-first century, contributing to its economic rebound. During the global financial crisis, improvements in terms of trade and subsequent increases in current account balance helped sustain economies of these resource-fertile countries. However, since the mid-2011, as the rising trend of commodity prices came to a halt, worries have been that the region's commodity exporters would experience mediocre growth and have their traditional economic problems (low savings/investment rate, low productivity, poor infrastructure) exposed. Meanwhile, surging resource nationalism and concerns about the manner in which Chinese NOCs operate have generated various sorts of criticisms toward China. While some allegations could be validated on a case-by-case basis, some are merely based on misconceptions and unsupported assumptions. In general, there are muddy waters with regard to implications of Chinese oil interests in Latin America.

In socioeconomic terms, attacks on China's involvement in Latin America's oil development often point to Chinese NOCs' employment practice and the way in which they deal with their oil resources acquired abroad. NOCs are denounced to hire only Chinese contracted workers and make little contribution to local employment. This might be true at the early stage of these companies' international footprints. But as they intensify their presence in Latin America, NOCs are more and more inclined to engaging locally as positive social players. Andes Petroleum Ecuador Ltd. of which CNPC holds 55% shares is reported to have employed over 453 local workers, over 95% of its total workforce since 2006.[53] Among them, about 100 have been promoted to managerial positions.

Due to its position as a net oil importer with ever-growing demand, China's entrance into Latin America's oil market is predicted to curtail oil available to other countries on the world market. While evidence is scarce about how much oil the Chinese government request NOCs to ship back home, custom statistics and news reports indicate that "most of this oil was sold on the open market to the highest bidder."[54] In 2013, Chinese NOCs produced an estimated 766.5 million barrels of oil overseas mainly through upstream M&A, of which 11% (84.3 mb) came from Latin America.[55] If hypothetically all oil entitlements from Latin America were to be brought back home, they would only make up 3.82% of the country's oil imports. Moreover, the 230,000 barriers per day of oil Chinese NOCs produced in Latin America only accounted for less than 2.41% of the region's daily production. In other words, the majority of Latin America's oil production does not go to China. Nevertheless, the Chinese government mandate to ease China's dependency on oil imports reflected in *the 12th Five-Year Plan for Energy Industry Development* might put pressure on NOCs to retrieve more foreign oil concessions since shipping home their entitlement oil can help evade fluctuations on the global oil market that are often encountered by oil imports. But even this final concern can be put at ease at least in the short run because of substantially lowered oil prices that give oil importers more buying power.

Other charges toward Chinese NOCs refer to the complex relationship they share with the Chinese government. Chinese state-owned oil oligarchs are suspected to use their political advantage to out-compete their Western counterparts, thus making the playing ground unfair. Although this assertion equates political stance with market competitiveness in an overly simplified way, it is not completely without merits. As of now, NOCs do not have the same level of penetration as major IOCs do in Latin America's oil market.

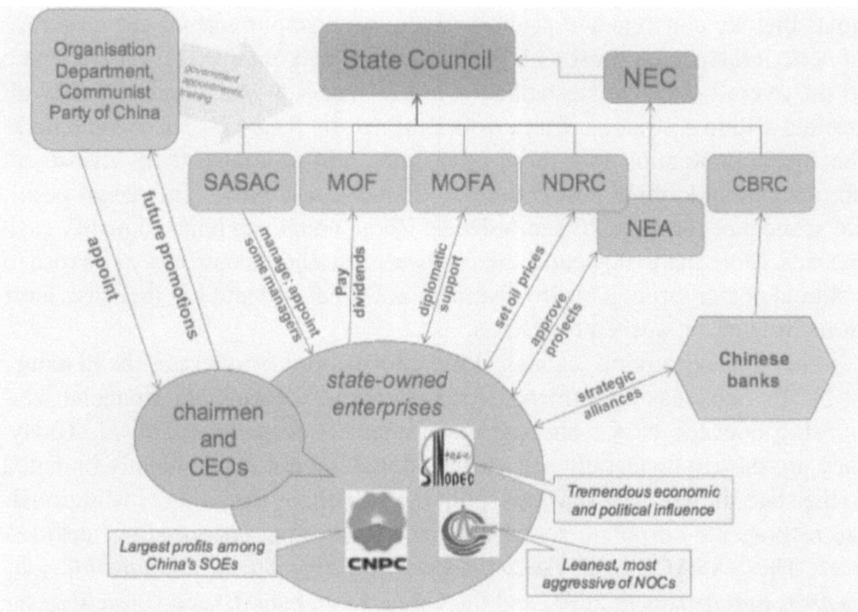

Notes: NEC = National Energy Commission; SASAC = State Assets Supervision and Administration Commission; MOF = Ministry of Finance; MOFA = Ministry of Foreign Affairs; NDRC = National Development and Reform Commission; NEA = National Energy Administration; CBRC = China Banking Regulatory Commission; SOE = state-owned enterprise.

Figure 4.2 NOCs' Complex Relationship with the Government. *Source*: Jiang and Sinton, "Overseas Investment," 25.

ExxonMobil, for instance, has a continuous business presence for about a century in Brazil and Colombia. Different from Chinese NOCs' focus on upstream M&A, ExxonMobil is also a strong participant in the downstream sector.[56] In South America, BP produced 30,000 barrels of oil liquids in 2003,[57] more than the Latin American oil production of all Chinese NOCs combined.

But as Chinese companies enter "a new phase of investing and operating abroad,"[58] they are quickly rising to global prominence. The level of competition NOCs will bring to the international oil market in the near future cannot be neglected. Further, since state-backed loan-for-oil deals carry political connotations, their beneficiaries—Chinese NOCs certainly, to some degree, gain leverage to compete with IOCs in acquiring overseas assets. Another "supera-market" advantage Chinese NOCs have is that as state-owned oil companies, especially the parent firms, they rarely have to respond to public shareholders. They are not required to pay a dividend to the government. Nevertheless, political leverage can prove to be a double-edged sword in some cases, adding another layer of complexity to the matter. For instance, dramatic changes in the global oil market can create profitability problems for beneficiaries of those loan-for-oil deals that settle on a given price.

Indeed, even though China's NOCs more often prioritize commercial goals that are concrete and profit-oriented, we cannot overlook the existence of state influence. Almost as a given, state-owned enterprises must respond to the overall objectives established in the state's 5-year plans and they all operate within a strategic framework outlined by SASAC. These facts tie to the fourth contention that these companies will politicize their investment abroad and undermine host countries' financial discipline. The recent political scandal of former CPC senior leader Zhou Yongkang tends to justify such worries. More than 100 people are implicated in Zhou's case and as he rose to political power through his involvement in China's oil and gas industry, most of them work or worked for NOCs.

Other than corruption scandals, little information corroborates the ill nature of NOCs' overseas investments. After corporate reforms, the financial relationship between NOCs and the government is *de facto* dissolved. Today, they are mostly financially self-sufficient and are not economically bounded to the state. In social dimensions, political guidelines can in fact institutionalize responsible corporate social behavior rather than encouraging recklessness. The SASAC issued *Guidelines on Fulfilling Social Responsibility by Central Enterprises* in 2008, and the China Exim bank issued *Guidelines for the Environmental and Social Impact Assessment of Loan Projects*.[59] Emphasizing that Chinese companies comply with all local rules and regulations, these policy guidelines offer official instructions for NOCs operating abroad. In Peru, CNPC has participated in the country's most recent (2004–2007) reconciliation of *Extractive Industry Transparency Initiative* (*EITI*) by disclosing its payments of oil resources.[60]

Lastly, China is accused to have cultivated the Dutch Disease and brought the resource curse to Latin America. There are rising concerns over Chinese oil investment regarding their negative consequences for environment and indigenous communities, as well as their bolstering of the region's left-leaning governments.[61] Meanwhile, different degrees of currency appreciations in Venezuela, Chile and Peru were also tied to difficulties in manufacturing exports, dangers of deindustrialization, and ultimately the Dutch Disease.[62]

Gallagher and Porzecanski, through studying industrial structure in Latin America and its bilateral trade with China, find only slight evidence supporting the contention of Dutch Disease. There has been no significant change in industrial structure of most commodity-exporting countries. Instead, the authors identify important complementarities between Chinese manufacturing exports and Latin American products, dismissing the proposition that China is outcompeting Latin American manufacturers on the world market. They further find that Latin American governments now have better fiscal mechanisms to transfer oil rents toward long-run development.[63]

Although China tends to reach loan-for-oil agreements with left-leaning governments in the region, it is rather diplomacy serving economic interests than the other way around. There is no indication that the Chinese government has any concrete role in supporting leftist regimes. China's focus at present is more about resource availability and regime stability instead of political rightfulness, as shown by a recent trend of Chinese NOCs moving their investment to resource-rich OECD countries. In Venezuela, Chinese oil investors are increasingly alarmed by the potential political instability and its possible attacks on Chinese oil interests.[64] With the long-hailed principle of "non-interference," the Chinese government is finding itself in a dilemma between protecting Chinese commercial and energy security interests and maintaining relationships with its strategic partners.

CONCLUSION

Predicted to become the world's largest oil consumer, China is committed to its global quest for oil. Over half of China's demand for oil is met by imports. This reality, combined with severe security and political setbacks in its traditional energy partners in the Middle East and Africa, results in an acute supply-demand conflict and requires China to seek new partners. Many Latin American countries, with their large untapped oil resources and relatively stable political environment, become a natural fit in China's search.

China's involvement in the region's oil business has expanded from upstream exploration and production to many mid- and down-stream activities, including engineering service, construction, refinery, transportation and storage. Its modes of cooperation also span from traditional merger and acquisition, joint ventures to government-backed loan-for-oil deals.

After two decades of industry restructuring, Chinese NOCs have corporatized and matured as competitive global players and China's main agents in oil cooperation with Latin America. In daily operations abroad, they act more like IOCs in setting their feet on commercial grounds. They expect to pump profits, gain technical and managerial know-how, improve reputation, and ultimately become internationally recognized enterprises. Assuming both commercial and political roles, they must reconcile these interests of their own and those of the Chinese government. In times of interest conflict, they are sometimes reported to prioritize their commercial concerns. Chinese NOCs are not puppets of the government, but they are not entirely independent from the state either. There is complex interdependence between NOCs and relevant government agencies. At the strategic level, NOCs must respond to government development plans and guidelines. After all, it was the state-launched reform that transformed identities and interests of Chinese NOCs.

There are mixed signals when it comes to assess the impacts of Chinese oil interests in Latin America. Although rapidly increasing their presence, Chinese NOCs, compared with IOCs are still not significant actors in Latin America's oil development. As they start to localize their workforce and engage in the local societies, these companies are merging with the global standards of social responsibility. With a limited production capacity, NOCs are not "locking up" oil resources, nor are they out-competing IOCs. Nonetheless, while it is obvious these companies are not state-run, it is still uncertain in which direction political influence and support can shape their behavior in the region. So far there has been little concrete evidence about China's contribution to emergence of the Dutch Disease and the resource curse. But since Chinese NOCs are relatively new players in Latin America, it is too soon to tell whether they are beneficial or harmful and further research is needed to study the evolving nature of their oil investments.

Under current developments in the global oil market where prices have been steadily falling since late 2014, China's appetite for oil has not been curtailed. Instead, Chinese demand for oil is projected to grow in 2015 and 2016. While the price slump can create debt issues for certain Chinese overseas investments (especially for those oil-for-loan deals in which prices are fixed), it should be highlighted that some Chinese loans to Latin American characterize floating prices and/or are purposed to support oil infrastructures. In this case, the effects of falling prices can be rather advantageous. In fact, with less money to spend on oil imports, Chinese NOCs can devote more capital to optimizing their overseas investments. Technical upgrading is urgently needed in Latin America's oil sector. Moreover, reduced demand for imports in other conventional markets (the North American oil market, for example) could possibly push Latin American oil producers closer to China's orbit. In this sense, an oil price downturn can bring opportunities for Chinese NOCs as they speed up securing their bids with more generous contract terms. During times of change, there might be more to gain for China in Latin America's energy business.

NOTES

1. "1" means "one plan," as the 2015–2019 China-Latin American Countries and Caribbean States Cooperation Plan; "3" means "three engines" of promoting comprehensive development, namely trade, investment, and financial cooperation; "6" means "six fields" of priority, referring to energy and resources, infrastructure construction, agriculture, manufacturing, scientific and technological innovation, and information technologies.
2. "Xi Jinping Attends China-Latin America and the Caribbean Summit and Delivers Keynote Speech, Comprehensively Expounding China's Policies and Propositions Toward Latin America, Announcing Establishment of China-Latin America Comprehensive Cooperative Partner," *Ministry of Foreign Affairs of the PRC*, July 18, 2014, accessed July 22, 2014, http://www.fmprc.gov.cn/mfa_eng/wjdt_665385/wshd_665389/t1176650.shtml.
3. Deisy Buitrago and Andrew Cawthorne. "Venezuela Gets $4 Billion from China in Exchange for Oil," *Petroleum World*, July 23, 2014, accessed July 24, 2014, http://www.petroleumworld.com/storyt14072301.htm.
4. "Short-Term Energy Outlook September 2014," US Energy Information Administration, accessed September 9, 2014, http://www.eia.gov/forecasts/steo/archives/sep14.pdf.
5. "China Is Now the World's Largest Net Importer of Petroleum and Other Liquid Fuels," US Energy Information Administration, last modified March 24, 2014, http://www.eia.gov/todayinenergy/detail.cfm?id=15531.
6. Julie Jiang and Chen Ding. "Update on Overseas Investments by China's National Oil Companies," *International Energy Agency* (2014): 36, accessed August 2, 2014, http://www.iea.org/publications/freepublications/publication/partner-country-series—update-on-overseas-investments-by-chinas-national-oil-companies.html.
7. Jiang and Chen, "Updates," 35.
8. "中国石油工程百年发展历程 (A Hundred Years of Chinese Oil Industry)," June 4th, 2009, *CNPC News Center*, accessed September 9th, 2014, http://center.cnpc.com.cn/bk/system/2009/06/04/001240529.shtml.
9. "A Hundred Years."
10. Trevor Houser, "The Roots of Chinese Oil Investment Abroad," *Asia Policy* 17 no.5 (2008): 144, accessed July 28, 2014, doi:10.1353/asp.2008.0009.
11. Houser, "The Roots," 144.

12. Shiyun Zhang and Huaiyu Yuan, "中国石油海外投资现状、问题及对策 (China's Overseas Petroleum Investment: Situation, Problems, and Policy Suggestions," *Natural Resource Economics of China* 11 (2005): 27.

13. Zhang and Yuan, "China's Overseas," 27.

14. Fangtang Wang, "60年新中国经济史记：行业本纪之能源业 (The PRC's 60 Years of Economic History: The Energy Industry)," *Ifeng Finance*, 2009, http://finance.ifeng.com/news/special/sytrq60th/.

15. Jin Zhang, *Catch-up and Competitiveness in China: The Case of Large Firms in the Oil Industry* (London: Routledge Cruzon, 2004), 101.

16. Xuchao Yan, 中国石油大重组 *(The Massive Restructuring of the Chinese Oil Industry)* (Beijing: Petroleum Industry and Trade, 1998), 1–3.

17. Chih-shiian Liou, "Bureaucratic Politics and Overseas Investment by Chinese State-Owned Oil Companies: Illusory Champions," *Asian Survey* 49, no. 4 (2008): 675–6.

18. Houser, "The Roots," 146.

19. Liou, "Bureaucratic Politics," 683–4.

20. Houser, "The Roots," 148.

21. Chietigj Bajpaee, "Chinese Energy Strategy In Latin America," *China Brief* 5, no. 14 (2005), http://www.jamestown.org/single/?no_cache=1&tx_ttnews%5Btt_news%5D=3870#.VBdD_y5dUpA.

22. "Mexico Overview," US Energy Information Administration, accessed September 10, 2014, http://www.eia.gov/countries/cab.cfm?fips=MX.

23. "Colombia Overview," US Energy Information Administration, accessed September 10, 2014, http://www.eia.gov/countries/cab.cfm?fips=CO.

24. "Venezuela Overview," US Energy Information Administration, accessed September 10, 2014, http://www.eia.gov/countries/cab.cfm?fips=VE.

25. "Major Events: 1991–2001," CNPC, accessed August 7, 2014, http://classic.cnpc.com.cn/en/aboutcnpc/companyprofile/history/MajorEvents/2001.shtml.

26. "Major Events."

27. Houser, "The Roots," 156.

28. Jiang and Chen, "Update," 7.

29. CNPCETRI, *Report on Domestic and Overseas Oil and Gas Industry Development 2012*, Beijing, 2013.

30. Jiang and Chen, "Update," 6.

31. Ibid.

32. "Short-term Energy."

33. KPMG China, *China's 12th Five-Year Plan : Energy*, 2011.

34. Julie Jiang and Jonathan Sinton, "Overseas Investments by Chinese National Oil Companies: Assessing the Drivers and Impacts," *International Energy Agency* (2011): 22, accessed July 20, 2013, http://www.iea.org/publications/freepublications/publication/overseas_china.pdf.

35. Data come from Jiang and Chen, "Update," and Jiang and Sinton, "Overseas."

36. Jian Gao, Dan Yang, and Xiucheng Dong. "贷款换石油:中外石油合作新模式 (Loan-for-Oil: New Mode of Foreign Oil Cooperation)," *International Business Cooperation*, no. 10 (2009), http://big.hi138.com/jingjixue/guojijingji/201001/209172.asp.

37. Gao, Yang, and Dong, "Loan-for-Oil."

38. Lily Kuo, "Venezuela Just Mortgaged Another $4 Billion of Its Oil Exports to China," *Quartz*, July 24, 2014. http://qz.com/239106/venezuela-just-mortgaged-another-4-billion-of-its-oil-exports-to-china/.

39. Jiang and Sinton, "Overseas," 14.

40. "中国石油在拉美 (CNPC in Latin America)," *CNPC*, accessed September 15, 2014.

41. "CNPC in Venezuela," *CNPC Worldwide*, accessed September 15, 2014, http://classic.cnpc.com.cn/en/cnpcworldwide/venezuela/.

42. Paddy Harris, "China Eyes Latin America Oil and Gas Potential," *Oil & Gas Technology*, May 30, 2014, http://www.oilandgastechnology.net/upstream-news/china-eyes-latin-america-oil-gas-potential.

43. Ivan Castro, "Nicaraguan Assembly OKs $40 Billion Chinese Canal to Rival Panama's," *Reuters*, June 13, 2013, http://www.reuters.com/article/2013/06/13/us-nicaragua-canal-idUSBRE95C1HI20130613.

44. "25 Biggest Oil Companies," *Forbes*, accessed August 9, 2014, http://www.forbes.com/pictures/mef45glfe/5-petrochina-4-4-million-barrels-per-day-3/.

45. Erica S. Downs, *Who's Afraid of China's Oil Companies?* (2013): 74.

46. Jiang and Sinton, "Overseas," 13.

47. Julie and Sinton, "Overseas," 13.

48. Liou, "Bureaucratic," 677.

49. Ibid, 679–82.

50. Downs, "Who's," 76.

51. Jiang and Chen, "Update," 8.

52. Houser, "The Roots," 153.

53. "CNPC," 11.

54. Houser, "The Roots," 162.

55. Jiang and Chen, "Update," 13–14.

56. "South America," ExxonMobil, accessed September 18, 2014, http://corporate.exxonmobil.com/en/company/worldwide-operations/locations.

57. BP, *Annual Report and Form 20-F 2013* (2013), 248.

58. Jiang and Chen, "Update," 33.

59. Adina Matisof, *Crude Beginnings: An Assessment of China National Petroleum Corporation's Environmental and Social Performance Abroad*, Friends of the Earth Report, San Francisco, February 2012, 18.

60. Matisof, "Crude," 23.

61. Jonathan Watts, "China's Exploitation of Latin American Natural Resources Raises Concern," *The Guardian*, March 26, 2013, http://www.theguardian.com/world/2013/mar/26/china-latin-america-resources-concern.

62. Rogelio Núñez, "La Enfermedad Holandesa Acecha a Varios Países Latinoamericanos," *Infolatam*, March 04, 2013, accessed September 18, 2014, http://www.infolatam.com/2013/03/04/la-enfermedad-holandesa-acecha-a-varios-paises-latinoamericanos/.

63. Kevin P. Gallagher, and Roberto Porzecanski, *China and the Latin America Commodities Boom: A Critical Assessment*, Political Economy Research Institute, Working Paper 192 (2009): 17–24.

64. Jiang and Chen, "Update," 34.

BIBLIOGRAPHY

Bajpaee, Chietigj. "Chinese Energy Strategy in Latin America." *China Brief* 5, no. 14 (2005). http://www.jamestown.org/single/?no_cache=1&tx_ttnews%5Btt_news%5D=3870#.VBdD_y5dUpA.

"Bolivia Approves Contract with China for $60 Mln Loan." *Ministry of Commerce People's Republic of China (MOFCOM)*, November 21, 2009. http://ng2.mofcom.gov.cn/article/chinanews/200911/20091106629189.shtml.

BP. *Annual Report and Form 20-F 2013*, 2013. http://www.bp.com/content/dam/bp/pdf/investors/BP_Annual_Report_and_Form_20F_2013.pdf.

Buitrago, Deisy, and Andrew Cawthorne. "Venezuela Gets $4 Billion from China in Exchange for Oil." *Petroleum World*, July 23, 2014. Accessed July 24, 2014. http://www.petroleumworld.com/storyt14072301.htm.

Castro, Ivan. "Nicaraguan Assembly OKs $40 Billion Chinese Canal to Rival Panama's." *Reuters*. June 13, 2013. http://www.reuters.com/article/2013/06/13/us-nicaragua-canal-idUSBRE95C1HI20130613.

CNPC. "Major Events: 1991–2001." Accessed August 07, 2014. http://classic.cnpc.com.cn/en/aboutcnpc/companyprofile/history/MajorEvents/2001.shtml.

"CNPC in Venezuela." *CNPC Worldwide*. Accessed September 15, 2014. http://classic.cnpc.com.cn/en/cnpcworldwide/venezuela/.

CNPCETRI. *Report on Domestic and Overseas Oil and Gas Industry Development 2012*. Beijing 2013.

Downs, Erica S. *Who's Afraid of China's Oil Companies?* (2013): 73–102.

ExxonMobil. "South America." Accessed September 18, 2014. http://corporate.exxonmobil.com/en/company/worldwide-operations/locations.

Forbes. "25 Biggest Oil Companies." Accessed August 9, 2014. http://www.forbes.com/pictures/mef45glfe/14-petrobras-2-6-million-barrels-per-day-3/.

Gallagher, Kevin P., Amos Irwin, and Katherine Koleski. *The New Banks in Town: Chinese Finance in Latin America*, Inter-American Dialogue. 2012. http://www.thedialogue.org/PublicationFiles/TheNewBanksinTown-FullTextnewversion.pdf.

Gallagher, Kevin P., and Roberto Porzecanski. *China and the Latin America Commodities Boom: A Critical Assessment*. Political Economy Research Institute. Working Paper 192, 2009.

Gao, Jian, Dan Yang, and Xiucheng Dong. "贷款换石油:中外石油合作新模式 (Loan-for-Oil: New Mode of Foreign Oil Cooperation)." *International Business Cooperation*, no. 10 (2009). http://big.hi138.com/jingjixue/guojijingji/201001/209172.asp.

Harris, Paddy. "China Eyes Latin America Oil and Gas Potential." *Oil & Gas Technology*, May 30, 2014. http://www.oilandgastechnology.net/upstream-news/china-eyes-latin-america-oil-gas-potential.

Houser, Trevor. "The Roots of Chinese Oil Investment Abroad." *Asia Policy* 17, no. 5 (2008): 141–66. Accessed July 28, 2014. doi:10.1353/asp.2008.0009.

Jiang, Julie, and Chen Ding. "Update on Overseas Investments by China's National Oil Companies." *International Energy Agency,* 2014. Accessed August 2, 2014. http://www.iea.org/publications/freepublications/publication/partner-country-series—update-on-overseas-investments-by-chinas-national-oil-companies.html.

Jiang, Julie, and Jonathan Sinton. "Overseas Investments by Chinese National Oil Companies: Assessing the Drivers and Impacts." International Energy Agency, 2011. Accessed July 20, 2013. http://www.iea.org/publications/freepublications/publication/overseas_china.pdf.

KPMG China. *China's 12th Five-Year Plan: Energy*, 2011.

Kuo, Lily. "Venezuela Just Mortgaged Another $4 Billion of Its Oil Exports to China." *Quartz*, July 24, 2014. http://qz.com/239106/venezuela-just-mortgaged-another-4-billion-of-its-oil-exports-to-china/.

Liou, Chih-shiian. "Bureaucratic Politics and Overseas Investment by Chinese State-Owned Oil Companies: Illusory Champions." *Asian Survey* 49, no. 4 (2008): 670–90.

Matisof, Adina. *Crude Beginnings: An Assessment of China National Petroleum Corporation's Environmental and Social Performance Abroad*. Friends of the Earth Report. San Francisco, February 2012.

Núñez, Rogelio. "La Enfermedad Holandesa Acecha a Varios Países Latinoamericanos." *Infolatam*, March 04, 2013. Accessed September 18, 2014. http://www.infolatam.com/2013/03/04/la-enfermedad-holandesa-acecha-a-varios-paises-latinoamericanos/.

"Petrobras Negotiates USD 10bn Loan with China Development Bank." *SeeNews Energy*, May 27, 2010. http://energy.seenews.com/news/petrobras-negotiates-usd-10bn-loan-with-china-development-bank-report-70772.

US Energy Information Administration. "Venezuela Overview." Last Modified June 20, 2014. http://www.eia.gov/countries/cab.cfm?fips=VE.

US Energy Information Administration. "Mexico Overview." Last Modified April 24, 2014. http://www.eia.gov/countries/cab.cfm?fips=MX.

US Energy Information Administration. "Colombia Overview." Last Modified January 7, 2014. http://www.eia.gov/countries/cab.cfm?fips=CO.

US Energy Information Administration. *STEO Archives*. Last Modified September 9, 2014. http://www.eia.gov/forecasts/steo/outlook.cfm#issues1996.

US Energy Information Administration. *Short-Term Energy Outlook September 2014*.

US Energy Information Administration. "China Is Now the World's Largest Net Importer of Petroleum and Other Liquid Fuels." March 24, 2014. Accessed September 9, 2014. http://www.eia.gov/todayinenergy/detail.cfm?id=15531.

Wang, Fangtang. "60年新中国经济史记：行业本纪之能源业 (The PRC's 60 Years of Economic History: The Energy Industry)." *Ifeng Finance*, 2009. http://finance.ifeng.com/news/special/sytrq60th/.

Watts, Jonathan. "China's Exploitation of Latin American Natural Resources Raises Concern." *The Guardian*, March 26, 2013. http://www.theguardian.com/world/2013/mar/26/china-latin-america-resources-concern.

"Xi Jinping Attends China-Latin America and the Caribbean Summit and Delivers Keynote Speech, Comprehensively Expounding China's Policies and Propositions Toward Latin America, Announcing Establishment of China-Latin America Comprehensive Cooperative Partner." *Ministry of Foreign Affairs of the PRC*, July 18, 2014. http://www.fmprc.gov.cn/mfa_eng/wjdt_665385/wshd_665389/t1176650.shtml.

Yan, Xuchao. 中国石油大重组 *(The Massive Restructuring of the Chinese Oil Industry)*. Beijing: Petroleum Industry and Trade, 1998.

Zhang, Jin. *Catch-up and Competitiveness in China: The Case of Large Firms in the Oil Industry*. London: Routledge Cruzon, 2004.

Zhang, Shiyun, and Huaiyu Yuan. "中国石油海外投资现状、问题及对策 (China's Overseas Petroleum Investment: Situation, Problems, and Policy Suggestions)." *Natural Resource Economics of China* 11 (2005): 27–29, 49.

"中国石油工程百年发展历程 (A Hundred Years of Chinese Oil Industry)." June 4th, 2009. *CNPC News Center*. Accessed September 9, 2014. http://center.cnpc.com.cn/bk/system/2009/06/04/001240529.shtml.

"中国石油在拉美 (CNPC in Latin America)." *CNPC*. Accessed September 15, 2014.

Part II

ENERGY AND EMERGING ECONOMICS IN THE 21st CENTURY

Chapter 5

Russian Energy Policy

Implications for Global Energy Security

Dina Moulioukova-Fernandez and Roger E. Kanet

Russia currently cannot be considered an influential player in the global energy market, despite its significant exports and the heavy reliance of Europe on those exports. We will argue that the domestic and international vulnerabilities of Russia in its energy sector make it rather a price taker and not an influential energy player on a global scale.[1] We believe that Russia could successfully address these issues through the institutions of global energy governance. Russia's preference, however, would be in the creation of a new energy regime, most likely under the umbrella of the BRICS, rather than participating in institutions constructed by the West.

We will develop our overall argument in four parts. In the first we provide a brief overview of what we consider to be Russia's core priorities and strategy and its vision of the future. We will further list domestic factors and limitations on Russia's energy policy in the context of its overall strategy and its use of energy for political purposes. In the third part, we briefly outline international factors that in our opinion shape Russia's behavior in its energy sector. Finally, we will pay particular attention to Russia's support for the construction of a new regime of energy governance, namely through collaboration with the BRICS.

RUSSIA'S STRATEGY AND PRIORITIES

The idea that an inherited "reputation for power" can constitute part of a state's overall portfolio of power resources is not new. The international relations scholar A.F.K. Organski noted that states "that have been great powers in the past may continue to trade upon their reputation for some time."[2] International prestige is a key asset in complementing, reinforcing and underpinning such a reputation. International prestige by itself is socially constructed as it is "a function of other actor's perceptions."[3] Such perceptions could be based, in part, on material factors. Some analysts argue, however, that the skill with which the state's leaders position prestige in the key networks of international life is equally important.[4]

In 2000, the Russian leadership set out its view of Russia's problems and the priorities for tackling them in a public document called "Russia at the Turn of the Millennium."[5] This clearly became the basis for a political and economic strategy that arguably predetermines Russia's behavior in the international arena. The basic claim for Great Power status was evident in this document that argued that "Russia was and will remain a great power. It is preconditioned by the inseparable characteristics of its geopolitical, economic and cultural existence. They determined the mentality of Russians and the policy of the government throughout the history of Russia and they cannot but do so now."[6]

The first of such priorities was, and continues to be, domestic consolidation. This view rests on the belief that domestic strength leads to greater international prestige, to respect and the willingness of other states to cooperate with Russia on an equal basis. For Russia, domestic power would be based on a "strong state" at home, which President Putin claimed is compatible with healthy democracy: "For Russians a strong state is not an anomaly to be discarded. Quite the contrary, they see it as the source and guarantee of order, and the initiator and the main driving force of change."[7]

Russia's idea of "healthy democracy" is tied to the socially constructed notion of "sovereign democracy." Martin Smith underlines three key and interrelated themes that underpin it.[8] The first is the importance of strengthening Russia, both domestically and internationally, in order to reduce the risks to its security. The second, closely related to the first, is the focus on Russia's sovereignty. The third is an affirmation that the Russian government would pursue its own course that is not to be dictated by others. It was apparent that Putin noted that the strength of Russia's sovereignty would increase with its ability to determine its own course both domestically and internationally. Putin's speech of 2005 should be taken in the context of his predecessor. While Russia was not perceived as fully sovereign in taking its decisions on the international arena during Yeltsin's presidency, it was under way during Putin's. Putin asserted: "It is precisely our values that determine our striving to enhance Russia's independence as a state and to strengthen its sovereignty. We are a free nation, and our place in the modern world—I wish to stress this particularly will be determined only by how strong and successful we are."[9] The core message of Putin's address was that Russia is acquiring "real" sovereignty, based on a new determination to pursue its own developmental path, constituted the essence of sovereign democracy.[10] It was designed, as Surkov put it, to give Russia back its status as a formidable power that is not independent on paper only.[11] Russia's concept of itself as a democratic state rests on the capability of prospering without undue reliance on other states. For Smith this entails limited engagement internationally, mainly to ensure that Russia's core interests are not seriously threatened.

DOMESTIC FACTORS

The Russian Federation is the world's second country in annual production of natural gas after the United States and is second only to Saudi Arabia in crude oil production.[12] It is hard to underestimate, therefore, the importance of Russia's energy sector for the global security of energy supply. Clearly oil and gas are considered key to Russia's return as an important player in the international arena. It is believed by some analysts that oil wealth will help it to regain its "superpower status."[13] It would seem, therefore, that abundance in Russia's energy sector would strengthen its claim to sovereignty and self-reliance. The reality, however, is far more complex.

First, energy could be considered as a major "value pump for its economy." Based on the figures provided by the Russian Statistics Agency, energy has been accounting for over 70% of its overall exports. Gaddy and Ickes note that Russia's political economy throughout its modern history has centered on the transfer of value from the resource sector to other parts of its economy. Resources continue to fuel Russia's economy and polity, as they did in Soviet times. They further argue that the current state of the Russian economy is structurally inefficient and impossible to modernize.[14] The economic structures bequeathed by central planning cannot compete in a market setting. The Russian fuel and energy sector serves as the generator and support of the Russian economy. In 2011, the non-oil sector current account deficit reached a record 13% of GDP, confirming the country's dependence on Russia's energy export sector.[15] We could argue that the performance of the Russian energy sector is crucial to support an otherwise unsustainable economy. Moreover, in addition to driving the Russian economy, this "value pump" finances the security net to millions of citizens who are employed by nonviable enterprises and depend on the state for heavily subsidized health and educational systems.

Second, although Russia's energy reserves have proven to be vast, so are the geological challenges that must be overcome to extract them.[16] Most of the newly discovered fields lie deeper than earlier finds. Most of the recently discovered fields are offshore and in permafrost zones (Sakhalin deposits and the Shtokman gas field in the Barents Sea). All these factors substantially influence the extraction costs and uncertainty about profitability of such explorations.

Third, we argue that the energy sector is one of the major contributing factors to Russia's "resource curse." The literature on the "resource curse" does not suggest that resource abundance is to blame for poor economic performance, conflict and authoritarianism. For example, the revenues from natural resources contributed to economic development in Canada. Instead, it suggests that it is the lack of appropriate institutions and political structures that is to blame. It could be argued, however, that the presence of resource wealth often undermines the creation of the right state structures and makes the state susceptible to the dangers associated with the volatility of global energy and financial markets. Thus, Russia finds itself in a "catch 22" situation, when it needs to promote diversification of its economy, while the only way this could be accomplished is through the delivery of the rents from energy sector.[17] As noted by Pavel Baev, the central paradox of this strategy of overcoming the resource curse is that the necessary volume of revenue in the energy sector can only be generated by massive new investment in the upstream production, so priority in revenue allocation effectively cannot be changed.[18]

Joe Hellman puts responsibility for Russia's uncompetitive resource-dominated economy on elite networks that eliminate competition and stall economic and political development in Russia. The Russian elites are characterized by rent-seeking behavior where "individuals seek returns from state-sanctioned monopoly rights."[19] As a result rent-seeking and authoritarianism provide "partial reform equilibrium" in Russia, according to Joel Hellman.[20] Partial reform equilibrium takes place when short-term winners, instead of supporting reforms, stall the economy in an equilibrium that generates concentrated rents for themselves with high costs on the rest of the society.[21] Despite the hope of its temporary nature, such equilibrium has proven to be internally coherent and stable in Russia—at least until the recent collapse of petroleum prices on the world market.

On the level of corporate governance, state companies started playing a more prominent role in the economy after Mr. Putin's rise to power. This has been a departure from the Yeltsin period, when state companies were losing ground and were sold off to private owners. Especially in the energy sector, under Putin, the role of such companies as Gazprom and Rosneft rose and the boards of these companies were joined by people very close to the President's circle. In 2004 the government took a decision to transfer a series of the largest Russian companies to the direct control of the government itself. These companies covered three major sectors of the economy[22] and the fuel and energy complex, along with electricity and atomic energy, was one of them. People from Putin's team began to systematically join the board of directors of these companies. Some of these officials have a visible public profile. For example then chief of staff, Dmitry Medvedev, became chairman of the board of Gazprom and former Prime Minister Sergei Kirienko went to the Russian nuclear energy state company, Rosatom. In 2012, Deputy Prime Minister Igor Sechin left the government and was appointed to the board that oversees all of Russia's state energy holdings he then reclaimed the chairmanship of Rosneft while remaining an advisor to President Putin. He also was appointed as a secretary of a presidential commission on energy, which supervises the energy industry in Russia.[23]

Therefore, Russia's economic system can be referred to as "state capitalism," where the state dominates markets. Some claim that such dominance is primarily for political gain.[24] Although business diplomacy does matter to Russia, it is usually through economic means that foreign objectives can be met by influencing other counties to align their policies with Russia. This has been argued to be the case in many instances with Russian energy policy.[25]

However, the connection is reciprocal; it can be argued that the presence of Russian officials on the boards of major companies allows them to use their position for the advancement of business interests. However, the number one priority for any business is to profit and seek markets for its goods. This priority may conflict with foreign policy imperatives of the state. Since influence in Russia flows in both directions, from the state to business and vice versa, the question rises whose interests are being ultimately served. It is unclear whether the board members would be loyal to the Kremlin or the corporate interests they are charged with overseeing.[26] It is a mistake, therefore, to argue that Russian state owned companies are driven purely by foreign policy imperatives.[27] In effect, the very nature of state capitalism makes the governance of the energy sector less efficient and prone to corruption.

Another relevant factor is Russia's desire to modernize and diversify its economy away from its dependence on resource exports. The paradoxical situation, as noted by Michael Bradshaw, is the need for Russia to continue to invest in its fossil fuel energy economy to generate the income that the country needs to finance its programs to diminish Russia's reliance on energy and volatility of energy markets. He further argues that the first half of the last decade may be seen now as a missed opportunity to apply energy revenues to modernize the country's infrastructure and promote diversification of its economy. It seems that Russia has successfully withstood the storm of global financial crisis. However, the costs of Western sanctions because of the Ukraine crisis and the dramatic fall in the global price of oil have already begun to create serious economic problems for Russian efforts at modernization.

In addition to domestic vulnerabilities caused by its energy sector, Russia supplies less than 10% of global oil or gas. With such a low market share it is hard for Russia achieve to "superpower" status in leveraging the global market. Peter Rutland summarizes this point by noting that only through collusion with other producers could there be any risk of serious influence by Russia in cornering the energy market. As a non-member of Organization of the Petroleum Exporting Countries (OPEC), Russia has not even attempted to coordinate its actions with the Organization and there has been little success in creating a natural gas version of OPEC. Rutland adds that overdependence on energy diminishes political stability and, in the absence of effective political institutions, usually means slower economic growth.[28]

THE INTERNATIONAL LEVEL AND GLOBAL GOVERNANCE

The volatility of energy markets is one of the major concerns in today's global economies. The geography of oil and gas production has shifted predominantly to non-OECD countries and states that have preferential treatment of their National Oil Companies over international oil companies. It is unfortunate that most of the current discussion of global energy security has centered on the security of supply, since for the producing states the security of demand and transit are equally important issues. In the case of Russia, there is a concern that Russia is overly reliant on exporting non-renewable resources. This puts Russia in a subordinate relationship with its customers and ultimately undermines its sovereignty. As was demonstrated during the long-term Russia-Ukrainian gas conflict, any potential energy weapon is a "double-edged sword." The "gas wars" of 2006 and 2009 showed that the relationships between energy producers and consumers are essentially symbiotic. Shutting off the supply of the gas to Ukraine was not only risky for Russia in economic terms, it also undermined Russia's reputation as a reliable supplier and encouraged European states to look for alternative sources of supply.[29]

The gas wars outlined a few negative aspects of such dependency. First, from a revenue standpoint, Russia's sales to Europe were reduced. Simon Pirani estimates these losses to have reached around $100 million per day.[30] More importantly, as a result of the crisis, Russia irrevocably damaged its more than 40-year reputation as a reliable energy supplier in the eyes of its European partners. The importance of the reputational factor in this conflict is difficult to underestimate, considering that energy revenues are vital for the Russian economy and that the European states represent its major customers. Moreover, we can argue that the negative impact of the dispute with Ukraine on Russia's reputation damaged both its immediate and its long-term business interests. First of all, it contributed to the intensification of the discourse inside the EU on the need for the diversification of its gas supplies with the goal of decreasing its dependence on Russia—for example by building the planned Nabucco pipeline.[31] As a result of such diversification, Russia could potentially experience a decrease of its share on European markets and face substantial losses in revenue. Secondly, the escalation of the conflict negatively affected the long-term strategic business interests of Gazprom in gaining access to additional customers.[32] According to Feklyunina, Gazprom had had some success in obtaining access to gas distributional systems in Southern Europe and the Netherlands. However, when it attempted to bid for the UK's largest distribution company, Centrica, it faced a wave of negative publicity that prevented it from successfully establishing its position in the UK.[33]

The institutional dimension, or lack of institutionalization—the balancing of supply and demand[34]—is another important international factor that affects Russia's energy security. There is a plethora of overlapping institutions and regimes that attempt to address the issue of energy market coordination. The G8+5 (Canada, France, Germany, Italy, Japan, Russia, UK, US) "+5" (Brazil, China, India, Mexico, South Africa), which was closely supported by the International Energy Agency (IEA) from 2005 to 2009, addressed some of the energy related issues at its various summits. The "+5," however, has never been dealt with on the basis of equality with the G8 and, when the G20 was born, G8+5 ceased to exist. However, the G20, which includes, in addition to G8+5, Argentina, Australia, Indonesia, Korea, Saudi Arabia and the European Union is a meeting place and not an agency. It requires strong supporting institutions to be the engine for effective energy cooperation.

The most influential organizations in the oil sector are arguably the IEA and OPEC founded to counterbalance and offset each other's influence. The IEA has the necessary institutional design and is considered by far the most substantial and influential body for international energy cooperation. It was established in 1970 by Organization for Economic Co-operation and Development (OECD) states as predominantly a consumer body. The IEA is a treaty organization and the provisions of its treaty confine its membership to countries which are already members of the OECD. Almost all OECD members (28 in all) now belong to IEA. When the IEA was founded, these countries accounted for the great preponderance of international oil demand. Today, IEA members account for only about half of world oil demand and this ratio is declining steadily.[35]

One of the weaknesses in the structure of global energy governance is the divide between developed and developing countries and the outdated format of the institutions involved in energy governance. As a result, some of the key players are excluded from consumer cooperation on energy security and supply. This partly is a question of formal mechanisms, but also concerns the mutual understanding and trust that is needed to handle difficult situations.[36]

RUSSIA AND WESTERN REGIMES

It could be argued that Russia's relations with the West emerged through two main stages. In the first, Russia attempted to construct its image as a Western state by aligning itself with Western powers through aspirations to participate in the regimes constructed by the West. This period was short-lived, in large part because of the weakness of Yeltsin's Russia and the Western perception of Russia as a liability rather than a strategically important partner. Such a perception resulted in the expansion of the West's sphere of influence by directly encroaching on Russia's neighborhood: the expansion of NATO, the EU and its launching of its soft power projects in Eurasia. The launching of the EU's Neighborhood Policy and Eastern Partnership that directly targeted former Soviet Union Republics (the sensitive areas of special interest for Russia), including the Association Agreement with Ukraine that sparked the current crisis in Ukraine, have been interpreted by Russia as direct threats to its interests.[37]

With the growth of oil prices and, consequently, the revenues that Russia received from them Russia became a different player than it was during its "pro-Western" post-Soviet period with its narrative for equality and respect. The watershed in Russian foreign policy[38] came with Putin's speech in Munich in 2007,[39] which announced Russia as a major international actor that was no longer willing to follow the lead of the West. It also asserted Russia's view of itself as a power pole in the international system. There soon followed the *Foreign Policy Concept of 2008* which suggested the existence of an alternative to the West value and development systems and competition between them.[40]

Did the West attempt to contain Russia's development as a Great Power? The short answer is "yes."[41] Ever since the dissolution of the Soviet Union, it can be argued, the United States and the European Union pursued policies aimed at drawing former republics of the Soviet Union into closer union with the West, while also attempting to limit Moscow's reestablishing a dominant position in post-Soviet space. Apart from the soothing rhetoric employed by the West, repeated direct encroachment on Russia's historic spheres of interests such as NATO's eastward expansion, US funded NGO political activities in Russia, a US-NATO military outpost in Georgia and missile defense installations near Russia, would naturally antagonize the state whose leadership is bound by the narrative of "Great Power" and "Strong State." As mentioned by Stephen Cohen the "struggle for Ukraine is yet another Chapter in the West's ongoing US-led march toward post-Soviet Russia."[42]

According to the tenets of regime theory, dominant powers create regimes that serve their interests and reflect their understanding of normative values.[43] Western states set up a variety of regimes intended to solve major regional problems. Among them, the G7 (later G8) was envisioned as the headquarters for the deliberation and resolution of major economic and political issues. The idea behind the establishment of the group was that decisions taken by the G8 would trickle down and be implemented by individual states; political mechanisms on the domestic level, as well as international organizations would help the process.[44]

In its attempt to join Western institutions in the immediate post-Soviet years' Russia invested considerable time and energy to join the then "G7" that led to the establishment of "G8" or "G7 + Russia," as it was referred to. By doing so, Russia aspired to join the Western regime and follow its main foreign policy directives. However, Russia was invited to participate in the G7 on "Western terms," as noted by Dmitri Trenin, with "door half open."[45] Talks that involved Russia in the context of the expanded G8 discussed predominantly non-economic issues, but did not include Russia in its discussions of financial decisions, thereby limiting the influence of Russia.[46] Russian analysts described this approach as the attempt by Western states to limit Russia's influence within the group, while utilizing Russia's external clout. This was done to champion the group's influence throughout the rest of the world. In conclusion, one can argue that Russia is not interested in membership in the "G8" with but limited participation. This is simply not acceptable, since Russia aspires to a more active role in world politics.

On the economic and geopolitical levels, the proposed Transatlantic Trade and Investment Partnership (TTIP) between the United States and the European Union presents a direct threat to Russia. On the economic level, the idea behind TTIP is to create a stronger economic bloc between the USA and the EU in the harmonization of safety requirements, regulations in financial markets, and the export of energy and raw materials from the US to the EU.[47] In this scenario the Partnership will establish a free trade and investment area that would generate $164 billion per year for the European Union and $131 billion for the US, with a total increase in trade volumes of 6% for the EU and 8% for the United States.[48] The economic benefits for these two countries seem to be obvious. The question remains whether this increased integration creates disadvantages for the states left outside of the Partnership, especially for Russia.

The most important issue for Russia in the TTIP relates to the energy sector, since energy exports comprise 75% of Russian sales to the EU.[49] If the TTIP eases access to the new gas generated by fracking from the United States to the European market, it will undoubtedly benefit both European consumers and American suppliers. At the same time, however, this new source of energy would substantially diminish the EU's dependence on natural gas from Russia, which has been a stated objective ever since the "gas wars" with Ukraine. This scenario would threaten Russia's macroeconomic and geopolitical interests. With the decrease of European demand, Russia would have to reorient its energy exports to other consumer markets, especially the ones in Asia.[50] One can argue that TTIP could be an important catalyzer in the changes of global energy landscape. At the conclusion of TTIP, natural gas from the United States may significantly increase and that will have serious geopolitical implications on the relations of the European states with Russia.

Some analysts underline the fact that the TTIP's aspirations are not purely economic. The relative economic decline of the transatlantic axis is expected to speed up with the rise of the emerging powers. Therefore, some see the true purpose of the TTIP as geopolitical. It aims to restore to the United States and the European Union the power to establish ground rules of the world economic order that other states would be forced to follow. It could be argued, therefore, that the TTIP is called upon to revive the West's economic and intellectual leadership with third countries being forced to adopt the rules imposed by the established powers. By implementing the Partnership the USA and the EU will become rule setters and regain the leading position in international trade and its development. The TTIP, therefore, could be seen as a part of the European and US reaction to their recent relative decline, as an instrument to regain leadership and greater influence in world affairs.[51] This situation would be similar to that in the period preceding the rise of the emerging powers, when developing or less developed countries were not able to defend their interests against the advanced countries in the international economic organizations constructed by the West.[52] At the same time the TTIP would lead to new protectionist standards against third countries such as China, India and Russia. This might trigger stronger cooperation among emerging economies in order to neutralize the negative consequences of the agreement. But, this would lead to further fragmentation of international trade and intensify the role of regional blocks.[53] To conclude, the Kremlin's calculations seem to favor alienating the West by building a cooperative bloc with emerging economies, in particular with China. It is important for Russia to understand with clarity what unites it with China. Moreover such calculations should take into consideration what such ties will bring for Russia in the long run.

The first argument in favor of closer ties with China is that they would provide a counterbalance in the international system to the unipolar domination of the United States. It would provide Russia and Putin's regime with a needed ally to serve as the legitimation of the current "Great Power" strategy. With China by its side, Russia feels able to promote itself as a global great power as one of the leaders in a post-Western world order.[54]

There is much that unites China and Russia normatively, even if they do not always share interests. For example, they share solidarity on issues such as sovereignty and territorial integrity, the need to counterbalance the

USA, etc.[55] Then there is the fact of China's insatiable appetite for energy, despite its being one of the leading countries in the development of alternative green energy technologies. China represents one of the largest energy consumer markets that could potentially help to diversify the consumption of Russia's energy exports.

However, a Russian decision to shift its economic relations away from the West and to expand its energy deliveries to China might not be as beneficial as expected. China's development of alternative energy technologies, such as shale gas and fracking, would undeniably increase China's leverage and negotiating strength. As a result, one might argue that Russia could lose more than it gained from such a shift in its energy exports by placing it in a subordinate position with Beijing. Given the fact that China is already challenging Russia's dominant role in Central Asia, such a strengthening of China's position vis-à-vis Russia represents a new challenge. In particular, a growing imbalance of power could result in the eventual loss of *de facto* sovereignty over some parts of Siberia and Russia's Far East. The Chinese could come to dominate Eurasia economically, exploiting its natural resources in a near-colonial relationship. On a more general level, Russia's aspirations of center of power would be reduced to a subordinate player in China-dominated world.[56]

RUSSIA'S ROLE AS A LEADER IN A MULTIPOLAR WORLD

One might argue that, due to the nature of Russia's network regime[57] and its discourse of "Great Power" that it has constructed, the projection of domestic strength vis-à-vis the external "other" is inevitable. It seems that at this point, the best bet would be for Russia to consolidate power domestically in response to potential challenges from the West. Such action could be supported domestically by the rhetoric established by the recent events in Ukraine.

What might Russia's calculations for the future be? Could we argue that the developments in Ukraine and Russia's role in them have once again supported Russia's priority in creating a new global order, rather than efforts to fit into regimes constructed by and dominated by the West? What awaits Russia if it chooses emerging economies as its primary allies?

Despite the very "diverse" characteristics of the states referred to by the acronym BRICS, there are unifying factors that can promote coherence in policy responses from Brazil, Russia, India, China and South Africa. There are two major benefits of constructing the BRICS as an international regime.

First, the BRICS as an international regime would provide an opportunity to address pressing issues that members face and can resolve within the block, such as energy security. Second, building on the example of OPEC as an international regime that has historically unified different actors with often divergent agendas, BRICS could serve as a platform to amplify the bargaining power of its member states with regimes and states outside of BRICS. In particular, they might further the creation of a multipolar world, along Russia's preferred lines, and obtain concessions from the West and the institutions it constructed for furthering its economic and geopolitical agenda.

It seems to be a consensus among some scholars that BRICS countries "lack coherence"[58] and "they do not share common political interests and are not a natural trading bloc"[59] and overall have little in common.[60] Armijo asks whether the notion of BRICS is a "mirage" that "appears to be forced."[61] As Heine notes, the BRICS countries consist of democracies and non-democracies hailing from four different continents, of very different size and economic performance.[62]

Yet, Roberts and Armijo list ten commonalities that in their opinion unite the BRICS and can potentially support their acting as a viable bloc.[63] Among the unifying features, they list several factors: large populations that provide demographic dominance in their respective regions, the rate of economic growth that makes members the largest economies outside of the OECD with the exception of South Africa, and the absence of special relations between BRICS states and other major powers, especially military and political alliances, that could potentially hinder their affiliation with BRICS as an international regime. Importantly, Brazil, Russia, India, China and South Africa have a long history of questioning or directly contesting the supremacy of the West and share the conviction that the crucial step in their joint strategy should be the demand for more influence and decision-making power in international economic and financial institutions. Such influence would push toward a more multipolar international system that would consequently enhance BRICS' role in world affairs. As Sadik Unay notes, in the emerging system of multipolarity, major players such as China and Russia have used their economic influence over Western politico-economic interests as a diplomatic tool to accelerate the transformation of global governance structures to a more balanced status quo. This perception transformed BRICS from an international investment strategy to a widely adopted symbol for the alleged decline of Western hegemony.[64] The question arises of the possibility of BRICS evolving into an international regime that might challenge the existing regimes dominated by the West.

Faced with criticism of the complexity of the definition of "regime," Keohane came up with a "lean" definition of the concept, as follows: "Regimes are institutions with explicit rules, agreed upon by governments that pertain to particular sets of issues in international relations."[65] Therefore, two important points can be made: first, that international regimes are directly related to international institutions and should be studied as such; and second, that international regime and "international organization" are not synonymous, even though in many cases regimes will be accompanied by organizations designed to support them. The major difference between them is that organizations can respond to events, whereas regimes are considered to be no more than set of rules, norms and principles and do not possess the capacity to act. However, many of the functions prescribed to regimes could not be carried without some organizational structure that is embedded in them.[66]

THE DECLINE OF DOMINANT POWERS

In international relations, states can be classified according to whether they are superpowers, great powers, regional powers or middle powers. Such a classification is based primarily on a state's military and economic capabilities, the geographic scope of its influence and the acknowledgement by other states of its power status. William Fox defined superpower as the great power that has a great mobility of power.[67] According to Fox, superpowers deploy different instruments of power, such as military, political, economic and cultural to promote their interests and ideological preferences throughout the world.

Unlike superpowers, great powers do not possess overwhelming dominance in all areas of power and may not have the capacity to bear the costs of regime maintenance. Not unlike superpowers, great powers are global players that have the interest of projecting influence beyond their immediate region and have capacity and desire to exercise influence in global issues. Regional powers are considered to be middle powers: to be somewhere in the middle of great powers and other states.

According to some analysts, based on this typology, the USA is the only state that could be considered as a superpower today. The European Union and Japan continue as global economic powers and have a substantial regional and global influence. China is considered as a rising great power, the second to the United States. Finally, Russia, India and Brazil—and to a lesser extent, South Africa—dominate their regions and have aspirations to great power status.[68] As a potential new regime, the BRICS comes in two major guises; on the one hand as a five-faced symbol of the growing weight of the emerging markets in the global economy and, secondly, as a political institution created to redress the current balance in world affairs that favors the largest non-Western powers.[69]

From the political point of view members of the BRICS bloc (Brazil, Russia, India, China and South Africa) these major regional powers have historical legacies based on the contestation of Western hegemony in the international system through communist, socialist, or non-aligned discourses. As such, they could easily associate themselves with the dissatisfaction that has stressed built-in inequalities that have been established in the modern West-centered regimes and exigencies of global capitalism exercised through multinational corporations.[70] Yet, they have been seldom analyzed as a group because of their timid joint actions. Some analysts compare them to the Non-Alignment Movement (NAM), since both are groups of emerging states that challenge the global pre-eminence of the West.[71] The comparison is not fully accurate, however, since NAM was a large heterogeneous group, similar to the present emerging world of which the BRICS represents a vanguard.[72]

Despite the differences among the members, this grouping of emerging economies is constructed on the basis of some common denominators. The first, as we have noted, is that BRICS countries are substantial in terms of their population and dominate their perspective regions in terms of demographic potential. Second, these countries constitute the largest economic bloc outside the industrialized countries in OECD with impressive economic growth figures and large domestic markets with considerable expansion capacity. Yet they are considered as developing countries because of their continuing developmental needs, informal economic sectors, etc. They generally adopt autonomous positions and stay distanced from Western-dominated politico-military frameworks, while also avoiding special security relations with Western powers exemplified by the relations between the United States and Japan for example.[73]

The construction of BRICS, as well as the deterioration of Russia's relations with the West, can be analyzed as a result of Russia's experiences in dealing with the West. Initially Russia attempted to join the states at the core of the international system by attempting to construct socially its image as a Western state and by aligning itself with Western powers. As we noted earlier, this period was short-lived and, from Moscow's perspective, was used by the West to expand its sphere of influence in Central and Eastern Europe through expanding membership and through the Eastern Partnership and the proposed Association Agreement with Ukraine. By 2005–2007 the Russian leadership rejected further tutelage by the West and announced its intention to pursue its interests regardless of Western criticism.[74] The power of five emerging economies in the BRICS could be helpful for Russia in pushing against Western domination through the institutions of global economic governance that play such an important role in establishing a broad international agenda. During the last BRICS summit on July 15, 2014, the group of emerging economies signed a long-anticipated document to create a $100 billion BRICS Development Bank and a reserve currency pool worth more than another $100 billion as an alternative to the existing US-dominated World Bank and the International Monetary Fund. The Bank has been set up to foster greater financial and development cooperation among the five emerging markets by countering the influence of Western-based lending institutions and the dollar. Documents on cooperation between BRICS export credit agencies and an agreement of cooperation in innovation were also signed.[75] Writing in 2012, Putin stressed that the BRICS were a "striking symbol of the transition from a unipolar world dominated by the United States to a more just world order."[76]

However, BRICS' future as a geopolitical bloc is not certain.[77] BRICS' members share some important positions, such as the defense of sovereignty, non-interference in internal affairs and a common desire to rebalance power within international institutions in their favor, or, the creation of non-West-centered institutional regimes. But, there are some challenges and contradictions inside the bloc. India perceives China as one of its security challenges; Brazil, India and South Africa are democracies, while China and Russia are authoritarian states.

Despite such inconsistencies within the BRICS group, Russia continues to use it as an instrument to advance its global agenda. Moreover, Russia has played the leading role in organizing the BRICS. Out of the five BRICS members, Russia is the only state with a recent history of global superpower status and wants to reassert itself as a global player. Analysts are divided on the importance of BRICS to Russia. Some believe that the importance of the BRICS does not extend beyond the issues of global economic governance.[78] Others consider the BRICS as the institutionalization and development of widespread awareness within global public opinion concerning the idea of a multipolar economic and political order. This could be considered one of Moscow's most successful international initiatives in recent decades.[79]

CONCLUSION

The current standoff of Russia with the West over Ukraine provides more questions than answers about Russia's visions as an international actor and about the future place of Russia in the global energy market. In this chapter, we demonstrate the complexity of Russia as an international actor. We have tracked the deterioration of Russia's relations with the West and its growing competition with both the EU and the United States and NATO for influence in post-Soviet Eurasia. We have noted the importance of Western control over major international economic regimes, as well as Russia's willingness to use economic blackmail in its relations with post-Soviet states, even when that has led to Russia's inability to meet its contractual obligations in supplying natural gas to EU member states. We have pointed, as well, to the concerns in Europe about the longer-term impact of the region's heavy dependence on

Russian energy and the inchoate plans to reduce that dependence. Finally, we have noted Russia's major efforts to find alternative markets for its energy exports, especially in China, and the prospects of BRICS, stimulated by Russian leadership, to establish an international regime that may challenge some of the rules of the existing Western-dominated economic regimes.

Russia's role in the future supply of energy is not at all clear. This derives, first of all, from the complexity of Russian decision-making in the area of energy exports. It will be influenced, as well, by the glut in global energy markets resulting from both a slowdown in demand from countries such as China and the explosion of new energy sources in North America and elsewhere. What is clear is the fact that the Russian Federation is likely to find the importance of energy exports—as both an economic and a geopolitical tool—of reduced importance, at least in the foreseeable future.

NOTES

1. As we complete the revision of this chapter in mid-December 2014, the price of petroleum has dropped to slightly above $60 per barrel, down from well over $100 in 2013.
2. A.F.K. Organski, *World Politics*. New York: Knopf, 1960.
3. Ibid.
4. Martin Smith, *Power in the Changing Global Order.* Cambridge: Polity, 2012.
5. An English-language translation of the document entitled "Russia at the Turn of the Millenium" can be found in Richard Sakwa, *Putin, Russia's Choice* (2nd edn.), Abigdon: Routledge, 2008.
6. Ibid.
7. Ibid.
8. Smith, *Power in the Changing Global Order.* See, also Graeme Herd (2009) "Russia's Sovereign Democracy: Instrumentalization, Interests and Identity," in Roger E. Kanet, ed., *A Resurgent Russia and the West: The European Union, NATO, and Beyond.* Dordrecht, The Netherlands: Republic of Letters Publishing, pp. 3–28.
9. *Rossiiskaya Gazeta,* 26 April 2005.
10. Smith, *Power in the Changing Global Order.*
11. *Izvestiya,* 13 September 2006.
12. Vladimir Basov, "Top 10 Oil and Gas Producing Countries in 2012," *Mining. Com*, 14 June 2013. http://www.mining.com/top-10-oil-and-gas-producing-countries-in-2012-24585/
13. Peter Rutland, "Russia as an Energy Superpower" *New Political Economy,* vol. 13, no. 2, June 2008.
14. Clifford Gaddy and Barry Ickes, "Russia's Virtual Economy." Washington, D.C.: Brookings Institution Press. 2002.
15. "Russian Economic Report Moderating Risks, Bolstering Growth," *The World Bank.* 2012. http://www.worldbank.org/content/dam/Worldbank/document/rer-27-march2012-eng.pdf.

16. Rutland, "Russia as an Energy Superpower."

17. Michael Bradshaw, "Russian Energy Dilemmas: Energy Security, Globalization and Climate Change," in Pami Aalto, ed., *Russia's Energy Policies National, Interregional and Global Levels,* Cheltenham: Edward Elgar, 2012.

18. Pavel Baev, "Russia Abandons Energy Superpower Idea but Lacks Energy for Modernization," *Strategy Analysis* (2010), 34(6).

19. For a recent assessment of the nature of the current regime and its inherent weaknesses, see Lilia Shevtsova. *Russia XXI: The Logic of Suicide and Rebirth.* Moscow: Moscow Center, Carnegie Endowment for International Peace, 2013. http://carnegieendowment.org/files/RussiaXXI_web_Shevtsova.pdf.

20. Joel Hellman, "Winners Take All: The Politics of Partial Reform in Postcommunist Transitions," *World Politics* (1998), vol. 50, no. 2.

21. For a brief assessment of part of the problem see Paul Krugman, "Putin's Bubble Bursts," *New York Times,* 19 December 2014. http://www.nytimes.com/2014/12/19/opinion/paul-krugman-putins-bubble-bursts.html.

22. Hellman, "Winners Take All."

23. Ewa Paszyc, "The 'Energy Tandem': Putin and Sechin Control the Russian energy sector" *EastWeek* 22, June 20, 2012. For a comprehensive discussion of corruption and the role of personal networks in Russia see Karen Dawisha, Putin's *Kleptocracy: Who Owns Russia?* New York: Simon & Schuster, 2014.

24. Ian Bremmer, *The End of Free Market: Who Wins the War between State and Corporations?* New York: Portfolio, 2010.

25. Andrei Tsygankov, "If not by Tanks then by Banks? The Role of Soft Power in Putin's Foreign Policy," *Europe-Asia Studies* (2006), vol. 58, no. 7. See, also, Bertil Nygren, *The Rebuilding of Greater Russia: Putin's Foreign Policy towards the CIS Countries.* Abingdon, UK: Routledge, 2008.

26. Jeffrey Mankoff, "Russian Foreign Policy and the United States After Putin," *Problems of Post-Communism,* (2008), vol. 56, no. 4, July/August 2008.

27. Ian Bremmer, "Thinking Beyond States," *National Interest* (2006*),* vol. 83.

28. Peter Rutland, "Russia as an Energy Superpower," *New Political Economy,* (2008), vol. 13, no.2.

29. Dina Moulioukova and Roger E. Kanet, "Decoding Russia's Energy Security. Perceptions Matter," in Bruce Bagley and Rémi Piet, eds., *Energy Security and Environmental Policy in the Western Hemisphere.* New York: Lexington Books, 2015, in press.

30. Simon Pirani, Jonathan Stern and Katja Yafimava, "The April 2010 Russo-Ukrainian Gas Agreement and its Implications for Europe." *The Oxford Institute for Energy Studies.* NG 42. June 2010.

31. Rutland, "Russia as an Energy Superpower." The development of this pipeline has since then been abandoned see Loskot-Strachota, Agata, and Janek Losocki (2013) "End of Nabucco—End of Southern Gas Corridor?" *Energy Post,* 27 June. http://www.energypost.eu/end-of-nabucco-end-of-southern-gas-corridor/

32. R. Ericson, "Eurasian Natural Gas Pipelines: The Political Economy of Network Interdependence," *Eurasian Geography and Economics,* vol. 50, no. 1 (2009).

33. Valentina Feklyunina, "Russia's International Images and Its Energy Policy. An Unreliable Supplier? "*Europe Asia Studies* (2012), vol. 64, no. 3, pp. 449–469.

34. Bradshaw, "Russian Energy Dilemmas."

35. Neil Hirst and Anthony Froggatt, "The Reform of Global Energy Governance," *Grantham Institute for Climate Change Discussion Paper No 3,* December 2012. http://www.chathamhouse.org/sites/files/chathamhouse/public/Research/Energy,%20 Environment%20and%20Development/1212granthamreport_energygovernance.pdf

36. Ibid.

37. John Mearsheimer, framing his argument within an explicitly structural realist perspective, blames the West for challenging Russia and, thereby, causing the current crisis. See his article "Why the Ukraine Crisis is the West's Fault: The Liberal Delusions that Provoked Putin," *Foreign Affairs,* vol. 93, no. 5, (2014), pp. 77–89. Jack Matlock, Jr., former U.S. ambassador to the Soviet Union, makes a similar point when he states: "The sad fact is that the cycle of dismissive actions by the United States met by overreactions by Russia has so poisoned the relationship that the sort of quiet diplomacy used to end the Cold War was impossible when the crisis in Ukraine burst upon the world's consciousness." See "Who is the bully? The U.S. has treated Russia like a loser since the end of the Cold War" *The Washington Post,* 21 November 2014. http://www.washingtonpost.com/opinions/who-is-the-bully-the-united-states-has-treated-russia-like-a-loser-since-the-cold-war/2014/03/14/b0868882-aa06-11e3-8599-ce7295b6851c_story.html (accessed 21 November 2014).

38. Roger Kanet, "The Failed Western Challenge to Russia's Revival in Eurasia," *International Politics,* (2015, in press).

39. Vladimir Putin, "Putin Slams US for Making World More Dangerous," *DW—World.DE Deutsche Welle.* 10 February 2007. http://www.dw6 world.de/dw/article/0,2144,2343749,00.html

40. President of Russia, *Foreign Policy Concept of the Russian Federation.* 12 July 2008. http://archive.kremlin.ru/eng/text/docs/2008/07/204750.shtml

41. Kanet, "The Failed Western Challenge to Russia."

42. Stephen F. Cohen, "Distorting Russia: How the American Media Misrepresent Putin, Sochi and Ukraine," *The Nation,* 11 February 2014. (3 March printed edition). http://www.thenation.com/article/178344/distorting-russia.

43. See, for example, Duncan Snidal, "The Limits of Hegemonic Stability Theory," *International Organization,* vol. 39, no. 4 (1985), pp. 579–614.

44. Alexander Lukin, "Russia's Identity Dilemmas: BRICS, the G8 and the Shanghai Cooperation Organization," in Francis A. Kornegay and Narnia Bohler-Muller, eds., *Laying the BRICS of a New Global Order: From Yekaterinburg 2009 to eThekwini 2013* Pretoria, South Africa: Africa Institute of South Africa, 2013, pp. 85–101.

45. Dmitri Trenin, "Russia Leaves the West," *Foreign Affairs,* Jul/Aug 2006, vol. 85, no. 4, p. 87

46. Lukin, "Russia's Identity Dilemmas."

47. Antonio De Lecea, "The EU-US Trade and Investment Before and After TTIP: What can Latin America Expect from It?" unpublished paper presented at the Conference Before and After the TTIP: Transatlantic Trade and Investment Partnership Between the European Union and the United States on 28 February 2014 at the University of Miami.

48. "Reducing Transatlantic Barriers to Trade and Investment an Economic Assessment." *Center for Economic Policy Research.* March 2013. http://trade.ec.europa.eu/doclib/docs/2013/march/tradoc_150737.pdf

49. Tamas Novak, "Implications of the TTIP on the Global Economic Integration of Central and Eastern Europe" unpublished paper presented at the Conference Before and After the TTIP: Transatlantic Trade and Investment Partnership Between the European Union and the United States on 29 February 2014 at the University of Miami

50. In fact, this has already begun to occur, as evidenced by recent agreements between Russia and China for the construction of new pipelines and the dramatic expansion of Soviet deliveries of natural gas to China. See "Gas agreement seen firming China-Russia energy ties," *Oil and Gas Journal,* 25 November 2014. http://www.ogj.com/articles/2014/11/gas-agreement-seen-firming-china-russia-energy-ties.html. See also, Novak, "Implications of the TTIP on the Global Economic Integration of Central and Eastern Europe."

51. Federico Steinberg, "US-EU Trade Negotiations: What is at Stake," unpublished paper presented at the Conference Before and After the TTIP: Transatlantic Trade and Investment Partnership Between the European Union and the United States on 28 February 2014 at the University of Miami.

52. Novak, "Implications of the TTIP on the Global Economic Integration of Central and Eastern Europe."

53. Ibid.

54. Bobo Lo, "A partnership of convenience," *The International Herald Tribune,* 8 June 2012.

55. Graeme Herd, "China, Russia and Eurasia: Neo-Tributary Relations and Geopolitical Realities?" In Matthew Sussex and Roger E. Kanet, eds., *Power, Politics and Confrontation in Eurasia: Foreign Policy in a Contested Area* Houndmills, UK: Palgrave Macmillan (2015, in press).

56. Lo, "A Partnership of Convenience."

57. Lilia Shevtsova describes Putin's regime as a system of government under which power is concentrated in the hands of a leader who relies on the bureaucracy, security forces, and big business, what some authors refer to as a "network state"—a system of government and private networks that control the economic and political system. See Lilia Shevtsova, *Russia Lost in Transition: The Yeltsin and Putin Legacies.* Washington, D.C.: Carnegie Endowment for International Peace, 2007, p. 41. See, also, Vadim Moshes and Arkady Kononenko, Russia as a Network State. Houndmills, UK: Palgrave Macmillan, 2011.

58. "The BRICS: The Trillion-dollar Club," *The Economist*, 15 April 2010. http://www.economist.com/node/15912964

59. Aseema Sinha and Jon P Dorschner, "India: Rising Power or a Mere Revolution of Rising Expectations?" *Journal: Polity*, (2009), vol. 42, no. 1, p. 88.

60. P.A. Jayan, "BRICS: Advancing Cooperation and Strengthening Regionalism" *India Quarterly* (2012), vol. 68.

61. Leslie Elliott Armijo, "The BRICS Countries (Brazil, Russia, India, and China) as Analytical Category: Mirage or Insight?" *Asian Perspective*, vol. 31, no. 4, 2007, pp. 7–42.

62. Jorge Heine, "A Tale of Two Very Different Summits, *The Hindu,* 24 April 2012.

63. Cynthia Roberts and Leslie Elliot Armijo, "The Emerging Powers and Global Governance: Why the BRICS Matter," in Robert Looney, ed., Handbook of Emerging Economies. New York: Routledge, 2014. See also, Suresh P. Singh and Memory Dube, "BRICS Guide 8: Major Strengths and Commonalities of BRICS," *GegAfrica*, 11 February, 2013. http://www.gegafrica.org/memory-dube/brics-guide-8-major-strengths-and-commonalities-of-brics.

64. Sadik Unay, "Reality or Mirage?: BRICS and the Making of Multipolarity in the Global Political Economy" *Insight Turkey,* (2013), vol. 15, no. 3, pp. 77–94

65. Robert Keohane, *After Hegemony: Cooperation and Discord in the World Political Economy.* Princeton: Princeton University Press, 1984, p. 57.

66. Andreas Hasenclever, Peter Mayer and Volker Rittberger, *Theories of International Regimes.* Cambridge, UK: Cambridge University Press, 1997, p. 11.

67. William Fox, "The Super-powers: Then and Now," *International Journal,* (1980), vol. 35, no. 3, Summer 1980.

68. Vidya Nadkarni, "Introduction" in Vidya Nadkarni and Norma C. Noonan, eds., *Emerging Powers in a Comparative Perspective the Political and Economic Rise of the BRIC Countries,* New York, London, New Delhi, Sydney: Bloomsbury, 2013, pp. 1–23.

69. Thomas Graham, "Russia, BRICS, and the United States," *Yale Journal of International Affairs*, Winter 2013.

70. Unay, "Reality or Mirror?"

71. Theodor Tudoroiu, "Conceptualizing BRICS: OPEC as a Mirror" *Asian Journal of Political Science*, (April 2012), vol. 20, no. 1, April 2012, pp. 23–45.

72. Ibid.

73. Unay "Reality or Mirror?"

74. In many respects the development of the concept of "sovereign democracy" was an indication of Russia's refusal to accept Western norms. For an extended discussion of the concept see Herd, "Russia's Sovereign Democracy."

75. "BRICS establish $100bn bank and currency reserves to cut out Western dominance." RT.com. 15 July 2014.

76. Vladimir Putin, "Vladimir Putin on Foreign Policy: Russia and the Changing World," *Valdai Discussion Club,* 27 February 2012.

77. Thomas Graham, "Russia, BRICS, and the United States," *Yale Journal of International Affairs,* Winter 2013.

78. Ibid.

79. Unay, "Reality or Mirror?"

Chapter 6

China and Latin America

The Case of Oil for Loans Agreements with Ecuador

Nashira Chavez

At the turn of the millennium, the world political economy has been driven by the accelerated growth of industrializing centers in Asia and, the global south[1] in general, and their demand for commodities. This trend has altered transnational economic networks as cross-border relations gravitate toward different states and markets other than industrialized countries. Energy is at the heart of these new relations. The fact is that energy security and diplomacy cross multiple issues in international politics; frequently energy security touches upon geopolitics, resource competition, dependence and great power relations. The emerging East-South relationship is a major current event linked directly and indirectly to energy markets.

China's quest for oil in particular has ignited a controversial debate over the nature of its objectives in the international oil market and the growing engagement of its national companies. Yet, less is known about how economic instruments serve Chinese foreign economic policy, how target countries reconcile such interests to their objectives, and how resource-rich countries leverage economic tools. This work discusses these distinct key areas through the lenses of China-Ecuador oil relations and offers three insights into the current study of oil politics between emerging economies.

The first section provides an overview of three main analytical assessments of China's energy interests abroad: the realist oil security assumption, a developmental profit-oriented approach, and a long-term soft power effort. Next, the work presents the conceptual links between oil and economic statecraft and argues that evidence demonstrates the aforementioned scenarios, which coexists among states international energy relations. Thirdly, I explore more closely the characteristics of economic oil statecraft between China and Ecuador from 2006 to 2013. I will offer some preliminary conclusions on the link between oil and politics and the different starting points of bargaining power and economic statecraft within emerging East-South relations.

VIEWS FROM AFAR

Beginning in 2002, Chinese oil investments took hold oil fields across continents. From Canada to Thailand and Kazakhstan, the blossoming of investments in upstream and downstream operations has caught the attention of business pundits and great power political strategists, raising a wide range of significant concern, optimism, and even uncertainty about its implications for global order. Until a couple of decades ago, international economic flow were reduced to Great Power relations and subordinated by military issues. Yet, as the twenty-first-century globalization progresses, both China itself and the topic of economic power are no longer at the border of foreign policy.

China's ties with developing countries moved to center stage as the recession and slow recovery of the West in 2009 coincided with the rapid growth of developing countries and their affluent economic activities became disconnected from traditional financial routes through New York or London. For Latin America in particular, this is issue is particularly important in a region traditionally considered US sphere of influence. At the same time, concerns over global economic crisis seems to be bringing back inquires about the return of the role state in economics. The redistribution of economic power, in relative terms, accompanied by flexible terms of engagement of a rising power, has revived interest in economic statecraft and the strategic use of economic instruments for political or business purposes.

The literature has brought to the limelight three perspectives on the expansion of China's oil diplomacy and the rest of the world.[2] One position proposes a realist explanation, marked by China's material needs to strengthen its oil capabilities as a result of its domestic demands and its great power characteristics. Oil stands alone when it comes to the foreign trade basket of a country by virtue of its strategic importance for the vitality, or at the very least, stability, of economic and even military programs.[3] A healthy economy requires a stable flow of oil to engineer growth, and in this light, the literature has paid considerable attention to China's scarce oil resources and its increasing demand, in tandem with its rapid industrialization and urbanization. This fact places China in a vulnerable position. Beginning in the 1990s, the country became a net importer and today is the largest importer and second consumer of oil in the world.[4] Concomitant with its sizeable industrialization, and despite registering a slower, yet still significant growth at 7%, China accounts for one-third of oil consumption growth in 2014.[5] In this context, Chinese foreign policy is driven by the unavailability of domestic reserves, by its outstanding annual growth, and by efforts to increase its position in the international oil market.

Appropriately, China does not leave oil to market forces alone. Domestically, Beijing highly regulates the production, distribution, and price of oil.[6] There are not proven reserves sufficient to keep pace with domestic demand, and as efforts to curtail coal use are underway, there is a lack of alternative fuels; the closest substitute to coal and oil, hydroelectric power, lags with a distant 6% share of the energy consumption mix.[7] A major initiative, for example, includes the construction of an oil storage facility to store more than 500 million barrels by 2020.[8]

Internationally, China's footsteps in the oil market seek to secure supply and clout in the oil order. Strategies seek to reduce its oil dependence through the diversification of imports, investment in fields overseas, participation in joint ventures, and acquisition of equity shares. In particular since 2008, China has expanded its oil supply through both overseas acquisition investments and loans-for-oil agreements in addition to the construction of an extensive pipeline networks with Russia, Kazakhstan, and Myanmar.[9] There is an intrinsic material interest to reduce its vulnerability to a deliberate oil crisis triggered by supplier nations and avoid the woes of market price fluctuation. Prospects of China's expanding role in the global economy equally spearhead strategic efforts to increase its oil market share and leverage in the commercialization and bargaining of energy with major oil producers.

Under this vision, the exclusive control of a strategic sector by state-owned companies, and its engagement abroad is an expression of economic statecraft. The operations of its state-owned oil corporations and the oil investments of its policy banks, China Export Import Bank (China Eximbank) and the China Development Bank (CDB) are hardly seen as integrated into the global economy. Instead, they are believed to be a more or less unified political body focused on bolstering economic instruments for political purposes, a perspective symptomatic of the "rising power" tag and power associated to a large, and in this case consumer, actor like China. Beijing's international economic power toolkit seems to be willing to withstand commercially undesirable terms to secure oil flows from resource-rich high-risk countries with its financial muscle channeling economic flexible inducements such as large low interest credits, subsidized financing, or payment in kind.[10] Any oil security goals is not a source of conflict as it targets the diversification, steady flow, and access to oil abroad and the use of military force or any kind of coercive capabilities to appropriate oil or control trade routes is largely absent.

The other side of the coin is led by a commercial rationale approach. China's dramatic economic outward engagement under its "Going Out" policy in the 1990s is a step ahead any offensive notions associated to its economic rise. Following its emblematic entry to the WTO under "Going out" in 1999, the government spearheaded the "Go Global" policy in 2000, a policy aiming at the internationalization of state-owned companies.[11] Any mercantilist notion is seen as a mischaracterization of operations abroad and instead claims transnational operations seek the expansion of market share, the mastery of oil technical skills, in the race to catch up with International Oil Companies, while Beijing seeks to unload the pressure of its outstanding reserves into outward foreign direct investments (OFDI). Further, evidence suggesting oil acquisitions are not automatically imported to China tones down claims of locking up oil supply. Two thirds of China's share of oil production was sold in the international market by the end of the last decade; there is no evidence that China's accumulation of oil imports it is reducing supply by the same amount as big country producers.[12] The main use of economic tools from this vision is consequently the search for profits, and a stake in exploration and production, the most profitable dimensions of the industry.[13]

Secure access to resources takes a back seat in a business rationale guided by profit-oriented calculus and commercial strategy to improve the opportunities of Chinese firms abroad.[14] In this case, an enterprise orientation predominates without precluding state guidance. Companies may have access to subsidized capital yet most oil is valued at market prices. China's oil engagements abroad are marked by the purposes of what Brautigam and Xiaoyang called "the developmental state," which uses acquisitions and loans as strategic tools for economic opportunities.[15] Along these lines, the internationalization of state-owned companies is seen as a natural extension of the globalization of the Chinese economy and workforce

There should be little fear that Chinese National Oil enterprises are an instrument of foreign policy. Corporate interests predominate against any political and monolithic entity vision. Erica Downs goes even further by asserting a "bottom up" organization in which the Chinese government is "hands off' oil operations as companies are the primary decision-makers and manager of resources.[16]

Lastly, the third view links oil diplomacy to Grand Strategy. China's position as a global player and in particular its growing expansion in Africa and Latin America raises suspicion about the nature of its intensions blurring the lines again between economics and politics.[17] Political considerations are far more involved, than in resource security, in advancing a Chinese long-term strategy as a global player. There is a lack of material evidence and capabilities directed toward offensive actions or conflict derived from oil accumulations and as such, China's intentions appear to be benign.[18] Many see the nature of Chinese engagement, however, as a cautious strategy mainly concerned with a balance relationship with the United States given its leverage in the oil market. While there is evidence that Latin America's increased oil exports to China have not trumpeted supply to the US—US imports most of its oil from the Western Hemisphere (54%)—the dealings raise alarm over Beijing's influence in what has been considered The U.S natural sphere of influence. The combination of burgeoning trade, climbing FDI figures, the presence of Chinese national companies, and the complementarity of their economies has set China among Latin America's main partners, bringing attention to how increasing links will affect norms. John Williamson, the architect of the Washington Consensus has explicitly raised concerns over how Chinese flexible contract terms may foreshadow the liberal economic principles underpinning Western order.[19]

China has found common ground with peripheral countries in their priorities as developing nations. Since 1953, Beijing has cemented an international strategy on the "Five Principles of Peaceful Coexistence"[20] and spearheads much of the statements of non-intervention and mutual benefit (win-win strategy) in its relations with Africa and Latin America. In times when the western rules are under scrutiny, following the economic crisis of 2009, China has taken a "de facto leader of the developing world" persona.[21]

Oil joins the portfolio of tools for economic statecraft including international aid, subsidized credits, and other generous assistance packages for infrastructure in light of its strategic role for the growth of the national economy growth and in turn, its economic lever of power for political purposes. Yet, economic tools can also have an economic purpose and Beijing's economic power provides a wide economic toolkit to guarantee both political leverage and economic advantage. In true, China's economic statecraft is often difficult to tease apart in the context of its closed regime and the secrecy cloud of its activities nevertheless it is possible to claim all these factors plays in the practical application of economic policy. The question remains, which is predominant?

OIL AND ECONOMIC STATECRAFT

Quincy Wright's statement, in his seminal work the *Study of International Relations,* argues that "the field of economics and politics overlap" presaged the struggle to draw a clear line between economics and politics in China's overseas oil exploration.[22] In common the three approaches assumed China's presence in resource-rich countries as a function of its increasing economic capabilities and its position as a rising power.

China is a difficult actor to read. Competing multidimensional identities shaped its behavior. Its vast actual power—territory, population, and military capabilities—coincides with a close domestic political system that simultaneously courts the international market by strategically accommodating its policies to a Western-oriented order. This compounded with the conflicted and alarming notions behind transitional power assumptions, enhances distrust over the capricious attributes of Chinese power

Power is a defining term in a major actor like China. A report from the World Bank shows China as having the fastest ascendance of all the major powers. Within a 30 years period, Beijing has multiplied its share of global output by seven, balloon its trade twenty second times, and has increased its share of global military spending threefold.[23] Its vast growth results today in renovated links with developing countries and the formation of deeper economic ties traditionally associated with major power politics . Yet, in the midst of economic reforms, in 1982, Deng Xiaoping eluded a military build up course for a strategy to "hide [its] capacities" and "bide [its] time".[24]

Yet, the current literature's attention on the asymmetrical disparities between China and small developing countries like Ecuador is fruitless and pose no major theoretical contribution. It is self-evident relative and absolute powers leans in China's favor. As such, its potential influence is overwhelming. At the same time, available economic resources do not mean effectiveness influence. Foreign policy as any policymaking involves a purposive behavior and this become more relevant as we recognize China's multidimensional goals and "win-win" manifesto.

Within China's economic power structure, policy efforts seem to take a myriad of forms and they might even be complementary. For China relations with developing countries, the distinction between tools of economic power is paramount as evidence suggests military force is not considered an option for now. Moreover, China for the most part has advanced positive economic instruments rather than negative sanctions.

The literature on China's relations with African and Latin America has identified four economic tools in China and developing countries oil diplomacy. First, there is China's international aid encompassing concessional lending toward infrastructure (including a share of payment of the debt in kind). The second one involves trade tools defined by preferential credits, partly subsidized, for the import of Chinese goods and tariff-free for African exports to China. A third tool is the purchase of resource of extractive operations or subsidized joint venture. Finally, mutual benefit loans consist of commercial-rate loans with a portion of debt payment in oil, also known as loans-for-oil.[25]

The various forms of specific economic techniques seem to serve a double purpose, to fulfill the needs of host countries while serving China's political, security and, developmental objectives. In this context, the case of Ecuador is instructive in two accounts. On one level, it identifies some of the different dimensions of economic engagement. At the same time, it empirically explains the nature and implications of such economic tools.

THE CASE OF ECUADOR

The oil industry is a central fixture in the Ecuadorian political landscape. It has been its main source of revenue, its platform to international politics through OPEC, the backdrop of the domestic power struggles, and a means to social reforms. The Ecuadorian oil boom in 1972 brought major economic and political changes. In light of the country's newfound wealth, the military was in a favorable position to take over the government and catalyze its long-term prospects to transform the economy and modernize its society.[26]

From then on, oil production has been a focus for military and democracies alike. While oil policies have diverged over time in the scale of participation of private companies, hydrocarbon resources are owned by the state. Resource nationalism seeks to maximize income and tap on oil revenues to finance public spending, nurture social reforms—many times to appease counter-mobilization—and gain some independence for its small, commodity driven economy.[27]

There is optimism for broader oil cooperation with a self-sufficient country like Ecuador. For once, the country is oil rich, there are not resource constraints, and as such offers great potential for energy efficiency and the maximization of revenues. Its strength resides in the availability of 8 billion proven reserves, which continue to increase year-over-year by 14. Despite the fact that oil accounts for 76% of Ecuador energy matrix consumption, more than half of its 527 thousand barrels produce per day is exportable surplus. The oil sector remains the main engine of growth accounting for 53% of total exports with a value fivefold greater than the succeeding export product (bananas), and also remains the main source of financing public spending.[28]

With all its associated fortune, the oil sector prompts deep concerns when it comes to policy and management efforts. Reports from the Inter- American Dialogue observe a stagnant production and reduced revenues. Today's 527,00 b/d is in fact a falling production from a 536,000 b/d peak record in 2006.[29] Poor performance—defined as the gap between increasing reserves and falling production—stemmed from lack of investment to upgrade

Table 6.1 Ecuadorian Oil by the Numbers

Proven Reserves	*8 Billion*
Daily production (Thousand barrels per day)	527
Oil Exports (Thousand barrels per day)	272
Oil Consumption (Thousand barrels per day)	255
Oil rents as % of GDP	19%
Oil as % total Exports	53%
Oil trade value	$13,411

Sources: U.S. Energy Information Administration, World Bank, and Economic Comission for Latin American and the Caribbean

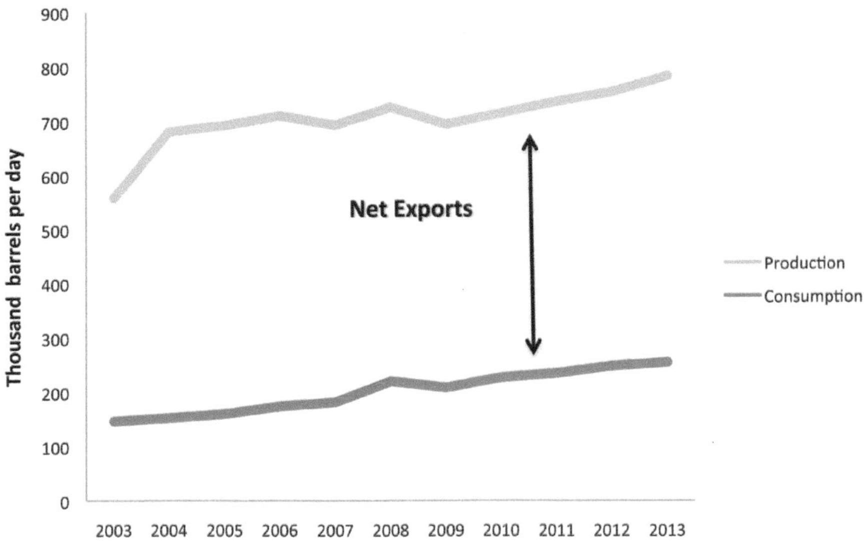

Graph 6.1 Ecuador's Total Oil Production and Consumption. *Source*: U.S Energy Information Administration.

technology for oil recovery, environmental standards, operational efficiency, and management of mature oil fields. A more productive course has also been hampered by the lack of foreign direct investment in the context of service contract for multinational companies.[30] Reforms in the hydrocarbon law in 20010 under president Correa prohibited companies to receive a percentage of profits for oil production replacing all operational agreements for fixed per barrel fees contracts in which the state earns 99% of oil income. In the mist of negotiation, private operations declined by 15% as some companies decided to exit and state-owned Petroecuador absorbed 73% of oil assets.[31] The lack of incentive for FDI and reinvestment of oil earnings appears to dent oil production despite the fact the state could triple its production. This, coupled with Ecuador's regressive fuel subsidy system and lack of refining capacity, suggests severe institutional and political factors affect the potential of the oil sector.

China's "Go out" strategy in Ecuador has played out under these domestic conditions. At the international level also, Ecuador's return to OPEC set a production quota of 520,000 b/d looming oil production even more. One type of economic tools is investments in specific acquisitions. China's two main national-owned companies China National Petroleum Corporation (CNPC) and China Petrochemical Corporation (Sinopec) have engaged in two direct stakes in oil blocks. The first and biggest of Chinese deals by CNPC subsidiary, Andes Petroleum, was signed in 2006 before renegotiations began in

2008, which consisted in buying the Canadian company EnCana assets for $ 1.42 billion.[32] Nevertheless, Chinese companies adhered to Ecuador renegotiations to get a better deal and remained in the country along with less than ten, mostly small, other companies.[33]

The purchase, while being an economic strategic asset, seems to follow normal commercial goals. EnCana assets were the third largest oil producer in Ecuador controlling three oil fields with an oil production of 75,000 b/d and 138 million barrels of proven reserve oil. The most interesting stake was probably its 36% share of the new Ecuadorian pipeline, Oleoducto de Crudos Pesados (OCP), with a capacity to transport 450,000 b/d, a number close to the national daily oil production. The other accord signed with Sinopec is marginal accounting for $500 million to develop two blocs. Since last renegotiations, there were no signs of new China biddings thus raising signs about the existence of less economic incentives toward the direct control of oil blocks and a rather more interest in stable oil flow through the oil-for-loan tool.

Recent speculations nevertheless suggest Chinese companies bidding in the 11th oil round are interested in the blocks surrounding Yasuní-ITT initiative, one of the largest oil reserves (amounting 796 million of barrels or 20% of Ecuadorian reserves). The ITT oil reserves lie under the Yasuni Park, one of the most diverse sites in the world and the core of a high-profile international Ecuadorian initiative to keep oil underground in return for half of the expected currency value from the international community.[34] At the same time of such international effort, leaked documents suggest the Ecuadorian government was negotiating a $ 1 billion loan on the condition Chinese companies take over the exploration of the ITT fields. As result, the document suggests main efforts have been under way to advance negotiations with Chinese companies—in August 2013, Correa announced the end of the ITT initiative due to insufficient monetary commitment to the trust (only $13 million)—to support China's potential $10 billion finance of a refinery on the coast (Ecuador imports refined products).[35] None of these efforts seem to have been planned with other oil multinationals giving priority to negotiations to Chinese companies. The results of the 11th round are pending and it remains to be seen its implications for Chinese participation in the oil sector yet questions remains over the "soft budget" power.

LOANS FOR OIL: A TOOL OF ECONOMIC STATECRAFT?

Oil-for-loans is the latest and innovative economic tool of engagement between China and resource-rich regions like Latin America. There is increasing interest in the study of the loans-for-oil tool, also known as mutual benefit loan, as the most apparent instrument of economic statecraft.

Most literature on the subject follows Chinese finance in Africa while studies on Latin America are recently emerging.[36] Yet, there are high stakes when it comes to Chinese loans in Latin America. The latest report by Gallagher, Irwin and Koleski estimate more than half of Chinese loans abroad went toward Latin America. By the end of 2012 figures amount to $85,804.60 million, 69% of this amount, 50,000 flowed into loans-for-oil mechanisms.[37]

Financial power and oil diplomacy intersect at the oil-for-loans strategy. Historically, oil nationalism has defined developmental models in Mexico, Brazil, Ecuador, and Venezuela to name a few. In the context of Ecuador's institutional framework and limited influence abroad, oil is a political expression of its sovereignty. Its flows, revenues, and, international demand have been conceptualized as resourceful tool for self-determination and a viability of its economy to the world. While stable oil supply is of paramount for China, the broader role permeates Beijing's long term attempts to increase its leverage in the oil international market and the politics of oil.[38] To this end, in and in an era where coercion is less tolerated by the international community, the possession of abundant financial assets vest an advantage for foreign policy goals.[39]

Table 6.2 Oil for Loans in between China and Latin America

Year	*Country*	*Debtor*	*Lender*	*Amount (USD billion)*	*Purpose*
2009	Brasil	Petrobras	CDB	10,000	Oil technology
2009	Ecuador	Petroecuador	Petrochina	1,000	Advance paymemnt for oil
2009	Venezuela	BANDES & PDVSA	CDB	4,000	Infrastructure
2010	Ecuador	Petroecuador	CDB	1,000	80% discretionary 20% oil-related
2010	Venezuela	BANDES & PDVSA	CDB	20,000	Infrastructure
2011	Ecuador	Petroecuador	Petrochina	1,000	Advance payment for oil
2011	Ecuador	Government	CDB	2,000	70% discretionary and 30% oil related
2011	Venezuela	PDVSA	CDB	4,000	Infrastructure
2011	Venezuela	PDVSA	ICBC	4,000	Housing
2012	Venezuela	BANDES & PDVSA	CDB	8,000	Infrastructure, agriculture, minning, and energy
	Total			55,000	

CDB: China Development Bank, ICBC: Industrial and Commercial Bank of China.
Source: Gallagher Kevin, Amos Arwin, and Katherine Koleski.(2013). "Un mejor trato: Análisis comparativo de los préstamos chinos en América Latina."

Oil for loans agreements between Ecuador and China have a clear purposive behavior, as they are oriented to issues of states' domain such as oil. Oil is a strategic resource and has instrumental objectives beyond its purchase: access of credit, apparent lack of conditional lending policies, expanding relations with another state actor, and an strategic alternative to other international actors such as western creditors. Its distinction also relies on the intervention of state-directed lending altering the way in which companies compete. China's ability to finance and many cases subsidized its oil companies grant access to heavy governmental intervention. The deals take place between state-state owned institutions. From China's side, the state-owned oil enterprise CNPC and Sinopec are tasked with the oil procedural transactions while both state-owned banks, China Development Bank and the Export-Import Bank of China, are clearly defined as "policy banks" by Beijing, focus on the financing side.[40]

On the Ecuadorian side, direct power has been vested to the Ministry of Finance to negotiate and sign the agreements. The state-owned oil company Petroecuador takes a secondary role in this case assuming the role. Preliminary interviews with former officials in the Ecuadorian oil industry highlight oil-for-loans agreements are managed by high-level officials close to the Executive and the Vice president office, the latter just recently in charge of strategic resources. Thus, the terms are managed by leading policy makers with the upmost secrecy, as none of the contracts are available for public information.

Since 2009 Ecuador has received 10 loans from China amounting to $9.9 billion. Eight of these agreements are on energy, 50% of which fall in the loans for oil category reaching $5 billion.[41]

Before 2008, there was no sign of China's financing muscle in the region; any Chinese loans were significantly below $1 billion and marginal disbursements by the Inter-American Development Bank (BID) and the World Bank (WB). The timing of the first Chinese loan to Ecuador followed the $3.2 billion debt default in 2008, raising assumptions Chinese funds have been used to replace access to Western credit.[42] Yet, other preliminary work suggests loans to Ecuador from BID and WB fell in 2005 before China entered the Latin America stage. Furthermore, such loans do not engage in the same domains. The terms of western credits concentrate in health and social programs while Chinese banks focus on development benefiting projects in infrastructure. Chinese financial practices nevertheless have risen attention to predatory finance in the way export credits and tied investments is used to secure oil deals.[43]

Ecuador and China's oil-for-loans are symptomatic of recent state-to-state energy deals trends focused on direct purchase and export credits, which consists on China providing cash advances repaid in oil sales to Chinese

Table 6.3 Oil for Loans in between China and Ecuador

Year	*Type*	*Lender*	*Amount (USD billion)*	*Purpose*	*Interest Rate*	*Premium per Barrel*
2009	Energy	Petrochina	1,000	Advance paymemnt for oil	7.25%	$ 1.25–1.30
2010	Energy	China EX-Im Bank	1,700	Coca-Codo-Sinclair hydroelectric dam		
2010	Energy	China Development Bank	1,000	80% discretionary 20% oil-related	6%	
2010	Energy	China EX-Im Bank	571	Sopladora hydroelectric dam		
2011	Energy	Petrochina	1,000	Advance payment for oil	6.9%	0.50
2011	Energy	China Development Bank	2,000	(Renewable energy development) 70% discretionary and 30% oil related	NA	
2012	Infrastructure	China EX-Im Bank	80	Road to Quito Airport		
2012	Other	China Development Bank	2,000	Finance 2013 budget deficit		
2013	Energy	China EX-Im Bank	312	Minas-San Francisco hydroelectric dam		
2013	Energy	Bank of China	299	Cañar and Naranjal dams		

CDB: China Development Bank, ICBC: Industrial and Commercial Bank of China.
Source: Gallagher Kevin, Amos Arwin, and Katherine Koleski.(2013). "Un mejor trato: Análisis comparativo de los préstamos chinos en América Latina."

oil corporations; a mechanisms that have been largely defined as a positive sanction in economic statecraft.[44] Transactions settle the volume of barrels shipped daily, an interest rate for the cash advance, dispute settlement terms, and some restrictions of oil resell.[45] Arrangements have been described by both parts as strictly commercial, harboring "mutual benefits" and "win-win" results.[46] Evidence suggests loans finance deficit in public spending, mostly toward infrastructure projects developed by the government.

A rising issue pinpoints to the efficiency and benefits of these loans in target countries like Ecuador. Is this really beneficial for Ecuador? What is its bargaining position? For now, analyses concentrate on the immediate outcomes with contradicting conclusions. Reports mainly from media outlets claim China has taken control of Ecuador's main export. A special report by Reuters claims as much as 83% of Ecuadorian oil production goes to Chinese corporations with figures expected to go as up as to 90% in the coming years.[47] For the smallest OPEC exporter, this could mean to have already committed its main source of revenue, a drastic shift for an industry that has been tightly controlled by the state. Ecuador as a result has only a 10% of its oil production to maneuver outside to Chinese commitment according to government officials. While the price of oil barrels allocated to China are valued at the market price the day they are shipped, loan transactions undermine the possibility to readjust supply flows and sell it to the highest bidder.[48]

In particular, since the first deal in 2009 the terms of the loans have evolved in China's favor. For example, the first contract guaranteed a premium for barrels of oil and prohibited reselling the oil to Ecuador's natural market, Peru, Colombia, and Chile. The following contract in 2010 cut the premium, approved the reselling of the oil in any market, and conditioned investment toward infrastructure projects. (OPEC countries prohibit reselling of their oil.) By the latest $1 billion loan in 2011, document leaks revealed a "Four Party Agreement" letter in which the Ecuadorian government grants Chinese companies a "sovereignty immunity waiver" to seize Ecuadorian oil from Ecuador's international partners if the country fail to repaid the debt.[49]

On the other hand, recent works by Gallagher, Irwin and Koleski argue oil for loans-agreements seem to benefit resource-rich Latin American countries such as Venezuela, Brazil, and Ecuador.[50] First, its interests are high but similar to what Ecuador would have to pay to international creditor as a high-risk country. Second, the volumes of oil shipment encompass more barrels than the ones needed it to repay the loans in order to obtain oil revenues. Calculations by Gallagher et al. estimate 79% of oil income returns to PetroEcuador after debt payments.[51]

In a wider spectrum oil commitments through new mechanisms such as oil-for-loans are potentially undermining the orthodox oil order and market rules. In terms of bargaining power, it means Ecuador, as other countries, may be relying on a single actor and enclosing its strategic asset to only one bidder. While is not new for Ecuador's oil production to depend on single customer, the United States was traditionally the main destination of oil exports, advanced credit terms seem to locked up future oil production.[52] The oil-for-loans agreements have been harshly criticized claiming China is securing oil at a low value and reducing the risks of lack of payment from Ecuador by guaranteeing oil supply. In all, conflicting arguments reveal the scant information on the oil-for-loans as a potential mechanism of positive inducement of economic statecraft and tied to other agreements.

CONCLUSION

As an importer, distributer and producer of oil, China will be a crucial actor in the oil market for years to come. Recently, China has increased its interdependence with Africa and Latin America and has become a relevant actor in resource diplomacy and international finance. Nevertheless, any application of the three main theoretical explanation of its oil quest should be taken with caution. Drastic security concerns are less likely; as a major importer, it is very unlikely that China can unilaterally withhold the flow of oil. Commercially speaking China remains an oil price taker and soft power continues to be an on going process. It is also apparent that at the empirical level, security supply, commercial rationale and soft power coexist. Future studies, must focus on identifying different economic mechanisms that inform us about policy preferences and the actual effect of economic and oil power beyond discussion of self-evident asymmetrical capabilities. This is particularly relevant when it comes to bilateral negotiations between states and the latitude of using these tools beyond the purchase of oil.

Contradicting and scant information on the Latin American experience, and Ecuador in particular, leaves many issues for further analysis. In particular the issue how Latin American countries leverage in this "win-win" framework and apparent policy of non-interference. It is clear China's interest in oil-for-loans agreements pursues both policy and profit objectives, as the magnitude of its capabilities allows greater leverage to address a multidimensional agenda. Yet it remains unclear the degree to which these agreements offer economic benefits to commodity-dependent economies which leads to suspect, for the supplier countries, politics weight heavily in the equation. Ecuador is feeling the negative impact of plummeting oil prices, as the margin of revenues has dropped by 47% and oil debt payment per barrel remains constant. The drop in oil crisis concomitant with China's slowing economy, may easily blame widening economic deficits of rentier countries on the old commodity curse narrative. To be true, the combination of both oil dependent economy and the volatility of oil prices do foresight acute vulnerability and shock waves. The other side of the story, however, has more to do with governmental preferences and their strategic use of their limited economic power to build extra-hemispheric relations or domestic support through this kind of policy. It is thus relevant that current analysis of oil politics asses oil bargaining beyond immediate results and scrutinized new oil practices in terms of the multiplicity of objectives and as an intermediate or instrumental factor to impact current oil competition (order), potential new forms of dependence, transparency issues, and alternative developmental arrangements.

NOTES

1. I am using the terms "global south" as used by the World Bank Institute to describe the increasing flows between developing countries, comprising countries in Africa, Latin America and most of Asia. Most of these countries are geographically located in the Southern Hemisphere but a comprehensive classification comprises medium and low-income countries (between 1,035 and 4,085 of Gross National Income per capita). According to the American University Center for the Global South, 157 out of 184 recognized states comprise the Global South. See for example, http://wbi.worldbank.org/wbi/devoutreach/article/525/rise-global-south-and-its-impact-south-south-cooperation http://www1.american.edu/academic.depts/acainst/cgs/about.html.

2. Deborah Brautigam and Tang Xiaoyang, "Economic Statecraft in China's New Overseas Special Economic Zones: Soft Power, Business or Resource Security?" *International Affairs* 88, no. 4 (2012). Michal Meidan, Philip Andrews-Speed, and Ma Xin, "Shaping China's Energy Policy: Actors and Processes," *Journal of Contemporary China* 18, no. 61 (2009); Suisheng Zhao, *China's Search for Energy Security: Domestic Sources and International Implications* (New York: Routledge, 2014).

3. Roland Dannreuther, "China and Global Oil: Vulnerability and Opportunity," *International Affairs* 87, no. 6 (2011); Christian Constantin, "Understanding China's Energy Security," *World Political Science Review* 3, no. 3 (2007).

4. U.S. Energy Information Administration, "China: Energy Report," (Washington D.C.: U.S. Energy Information Admnistartion, 2014), http://www.eia.gov/countries/cab.cfm?fips=ch.

5. Ibid.

6. Philip Andrews-Speed and Roland Dannreuther, *China, Oil and Global Politics*, (New York: Routledge, 2011).

7. Energy Information Administration, "China: Energy Report."

8. Ibid.

9. "The Pipelines Feeding China's Burgeoning Economy," *Pipelines International* 2011, https://www.google.com.br/search?q=The+Pipelines+Feeding+China%E2%80%99s+Burgeoning+Economy&rlz=1C5CHFA_enUS503US503&oq=The+Pipelines+Feeding+China%E2%80%99s+Burgeoning+Economy&aqs=chrome.69i57.723j0j4&sourceid=chrome&es_sm=91&ie=UTF-8.

10. Friedrich Wu and Koh De Wei, "From Financial Assets to Financial Statecraft: The Case of China and Emerging Economies of Africa and Latin America," *Journal of Contemporary China*, 23, no. 89 (2014).

11. Ibid.

12. Erica S. Downs, "China's Quest for Overseas Oil," *Far Eastern Economic Review*, 170, no. 7 (2007).

13. Ibid.

14. Brautigam and Xiaoyang, "Economic Statecraft in China's New Overseas Special Economic Zones: Soft Power, Business or Resource Security?"

15. Ibid.

16. Downs, "China's Quest for Overseas Oil."

17. Bates Gill and Yanzhong Huang, "Sources and Limits of Chinese 'Soft Power'," *Survival* 48, no. 2 (2006); Library of Congress Congressional Research Service, "China's Foreign Policy and 'Soft Power' in South America, Asia, and Africa," (Washington D.C. : Library of Congress, 2008).

18. Juan Gabriel Tokatlian, "A View from Latin America," in *China's Expansion into the Western Hemisphere. Implications for Latin America and the United States*, eds. Riordan Roett and Guadalupe Paz (Washington D.C.: The Brookings Institution, 2008).

19. John Williamson, "Is the" Beijing Consensus" Now Dominant?" *Asia Policy* 13, no. 1 (2012).

20. (1) Territorial integrity and sovereignty (2) non-violence; (3) non-interference in internal affairs (4) equality and mutual benefit and (5) peaceful coexistence.

21. Ian Taylor, "China's Oil Diplomacy in Africa," *International Affairs* 82, no. 5 (2006).

22. Quincy Wright, *The Study of International Relations*, The Century Political Science Series (New York,: Appleton-Century-Crofts, 1955).

23. Daniel Kilman, "Is China the Fastest-Rising Power in History?," *Foreign Policy*, May 16, 2014–2014.

24. Ibid.

25. Brautigam and Xiaoyang, "Economic Statecraft in China's New Overseas Special Economic Zones: Soft Power, Business or Resource Security? Kevin Gallagher, Amos Irwin, and Katherine Koleski, *¿Un Mejor Trato?: Analisis Comparativo De Los Prestamos Chinos En America Latina* (2013).

26. George Philip, *Oil and Politics in Latin America: Nationalist Movements and State Companies* (Cambridge: Cambridge University Press, 1982).

27. Ibid.

28. CEPALSTAT, Economic Commission For Latin America and the Caribbean, 2014, http://estadisticas.cepal.org/cepalstat/WEB_CEPALSTAT/Portada.asp?idioma=i.

29. Gerardo Arriaga, "Energy in Latin America 2010–2020: Headed Fo a Lost Decade?" in *The Americas in Motion: Looking Ahead 30th Anniversary Sol M. Linowitz Forum Commemorative Volume* (Washington D. C.: Inter-American Dialogue, 2013).

30. Business Monitor, "Ecuador: Oil & Gas Report," (London: Business Monitors, 2012). http://store.businessmonitor.com/ecuador-oil-gas-report.html.

31. Ibid.

32. Andres Schipani, "Ecuador Punts Its Oil Bidding Round but China May Be Only Taker," *Financial Times*, April 24, 2013. http://blogs.ft.com/beyond-brics/2013/04/24/ecuador-punts-its-oil-bidding-round-but-china-may-be-only-taker/.

33. Monitor, "Ecuador: Oil & Gas Report."

34. "Ecuadorians Fear Their Country Is Being Taken over by China," *Business Insider*, August 21, 2012, http://www.businessinsider.com/ecuadorians-fear-their-country-is-being-taken-over-by-china-2012-8 .

35. Amazon Watch, "Beijing, Banks and Barrels: China and Oil in the Ecuadorian Amazon," (Washington D.C.: Amazon Watch, 2014, http://amazonwatch.org/assets/files/2014-beijing-banks-and-barrels.pdf.

36. Brautigam and Xiaoyang, "Economic Statecraft in China's New Overseas Special Economic Zones: Soft Power, Business or Resource Security?"; Taylor, "China's Oil Diplomacy in Africa."

37. Gallagher, Irwin, and Koleski, *¿Un Mejor Trato?: Analisis Comparativo De Los Prestamos Chinos En America Latina.*

38. Taylor, "China's Oil Diplomacy in Africa."

39. David Allen Baldwin, *Economic Statecraft* (Princeton: Princeton University Press, 1985).

40. Hoy, "Más Deudas Con La China," *Hoy*, July 16, 2010, http://www.explored.com.ec/noticias-ecuador/mas-deudas-con-la-china-419176.html.

41. Gallagher, Irwin, and Koleski, *¿Un Mejor Trato?: Analisis Comparativo De Los Prestamos Chinos En America Latina.*

42. Kevin Gallagher, Katherine Koleski, and Amos Irwin, *The New Banks in Town: Chinese Finance in Latin America* (Washington D.C.: Inter-American Dialogue, 2012), http://www.ase.tufts.edu/gdae/policy_research/NewBanks.html.

43. Peter Evans and Erica Downs, "Untangling China's Quest for Oil through State-Backed Financial Deals," (Washington D.C.: The Brrokings Institution, 2006), http://www.brookings.edu/research/papers/2006/05/china-downs.

44. Joshua Schneyer and Nicolas Medina, "China's Power Play in America's Backyard," *Reuters, Nov. 26,* 2013, http://uk.reuters.com/article/2013/11/26/china-ecuador-taurus-idUKL2N0IT2BH20131126; Baldwin, *Economic Statecraft*.

45. Schneyer and Medina, "China's Power Play in America's Backyard."

46. Hoy, "Ecuador Lidera Inversión China En Latinoamérica," *Hoy*, Sept. 20, 2010, http://www.explored.com.ec/noticias-ecuador/ecuador-lidera-inversion-china-en-latinoamerica-431041.html.

47. Schneyer and Medina, "China's Power Play in America's Backyard."

48. Ibid.

49. Watch, "Beijing, Banks and Barrels: China and Oil in the Ecuadorian Amazon."

50. Gallagher, Irwin, and Koleski, *¿Un Mejor Trato?: Analisis Comparativo De Los Prestamos Chinos En America Latina.*

51. Ibid.

52. U.S. Energy Information Administration, "Ecuador: Energy Report," (Washington D.C.: U.S. Energy Information Admnistartion, 2014), http://www.eia.gov/countries/country-data.cfm?fips=ec.

BIBLIOGRAPHY

Andrews-Speed, Philip, and Roland Dannreuther. *China, Oil and Global Politics.* New York: Routledge, 2011.

Arriaga, Gerardo. "Energy in Latin America 2010–2020 Headed for a Lost Decade?" In *He Americas in Motion: Looking Ahead.* Washington D.C.: Inter-American Dialogue, 2013.

Baldwin, David Allen. *Economic Statecraft.* Princeton: Princeton University Press, 1985.

Brautigam, Deborah, and Tang Xiaoyang. "Economic Statecraft in China's New Overseas Special Economic Zones: Soft Power, Business or Resource Security?" *International Affairs* 88, no. 4 (2012): 799–816.

CEPALSTAT, Economic Commission For Latin America and the Caribbean, 2014, http://estadisticas.cepal.org/cepalstat/WEB_CEPALSTAT/Portada.asp?idioma=i.

Congressional Research Service, Library of Congress. "China's Foreign Policy and 'Soft Power' in South America, Asia, and Africa." 139. Washington D.C.: Library of Congress, 2008.

Constantin, Christian. "Understanding China's Energy Security." *World Political Science Review* 3, no. 3 (2007).

Dannreuther, Roland. "China and Global Oil: Vulnerability and Opportunity." *International Affairs* 87, no. 6 (2011): 1345–64.

Downs, Erica S. "China's Quest for Overseas Oil." *Far Eastern Economic Review*, 170, no. 7 (2007): 52–56.

"Ecuadorians Fear Their Country Is Being Taken over by China," *Business Insider*, August 21, 2012.

Energy Information Administration, U.S. "China: Energy Report." Washington D.C.: U.S. Energy Information Admnistartion, 2014.

———. "Ecuador: Energy Report." Washington D.C.: U.S. Energy Information Admnistartion, 2014.

Evans, Peter, and Erica Downs. "Untangling China's Quest for Oil through State-Backed Financial Deals." Washington D.C.: The Brookings Institution, 2006.

Gallagher, Kevin, Amos Irwin, and Katherine Koleski. *¿Un Mejor Trato?: Analisis Comparativo De Los Prestamos Chinos En America Latina.* Washington D.C.: Inter-American Dialogue 2013.

Gallagher, Kevin, Katherine Koleski, and Amos Irwin. *The New Banks in Town: Chinese Finance in Latin America.* Washington D.C.: Inter-American Dialogue, 2012.

Gill, Bates, and Yanzhong Huang. "Sources and Limits of Chinese 'Soft Power'." *Survival* 48, no. 2 (2006): 17–36.

Hoy. "Ecuador Lidera Inversión China En Latinoamérica." *Hoy*, September 20, 2010.

———. "Más Deudas Con La China." *Hoy*, July 16, 2010.

Ikenberry, G. John, David A. Lake, and Michael Mastanduno. *The State and American Foreign Economic Policy.* Ithaca: Cornell University Press, 1988.

International, Pipelines. "The Pipelines Feeding China's Burgeoning Economy." *Pipelines International*, 2011.

Kilman, Daniel. "Is China the Fastest-Rising Power in History?" *Foreign Policy*, May 16, 2014.

Meidan, Michal, Philip Andrews-Speed, and Ma Xin. "Shaping China's Energy Policy: Actors and Processes." *Journal of Contemporary China* 18, no. 61 (2009): 591–616.

Monitor, Business. "Ecuador: Oil & Gas Report." London, 2012.

Philip, George. *Oil and Politics in Latin America: Nationalist Movements and State Companies*. Cambridge: Cambridge University Press, 1982.

Schipani, Andres. "Ecuador Punts Its Oil Bidding Round but China May Be Only Taker." *Financial Times*, April 24, 2013.

Schneyer, Joshua, and Nicolas Medina. "China's Power Play in America's Backyard." 7. New York: Reuters, 2013.

Spegele, Brian. "China Opens up on Oil Reserves." *The Wall Street Journal*, November 20, 2014.

Taylor, Ian. "China's Oil Diplomacy in Africa." *International Affairs* 82, no. 5 (2006): 937–59.

Tokatlian, Juan Gabriel. "A View from Latin America." In *China's Expansion into the Western Hemisphere. Implications for Latin America and the United States*, edited by Riordan Roett and Guadalupe Paz. Washington D.C.: The Brookings Institution 2008.

Watch, Amazon. "Beijing, Banks and Barrels: China and Oil in the Ecuadorian Amazon." Washington D.C.: Amazon Watch, 2014.

Williamson, John. "Is the" Beijing Consensus" Now Dominant?" *Asia Policy* 13, no. 1 (2012): 1–16.

Wright, Quincy. *The Study of International Relations*. The Century Political Science Series. New York,: Appleton-Century-Crofts, 1955.

Wu, Friedrich, and Koh De Wei. "From Financial Assets to Financial Statecraft: The Case of China and Emerging Economies of Africa and Latin America." *Journal of Contemporary China*, 23, no. 89 (2014): 1–23.

Zhao, Suisheng. *China's Search for Energy Security: Domestic Sources and International Implications*. New York: Routledge, 2014.

Chapter 7

Economic Growth and the Role of India in the Global Energy Security System

Challenges Ahead

Diana Soller

India is a state of many paradoxes. It gained its independence in 1947 after a fierce struggle against British imperialism, but retained the Raj's major bureaucratic and military institutions. It is proud of its Hindu civilizational past, but it chose to build the state under constitutional secularism.[1] Its project of modernization was a composite of different ideologies of progress:[2] nationalism created through the independence movement as the vision of the Indian future; state centralized economy with social democratic redistribution (that never quite worked) that changed into a market oriented economy in the 1990s that never really waved the presence of the state;[3] and a pluralist democracy as the way to bind together a myriad of elements originated in different peoples' with different languages, cultures and religions.[4]

Internationally, right after the independence, India adopted an internationalist foreign policy known as nonalignment, that gave New Delhi based on the idea of a an independent foreign policy, free from great powers' constraints, which later was replaced by an alliance with the Soviet Union in the 1970s, and a return to an independent, however more pragmatic, foreign policy in the 1990s after the dismiss of the USSR. Since then, India refuses permanent alliances but is open to partnerships with diverse international actors, with preference for great and emerging powers and states with which it shares historical cordial relations.

Since the 2000s, New Delhi embarked on a path of fast growth, that created on India's elites a sense that despite all the remaining internal problems—inequality, poverty, bureaucratic inefficiency—India has the opportunity of raising into the status of international great power. However, to become a great power *de facto*, there are at least three conditions that New Delhi has to meet: enhancing internal development, maintaining the fast pace growth, and be recognized by its pears as an international responsible stakeholder. Despite the enthusiasm, Indian elites are well aware of the great challenges ahead.

To a certain extent, this brief description mirrors India's energy security policy. The three major problems facing New Delhi are (1) the lack of coordination among state agencies; especially since the beginning of the 2000s, India's fast growth creates further energy needs that requires planning ahead; (2) the competition with China, that has a similar energy profile but started its energy quest before India; and (3) the need to reconcile India's development and fast growth programme with the political will to negotiate a climate change regime according to high international standards. Although there is an almost unanimous claim that India's foreign policy should foster internal development,[5] pollution already has and global warming will have great negative impact on India in the future.

In what concerns energy security in a global perspective, the policies India is going to adopt will have impact in the international system in three ways: (1) India is a very large energy consumer state and tends to be even larger and very dependent on oil imports. As the dependence tends to grow up to more the 90%, India's quest for energy has the potential to create unbalances; (2) India's quest for resources is in direct competition with China. The way New Delhi and Beijing will handle their resource competition will deeply influence their relationship and the regional security and stability in Asia; and (3) India is one of the major international polluters; therefore its engagement in climate change international regimes—that will always be balanced with its process of industrialization and economic growth as suggested above—will have an impact in the reduction of the global CO2 footprint.

To understand the complex picture of India's energy security and its challenges, this paper is organized in five sections. It starts by articulating a theoretical approach necessary to shed light over the challenges of late developers in today's globalized world. This section will specify which the possible paths ahead are. The second outlines India's profile as an energy producer and consumer. This is important to determine the state of the art and possible scenarios for India's energy security strategy. The profile will allow the identification of security gaps and what decision makers are trying to do suppress them. Third, it will look at the historical background of energy security policies: it will show that India's energy security went through four phases: the first was characterized by diversification of sources in the market; the second by a dependency of the Soviet Union; the third to a readjustment to diversification of sources. The last one that started at the beginning of the 2000s, was induced but India's rapid growth and the need to find more reliable sources of energy supply. As we will see in section four, the attempts have so far produced meager results. The last section will equate India's energy security with the wider international system, through looking at the relationship between energy security and climate change. This section will highlight part of the external challenges that India has to face in this process of becoming energetically secure. They range from Western pressure to the creation of sources of cleaner energy and adaptation to an international environmental regime that does not favor India's immediate national interest; the need to develop more consequent cooperation with China which at this moment is the more important India's competitor for energy supply; and the need to deal with an hostile and unstable neighborhood which has been the greatest obstacle to India's acquisition of natural gas.

ENERGY SECURITY AND INTERNATIONAL RELATIONS THEORY: INDIA IN A WORLD OF SCARCE RESOURCES

Energy security—defined by the Indian government as the ability to "supply lifeline energy to all citizens as well as meet their effective demand for safe and convenient energy to satisfy various needs at affordable costs at all times with a prescribed confidence level considering shocks and disruptions that can be reasonably expected"[6]—comes with particular features in the twenty-first century.

To the classical competition for scarce resources and the increase of instability on geographical areas abundant in resources (i.e., Russia, the Middle East, Venezuela), one must add now the fact of the rise of new actors in the international system, that are at once developing states and candidates to the status of great powers. By definition, fast developing states require greater energy supplies both to enhance their industrialization and to respond to the demands of their growing middle class.[7] Simultaneously, achieving the status of great powers requires continued industrialization and the maintenance of fast growth rates. These two conditions, together with the fact that these countries do not produce nearly enough energy to fulfill their needs, make states like India and China very demanding—and unpredictable—energy seekers. In this section, this article looks for a way to theorize how these particular states behave in what concern their energy policies.

NEOREALISM AND NEOLIBERAL INSTITUTIONALISM DURING THE COLD WAR

During the Cold War structural realism and neoliberal institutionalism provided an answer to almost every question of international relations.[8] Mainly due to Kenneth Waltz seminal work *Theory of International Politics*,[9] and the rise of neoliberal institutionalism that developed a theory of interdependence and cooperation based on similar premises of anarchy, states' rationality and conditionality,[10] systemic theories became widely accepted.

They put forth a few predictions: as for realism, states were to act according to constraints and opportunities provided by the international system (anarchy was insurmountable); states would copy one another's good practices; and they would ally or try to increase their power capabilities to balance third parties, as others' international behavior was impossible to predict, and survival required a meticulous strategy of self-help. In some cases—liberals would add to realists' conclusions—states would cooperate if that would come with mutual benefits—international regimes were nothing but the result of the political will of same states—usually the most powerful ones in the system—to solve common problems or enhance their mutual benefits. Importantly, structural realism and neoliberal institutionalism based their predictions in strict state rationality—and rational behavior could only be identified with behaviors stated in their theories.

In sum, the international system was conceived as anarchic, the states were black boxes or billiard balls—meaning rational actors that behaved similarly because they have the same international aims[11] being self-help the top priority. Logically, in a world of scarce resources, developing states such as India, would build their foreign policy on energy security as a mix of zero-sum game of relative power—trying to assure access to resources that other states would not be able to reach—and absolute gains of cooperation, also necessary to assure lack of interference in energy transportation, transparency in relations with producer states and to avoid resources conflicts. In this view, any political tool would be acceptable (including armed conflict) as long as the Indian population would have comfortable and continued access to the necessary energy.

NEOLIBERAL INTERNATIONALISM REVISED AND GLOBALIZATION COSMOPOLITANISM AT THE END OF THE COLD WAR

The end of the Cold War came with a new reality: the United States became the sole superpower. Despite the realist claims that power politics would soon make a comeback with all its security dilemmas attached[12]—including the ones related with energy security—and the fact that anarchy still explained some regional security complexes,[13] the idea that the international system was turning into a more hierarchical form became widely accepted. The post-1990 international order could "not eliminate international bargaining" but displaced "brute force to the nature and the extent of the authority exercised by the dominant state."[14]

There were at least two versions on how this new international order would affect states' relations. One represented an evolution of the liberal international tradition[15] and the other was a more cosmopolitan approach to economic international relations, where its advocates applied modernization theory to international relations, especially globalization.[16] Both concluded the same through different explanations and methods: due to the international affairs' new hierarchy, driven by hegemony or by the laws of the market, states did not need to resort to war for scarce resources as there were mechanisms to access them through institutional (neoliberal institutionalism) or supranational (cosmopolitanism) relations. States like India would have their role and status in the twenty-first-century international order and their fair share of access to international resources, as long as they were willing to accept the new rules of the game.

From a neoliberal institutionalist perspective, what started to happen in the beginning of the 1990s was that zero-sum games could be replaced by win-win strategies that had the advantage of a more just distribution and a more peaceful international system. This less idealist approach to the "harmony of interests" theory[17] enounced by liberal institutionalists[18] was to be regulated by a hegemonic power, now global, as one of its provision of public goods' task[19] mostly through international institutions and regimes.[20] The American-led liberal order promoted "open markets, international institutions, cooperative security, democratic community, progressive change, collective problem solving, shared sovereignty [and] the rule of law."[21] Being India a democracy, it would be in optimal conditions to join the "hierarchical order with liberal characteristics"[22] and enjoy its security policies. That was offered both in theoretical and practical terms, especially in the nuclear agreement sign by the United States and India in 2006. More than the treaty itself, it was a demonstration that America would accept India among its allies. At least in theoretical terms, this should also apply to access to energy, that should also be regulated through international institutions.

The cosmopolitan view—also an interpretation of the "harmony of interests" in an economic perspective—was shared by enthusiasts of economic neoliberalism and cosmopolitanism alike. The common view was that the international system was going through a phase of transformation triggered and accelerated by the forces of globalization and the "new international division of labor."[23] The result would be an increase of wealth throughout the countries that engaged in market economy (neoliberalism) and/or the blurring of international inequalities (cosmopolitans).

Open markets would also have a positive impact in energy security as increased openness would provide wider possibilities of diversification of sources. In the overall, globalization was perceived as an opportunity for a wealthier and more convergent world. This optimistic view fitted the Indian case. After the collapse of the Soviet Union in 1991 and the almost simultaneous foreign debt crisis, India requested substantial international loans that were granted by international institutions in exchange for market reforms.[24] These two traditions perceived India's reforms as a path to greater growth and international integration in the US-led liberal order.[25]

However, this enthusiasm with order and globalization was curtailed but the 2008 economic crisis that undermined both the United States legitimacy as a global leader and the Washington Consensus.

IR THEORY, MODERNITY AND THE RISE OF DEVELOPING POWERS

The international economic crisis of 2008 had a double effect: to shed light over the fragilities of the optimist perceptions of the end of the 1990s and the 2000s and to acknowledge the economic rise of a number of states with different political characteristics of the traditional great powers, namely China, India and Brazil. Russia also seemed to be making its comeback in world affairs, but its trajectory as a resurgent power made it a different case from the others.

The theoretical hypothesis put forward so far—(1) the international system favors a race for scarce resources (2) punctuated by cooperation, (3) that became more likely after the end of the Cold War due to the globalization of the US hegemony and the globalization of world market—became insufficient in the presence of these new actors. To understand which of these hypotheses would stand, it became necessary to apply to the systemic predictions a reductionist theory.

A reductionist theory (an unfortunate name that does not do justice to its content) "is a theory about the behavior of the parts" (i.e., the states preferences also determine outcomes in the international system). From a reductionist perspective, "international outcomes are simply the sum of the results produced by the separate states, and the behavior of each of them through its internal characteristics."[26] This new perspective became necessary because states like India combine two elements that were not common in international relations until recently: they are rising states that can challenge the balance of power in the international system[27] but they are simultaneously developing states, i.e., their modernization process is still to be completed.[28] International relations theory is mainly about relations of great powers, with their bureaucratic, institutional and economic apparatuses fully developed. Being India in its transition phase it is pertinent to ask how modernization would impact its foreign policy and quest for energy security.

There are at least three modernization theories. Barrington Moore's that defends that relations among social classes originate different routes to modernity, determining model of development and regime type[29]—also very dependent on the colonial heritage, when there is one;[30] Samuel Huntington's that theorizes about the importance of states' capacity and institutional quality, tested in times of change that require the right blend of authority/legitimacy and flexibility to absorb new elements at each critical juncture;[31] and Martin Seymour Lipset's liberal modernization theory, a progressive view of history, that defends that industrialization leads to a logic of modernization as follows: states develop economically, their middle classes enlarge and they demand more freedom and constitutional guarantees—which leads to institutional reforms.[32]

All three theories point to stages of development: there is path dependency (colonialism heritage) and certain steps that each state has to follow from their traditional social status (sometimes referred as pre-modern) into a fully modern society. As modernity refers to political, economic and social changes—reflected later in institutional arrangements—it is essential that they reflect historical patterns and logically different outcomes. Even Lipset's more deterministic approach allows room for intervening variables—namely endogenous (regime type) exogenous (international system shocks) obstacles to the fruition of liberal modernity. They also commonly point the need of developing a middle a class that in many ways is the trigger for modernization, especially in countries with democratic regimes. As a democracy, India's government legitimacy is highly dependent on the stable delivery of public goods to the growing middle class.

More recently, Barry Buzan and George Lawson developed an analytical framework to analyze the modernization processes. It contains three elements: (1) industrialization—completed when there is full integration in the global market; (2) "rational state-building," "the process by which administrative and bureaucratic competence were accumulated and cage within national territories"; and (3) an "ideology of progress," providing the basis of social consensus and in first stages of modernization, how to proceed through modernization stages.[33] The authors also reinforced the idea expressed by other modernization theorists that colonization might have brought some forms of institutionalization to colonies—as mentioned before, India inherited the British bureaucracy—but also led, in some cases, to de-industrialization, in detriment of the extraction of resources. India is pointed out as an example of these practices.[34]

In what concerns India's modernization stage regarding these three indicators, the level of development is mixed. India has its own ideology of progress. It was developed in the 1920s and the 1930s for the nationalist movement that came to be agglomerated in the Congress Party, at once highly eclectic but united around a compromise toward progress, nationalism, secularism and democracy.[35]

In what concerns state-building, India has been showing a remarkable institutional resilience and flexibility in absorbing cyclical changes. The most important moments were the reenacting of democracy after the Emergency period (1975–1977)[36] and the adaptation to market economy in the 1990s. However, there are also some "battles half won." Two examples are difficulties in elimination of poverty and inequality,[37] or the number of Indians who work in the informal sector—93% in 2013.[38] This means that only 7% of the population pays labor taxes. These indicators reveal weak state capacity visible also on energy distribution. As we will see later, around 300 millions of Indians does not have access to electricity.

A set of reasons explain lack of state capacity. Besides the already mentioned colonial legacy, India still has to manage other deeply ingrained problems such as the heritage of hierarchy of its traditional society, bureaucratic inefficiency, power struggles between regional and central government, and the closeness of Indian elites.[39] In a recent account of India's foreign policy preferences, the authors noted that it is hard to pinpoint how much influence civil society has in the process of decision making.[40]

The industrialization measurement it harder to access: competitiveness in the market is still an important indicator, but there is much less consensus on the in what concern the economic model. However, three points should be made: first, India's economy is far from being fully liberalized. As a matter of fact, "India's openness to the world remains relatively modest." The major shift consists in moving away from socialism into "a warm embrace of state and business groups."[41] Second, to reinforce the previous point, the second wave of economic reforms never took place (partly because the Indian elites were satisfied with the fast growth results of the first wave). Finally, this has implications in energy security, as the market is highly liberalized and the Indian state frequently creates barriers to energy products traded by private companies.[42]

In sum, the international relations theory predicts a number of outcomes ranging from conflict as a consequence of the quest for scarce resources to cooperation in a more or less hierarchical or globalized order. However, the specificity of states like India, simultaneously modernizing and rising and the way its institutions will evolve will have a deep impact on the foreign policy choices regarding energy security.

INDIA'S PROFILE AS AN ENERGY PRODUCER AND CONSUMER

A successful national energy security policy today has to fill at least three requirements: the provision of the diverse forms of energy to the population, the assurance of medium-term access to energy according to economic growth needs, and to contribute to reducing the Carbone footprint in the planet. Accordingly, the way India will conduct its foreign policy of energy security will have major consequences globally.

India is the forth energy consumer country in the world, and its condition of developing state with a very dynamic economic and with "the needs of energy supplies continues to climb,"[43] will tend to create additional stress in the quest for finite resources. Furthermore, according to the World Health Organization of the United Nations, thirteen of the twenty most polluted cities in the world are in India—being New Delhi the most polluted of all.[44] Equally, the way India will handle its participation in climate change regimes will have global consequences.

Domestically, India is at a crossroad. It has to improve the problem of lack of state capacity (beaurocratic fragmentation in what concerns energy security), deficient infrastructure, and distribution problems. The per capita consumption in very low compared to other countries (490 kg of oil and 1780 WA of electricity) and 25% of the population does not have access to electricity. The 75% that does have experience regular energy cuts and electricity blackouts throughout the country. The pressure for augmenting capacity is not only due the shortages mentioned before it is also related to the exponential growth of the middle class, estimated to rise from 5% of the population (in 2007) to 41% until 2025.[45]

Equally important is the migration from the countryside into the cities. According to the last census (2011), the number of people living in cities raised 3.25% reaching 37.7 million. The numbers may seem negligible, but authors point to "urban distress" in many Indian regions.[46] This picture leads to a logical conclusion: in India, the "social distribution of energy" is not efficient, as "despite the claims of increased penetration [of energy in the society], in the case of disadvantaged sections (Scheduled Castes, Scheduled Tribes and Other Backward Classes), the number of households using biomass for lack of modern energy option has actually gone up in certain income categories."[47]

Still in the domestic realm, it is important to note that India produces 70% of the energy it consumes. It holds the third largest reserves of coal in the world and it produces a large amount of energy from biomass and waste, and a residual amount of nuclear energy. The energy production sector has grown in the last years, but not enough to keep up with the demand.[48] The same is valid for exports (especially cool): its recent increase does not compensate for India's external dependence—especially of oil—creating an energy trade deficit of US$190 billion in 2012–2013.[49]

Moreover, as the graphic below illustrates, only 4% of Indian energy is originated in renewable and nonpollutant sources. This creates a strong pressure both from the internal public opinion and the international community, both very critical of India's environmental records.

Internationally, India has two immediate problems: the most important is the dependence on external oil. As a "high energy-intensive economy" New Delhi is the largest net importer of crude oil and petroleum products.[50] Currently of crude oil and petroleum products external dependency reached 70% and is likely to rise to 92% in five years. As the graphic bellow shows production remained steady while consumption almost doubled in the last fifteen years.

Sixty two percent of the oil imports came from the Middle East—mainly from Saudi Arabia, Iraq and Iran—an especially unstable world region. The other significant importers, Venezuela and Nigeria, cyclically go through political crisis—recently Caracas was considered a security threat by the US government and it is enduring a long political and economic crisis. Interestingly, the imports from Russia are negligible, in sharp contrast to what use to happen during the Cold War. In the overall, the sources are diversified (there is no dependency on one country) but New Delhi was not able to reach more stable markets that have been upstaged by China.

The natural gas dependency is not as significant, ranging from 15% to 20%.[51] Its consumption produces other types of problems: most pipelines or pipeline projects pass through neighboring hostile countries or unstable regions. The one in Myanmar is already dominated by China's investment. The Iran-Pakistan-India pipeline project, while logical in terms of supply-demand (Iran is a large producer while India and Pakistan are large consumers) has been in check due to poor bilateral relations among the two subcontinent neighbors. The same applies to the project of Turkmenistan-Afghanistan-Pakistan-India; the latent conflict between Delhi and Islamabad and the great instability of Afghanistan—that became more evident since the terrorist attack in Peshawar in December 2014—have kept India's

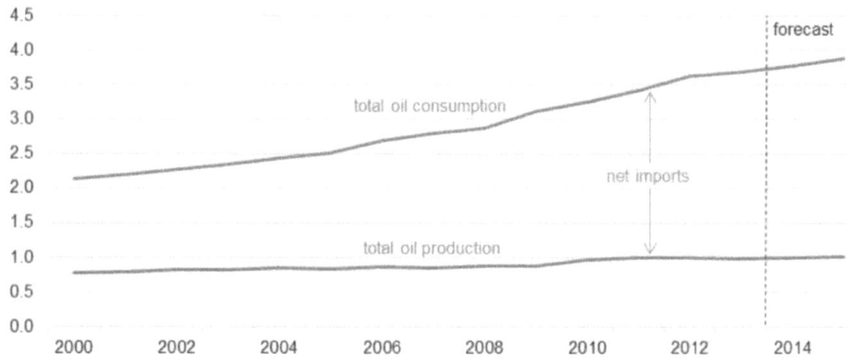

Figure 7.1 India petroleum and other liquids production and consumption, 2000–2015. *Source*: U.S. Energy Information Administration, International Energy Statistics and Short-Term Energy Outlook June 2014.

government undecided about it. The same applies to some possible route coming from Russia and Central Asia that has been postponed for years, and became more difficult since the war with Ukraine started.[52]

Despite these difficulties, India has so far been able to assure enough oil and gas imports through a using a strategy of "energy mix." This strategy is likely keep on being pursued in the future, but the amounts purchases required is increasing fast. According to the Indian Plan Commission, if India keeps its growth rates at 5% to 6% a year, by 2030 the energy demand will triple and the electricity demand will rise five to six times. Similarly, commercial energy requirements are expected to increase 6% a year.[53] In 15 years India energy mix will require 45% of oil and gas—still a significant dependency from outside providers.

FOREIGN POLICY AND ENERGY SECURITY: HISTORICAL BACKGROUND AND PRESENT

This article's introduction mentioned three phases of India's policy: the Nehruvian phase that lasted from Independence to 1964; Indira Gandhi's years (followed by a transition period from ranging from her death in 1984 to the beginning of the end of the Cold War) and the current period that started in the 1990s and it is ongoing. It also suggested that the 2000s are a decade of paramount importance, as they correspond to India's change in self-perception. At this juncture, Indian elites started to think about their country as a possible future great power.

These four periods roughly translate into the four phases of India's energy security policy.

1. Between 1947 and 1971, India based its energy security strategy in the diversification of sources. As policies are path dependent, the suppliers' choice was deeply influenced by India's colonial legacy: the providers were mainly Western oil companies. This diversification strategy started to change in the middle of the 1960s. India reinforced the centralized management of its industry and implemented a comprehensive Imported Substitution Industrialization plan, aiming self-sustainability.[54] This political choice drew India closer to the Soviet Union that started, along with Romania, to be one of India's chief energy providers. Accordingly, New Delhi centralized its oil acquisitions in the Indian Oil.[55]
2. The dependency of the USSR rose since 1971 when India signed a twenty years Peace, Friendship and Cooperation Treaty with the Soviet Union.[56] The USSR became the major energy provider of India and remained so for the next 20 years; Moscow provided energy at low prices and produced

enough to meet India's needs.[57] This period is framed by an overall discontinuity in India's foreign policy from autonomy and non-alignment to proximity with a great power.[58]

3. The 1990s were a lost decade to India. The collapse of the Soviet Union had a major impact in New Delhi for three reasons: first because the USSR was the major provider of vital elements for India's development including, as mentioned above, energy, but also military and economic aid; second because the Soviet Union provided the model of centralized economic development adopted by Nehru and maintained by Indira Gandhi, and that same model was the reason for the Soviet Union dismiss. India was left without references and in need to recreate itself economically. Third, and more importantly, the Soviet Union abruptly interrupted the provision of public goods to India, precipitating the New Delhi in a financial crisis of foreign debt.[59]

These events set India in the course of changing its economic and financial policy. India requested the IMF support and engaged in a comprehensive set of reform toward a more market oriented policy. These reforms were not entirely new: in the end of the 1980s Rajiv Gandhi had already started to forge new elite pacts with the Indian business elite and was gradually introducing market friendly measures. Some authors go as further as to say that this movement toward the markets started very slowly during Indira Gandhi's administration.[60] However, the debt crisis and the IMF intervention accelerated the process that is still to be concluded: the first round of reforms was very successful, as India's growth rate rose greatly, but the second round of reforms were never implemented, and there is no indication that it will be in the near future.[61]

It took India some time to adapt to the absence of the Soviet Union both in provision of energy as well as in the security of its transportation—as India's neighborhood is unstable and without Moscow to balance New Delhi's rivals' new problems rose. The Indian decision makers opted for diversifying energy providers, which has been considered a successful policy until the turn of the century. Despite the changes from multiple suppliers into dependency of the Soviet Union and again a turn to multiple suppliers, there is certain continuity among these three periods: the energy file was handled a domestic issue, and its management was spread through different ministries across the Indian administration—according to the final uses of energy. Moreover, the first and third periods were guided by the conviction that diversification was sufficient to secure provision of energy, according to India's needs. However, this idea was not going to last long. A decade later, India would declare its energy insecurity and try to change its policies on that regards. What changed between 1991 and the beginning of the 2000s?

2000S: WHEN ENERGY BECAME A FOREIGN POLICY ISSUE.

The energy security picture in India dramatically changed in the 2000s, as its leaders' realized that energy was in the verge of becoming an acute problem. Former Prime Minister Manmohan Singh affirmed in 2004 that "the quest for energy security is second only in our scheme of things to our quest for food security. India is dependent on imported energy and what goes in the world today the growing instability of supplies, give rise to new challenges (. . .) Energy security is of critical importance to India."[62]

The change is the official discourse as well as in the policies is due to three factors: (1) the growth of the middle class demanding better access to energy goods (and this new middle class is preponderant as it is formed by the new consumers that the India government wants to preserve and enlarge); (2) energy started to be seen as "a significant constraint on India's ability to reach and maintain 8 to 10% growth over the next several decades, and to take advantage of the so-called 'demographic dividend' will be the availability of energy resources";[63] (3) India's self-perception changed; as an aspiring great power, New Delhi has to be sure that its growth and power will not be "limited" by poor access to energy, as energy is an important "driver of competition" and India is currently "structurally dependent" on third parties, which perils its cherished "independency" in foreign policy decision making.[64] As such, energy security ceased to be a domestic concern to be a foreign policy issue.

This change of perception had two immediate results: it led to the launch of a number of separated policies in the name of the improvement of the overall energy security—being the chief policy the creation of a more centralized bureaucratic mechanism to decide about energy security. The second result was to reveal a grim picture for India's future energy security. New Delhi faces three structural problems yet to be solved: (1) bureaucratic fragmentation—that leads to the absence of a comprehensive energy security strategy; (2) the design of its economic institutions: India's energy market purchase is highly regulated which creates difficulties in the liberal energy market in which it is competing; and (3) perhaps the most difficult to solve, is the competition with China for energy resources. The Asian aspiring powers have a similar energy profile, but Beijing started its quest for energy earlier than New Delhi. Furthermore, China has more competitive conditions to offer, becoming the most likely winner in competitions for scarce resources.

BUREAUCRATIC FRAGMENTATION

Since the beginning of the 2000s the Indian government launched a number of initiatives to improve India's performance on energy security. The major bureaucratic changes were two—energy security came to be discussed and decided at the level of the Foreign Ministry, including the establishment of the Energy Security Unit of Foreign Ministry in 2007. The goal of this Unit was to make India a self-sufficient state in what concerns energy resources in 2030 (more recent books and reports look at the United States recent shift to energy sufficiency as an example), but at this critical juncture, this is a very difficult task, due to the already mentioned heavy dependency on imported oil. A significant part of it is used transportation fuel, very difficult to replace within a short period of time.

For the time being the Foreign Ministry has reinforced the strategy of diversification of foreign suppliers coupled with small investments on foreign oil fields and developing oils assets in states like Sudan, Russia, Libya, Nigeria, East Timor, Vietnam, Iran, Syria and Brazil—none of which are among India's most important suppliers.[65] The same strategy of diversification has been apply to natural gas purchase (bought mainly from Qatar and Iran) in 2004, that has known very little success on what concerns pipeline transportation for reasons already mentioned above. These investments are important but they still represent a very small percentage of India's future energy requirements.

Overall, the turn of the energy security file to foreign policy has been positive, although most of the measures had mild effects. India has yet to acquire two fundamental tools for energy security: a comprehensive strategy and a cooperative relationship between domestic actors (including states bureaucracies and public and private companies that trade in oil and natural gas). This is due to its history of fragmenting decision making in energy issues, and the too recent creation of the Energy Security Unit of Foreign Ministry.[66]

ECONOMIC DESIGN CONSTRAINTS

Other difficulty is related with the "highly regulated fuel prices for consumers, fuel subsidies that are shouldered by the government and state-owned upstream companies, and inconsistent energy sector reform currently hinder energy project investment. Some parts of the energy sector, chiefly coal production, remain relatively closed to private and foreign investment, while others, such as electric power, petroleum and other liquids, and natural gas have regulated price structures that discourage private investments."[67] Although there has been some liberalization of the energy markets in the end of the 2000s, its impact is yet to be seen,

Think tanks and scholars have been calling both for the privatization of state owned companies of energy production and distribution[68] and for the liberalization of India's market for energy to international companies.

However, India's energy sector has been organized through top down allocations that "undercut market forces and act as indirect price interventions."[69] And despite the fact that India's government has been encouraging private companies to invest in overseas asset acquisitions, neither these are significant nor the synergies between actors have been strong and very productive.[70]

The economic dimension of India's energy security, combined with the poor distribution across the population and the lack of infrastructure, has been one of the most criticized flaws on the country's quest for energy security. Think tanks, especially Indian based, have been calling for reform in this particular dimension.[71]

COMPETITION WITH CHINA

However, India's major energy security challenge might be competition with China. Since 1962 the bilateral relations between the two countries have been tense due the Sino-Indian war over the Himalayan border. The war had such an impact on India that "in the aftermath of the 1962 war [Patel's realpolitik view] became the representative mainstream Indian thinking *on China*."[72] Since then, the relations between these countries vary from cordiality—India sides China often in multilateral *fora* and both states are active members of the BRICS—to rivalry, especially in the bilateral and regional realm.[73]

The geopolitics of energy security reflects this reality. The two Asian giants have similar energy profiles and energy security policies: "their approach towards energy predicament remains rather traditional in so far as it is largely state-centric, supply-side biased, mainly reliant on oil, and tends to privilege self-sufficiency. It is towards and aggressive pursuit of energy resources, particularly oil, across the globe that China and India seem to have focused their diplomatic energies in recent years (. . .)."[74]

Competition, even conflict, should be the predictable outcome. Nevertheless, the two countries signed an agreement "to work together to secure energy resources without unnecessarily bidding up the price of those resources, thereby agreeing to a consumer cartel representing 2.3 billion potential consumers. Together, their combined markets and purchasing power offers an extremely attractive package to energy-producing states, especially the one that face western pressure over their human rights records or the nature of their political institutions."[75] The agreement should create a win-win strategy through reduction of costs, distribution of assets among signatories and transparency in policies.

However China had already overshadowed India before the agreement was: it has secure oil provision contracts with the Central Asian countries, and very strong economic ties with African and Latin American countries.[76]

Beijing started the energy quest before India, with a more organized energy security strategy, and a freer hand to offer to the sellers. "The Chinese companies not only enjoy a head start over their Indian rivals, but they also have deeper pockets."[77] As such, the durability and efficiency of the agreement is still to be proven.

In 2008, a scholar wrote a very pessimistic assessment on India energy security. He stated that "the Indian government has only recently woken up to the challenge of managing the nations' energy security with the realization that it has already fallen behind other major players, such as China. Despite this, India continues to lack an overarching energy strategy because of the lack of consensus on the crucial choices that the nation needs to make in the domestic political as well as global context."[78] Since then, the state of affairs has not improved considerably. Besides the need of keep pursuing the fairly successful strategy of diversification of sources, India needs to create better infra-structure, dramatically improve distribution of energy, adapt its economic apparatus to the energy market, create a more coordinated and transparent strategy for energy security, and engage in a quest of scarce resources with a number of countries, especially China, in a regional geopolitical environment that does not favor its position.

INDIA AND CLIMATE CHANGE: BETWEEN NATIONAL INTEREST AND INTERNATIONALIST ASPIRATIONS

Interestingly, and despite the already mentioned urgent need to become more secure energetically, India's decision makers and especially India's civil society have major concerns about environmental issues. The motives that move the later—especially environmental activists, very preeminent in the subcontinent—are not necessarily the same that the ones that motivate the later. However, there are three fundamental reasons that turned Indian elites' attention to these problems: (1) public opinion pressure; (2) India's very high pollution levels, especially in the major cities (according to the World Health Organization of the United Nations, New Delhi is the most polluted city in the world) that create public health problems; (3) the elites' political will to negotiate a climate change agreement that will take into account India's national interest and its political will to contribute to a better planetary environment. In other words, India knows that "accepting international commitments would be a way of jeopardizing domestic development" but it is willing to accept "common but differentiated responsibility"[79] (Patrick 2010, 48, 51).

As the two first reasons were already approached in section 3, this section will concentrate on the third motive, international climate change policies.

It is well known that India joined the BASIC countries—a group headed by China, India and Brazil—to block the international climate change agreement at the Copenhagen Summit in 2009. Three reasons explain this position: first, the leaders of southern states led this process with a fundamental goal: to symbolically show the Western leaders, who took the initiative to join together and draft an agreement that others were suppose to follow, that they were no longer rules takers. Although the BASIC was a "tactical agreement" among states with divergent interests in many other areas, it was also a way of making clear that southern countries are able to oppose Western impositions (Jones 2014, 66).[80] Furthermore, as a scholar noted, the Copenhagen Conference is a clear example of "clash" of "different understandings of global responsibilities (. . .)" (Pu 2012, 3).[81]

Second, rising powers continuously claim that "150 years of industrialization in the West have inflicted nearly all the damage (. . .). They insist that Americans and Europeans have no right to expect developing countries to sharply limit their growth to clean up the rich world's mess" (Bremmer 2012, 10).[82] This might is true. However, it is impossible to reach a global environmental solution if rising powers, especially China and India, do not accept a share of the burden. And although the idea of shared by differentiated responsibilities come to be accepted by all the intervenient parties, there is still no agreement about how to distributed everyone's fair share.

Third, the parties were not capable to reach common ground as climate change agreements have very costly consequences. As such, despite states' recognition that the environment problems must be urgently addressed, the tendency has been to move from multilateral agreements into "aggregate climate governance," that is, local/regional coordination of environmental efforts or countries' individual plan to address these problems. As many other countries, India has launch its own National Action Plan on Climate Change in 2008.

Besides the obvious difficulties—especially the tension between economic development and concerns with reducing the CO_2 footprint—India has three additional problems: one is related to the difficulties to access clean energy and the difficulties in access technologies to reduce pollution at a relatively affordable price. Most of it is patented and manufactured by private industries and are expensive systems. According to the Center of Policy Research, India has weather conditions to generate 28% of its electricity through sun and wind in 2030, if it would invest US$ 100 billion (Goldberg 2015).[83] This could also represent a big boost in Indian economy, but Indian leaders have routinely explained their reluctance in making these investments with India's still very large numbers of citizens living below the poverty line.

Recently Narendra Modi gave the same explanation to Barack Obama in the official visit paid by the United States to Delhi, last January. This event sheds a light on India's second problem: the mounting pressure of other great powers, especially the United States and China.

Despite the friendly atmosphere of the visit, it did not escape the media that the American President took the opportunity to pressure India into contributing to reducing the greenhouse emissions. A large part of its farewell address has about the centrality of climate change on the international society and how India, due to its size and pollution level, must get further involved in the solution (Baker and Barry 2015).[84] It is predictable an increase on the diplomatic pressure toward India, as date of the Paris Summit that will take place in November 2015 gets closer.

The other source of pressure is China. Despite aligning with India in the 2009 Copenhagen Conference, China recently broke a deal with the United States considering both countries contribution to reduce the CO2 pressure in the planet. In November, during Barack Obama's official visit to China, the American president and his Chinese homologous Xi Jinping had announced a climate change agreement where the United States is obliged to reduce CO_2 emissions in a range of 26% to 28% until 2025, while Beijing will make an effort to reach picking of CO_2 emissions before 2030 by replacing fuel fossils use in primary energy consumption to around 20% (White House 2014).[85]

The Indian experts believed the agreement was bad for the world (as it sets lower standards than the ones envisaged in the failed Copenhagen Summit of 2009) but it was also bad for India. This agreement meant less room to negotiate in the upcoming United Nation Climate Conference making more difficult to sustain India's position on shared but just responsibility, and it put India behind in long standing and mentioned and mentioned above claim: the transfer of technology to curtail pollution. Moreover, the United States-China climate change agreement was negotiated without the knowledge of India's elites, creating a discomfort situation: New Delhi stood behind Beijing in the Climate Change Summit of 2009 only to find itself isolated five years later. Again, in climate change, China positioned itself ahead of India, with consequences still to be known.

FINAL NOTES

As this article tried to make clear, India is at a critical moment in what concerns its energy security. During the 2000s, New Delhi's fast economic growth dramatically changed the country's energetic needs and self-perception in the following logic: to become a great power India needs to keep growing at a fast pace. But to do so it has to guarantee energy supplies that will guarantee this growth—in terms to industrial production—and to fill the need of a growing middle class of consumers that demand an increase on the quantity and reliability of energy supply.

To keep up with this major goal, India has to solve at least three fundamental problems: (1) the lack of state capacity (that translates into very weak infra-structure and unequal distribution among the Indian population) and bureaucratic fragmentation (that leads to difficulties in creating a comprehensive strategy of energy security); (2) the need to adapt its economy to the energy market (India has a number of protectionist features that preclude easier market exchanges with third parties in the current liberal energy market); (3) to address competition with China (that has a very similar energy profile but a more efficient and centralized state bureaucracy and started its energy quest and climate change strategy earlier than India, which gave Beijing a clear competitive advantage). None of these problems are assuaged by India's weak position on climate change international negotiations and its need to integrate it in its path to development.

The theoretical section suggested that internationally India has two ways of addressing its quest of energy security—through fierce competition for scarce resources, which it has tried to avoid due to pragmatic (low military capability) and ideological (self-image as a peaceful state) reasons, or through cooperation—either in the framework of the liberal international order, or by bilateral and multilateral agreements with other nations (India seems to prefer the later due to its traditional non entangling alliances policy that dates from the Independence and was reinforced by the 1990s experience of the collapse of its only ally).

Possibly, India is going to choose a mix of all this elements. But the way the mix is going to look like is going to depend on the evolution of two elements of India's modernization process: the economic system and the decision making of elites. As these decisions are yet to be made, it is difficult to predict India's path toward energy security. However, due to India's size and ambitions, it is already clear that it is going to have a strong impact on the international system energy security.

NOTES

1. R. Guha, *India after Gandhi: The History of the World Largest Democracy* (New York: Harper Collins, 2007), 110.
2. S. Khilnani, *The Idea of India* (New York: Farrar, Straus & Giroux, 1999), 9.
3. A. Kohli, *Poverty amid Plenty in the New India* (Cambridge: Cambridge University Press, 2012), 36.
4. M. Pardesi, "Review Essay: Understanding the Rise of India." *India Review*, 6 (3), (2007): 209–231, 2010.

5. S. Khilnani et al, *Non-Alignment 2.0: A Foreign Policy and Strategy for India in the 21st Century*, (Gurgaon: Penguin Books, 2013), x, S. Tharoor, *Pax Indica: India and The World of the Twenty First Century* (London: Alan Lane, 2012), 5.

6. M. Pardesi and S. Ganguly, "India and Energy Security: A Foreign Policy Priority," in *Indian Foreign Policy in a Unipolar World*, ed. Harsh V. Pant (London: Rutledge, 2009), 103.

7. Pardesi and Ganguly, "India and Energy Security," 99.

8. S. Walt. "International Relations: One World, Many Theories." *Foreign Policy*, 110 (1998): 29–32, 34–46, 31.

9. K. Waltz, *Theory of International Politics* (New York: Columbia University Press, 1979, 2010) and K. Waltz, "Reductionist and Systemic Theories," in *Neorealism and Its Critics*, ed. Robert O. Keohane, 47–69 (New York: Columbia University Press, 1986), 53.

10. R. Keohane, *After Hegemony: Cooperation and Discord in the World Political Economy* (Princeton: Princeton University Press, 1984, 2005), 8–9 and R. Keohane, "Realism, Neorealism and the Study of World Politics" in *Neorealsim and Its Critics*, by Robert O. Keohane, 1–26 (New York: Columbia University Press, 1986), 19.

11. J. Mearsheimer, *The Tragedy of Great Power Politics* (New York: W. W. Norton & Company, 2001), 31.

12. C. Layne, "The Unipolar Illusion: Why New Great Powers Will Rise," in *International Security* 17 (4), (1993): 5–51, 7.

13. B. Buzan and O. Waever, *Regions and Powers: The Structure of International Security* (Cambridge: Cambridge University Press, 2013), 49.

14. D. Lake, *Hierarchy in International Relations* (Ithaca: Cronell University Press, 2009), 15.

15. See J. Ikenberry, *After Victory: Institutions, Strategic Restraint, and the Rebuilding of Order After Major Wars* (Princeton: Princeton University Press, 2001).

16. See D. Held et al, *Global Transformations: Politics, Economics and Culture* (Stanford: Stanford University Press, 1999).

17. E. H. Caar, *The Twenty Years' Crisis: 1919–1939* (New York: Harper & Row, 1964), 50.

18. See R. Keohane abd J. Nye, *Power and Interdenpendence* (New York: Harper Collins, 1977, 1989) and Keohane, *After Hegemony*.

19. See Keohane, *After Hegemony*.

20. See Ikenberry, *After Victory*.

21. J. Ikenberry, *Liberal Leviathan: The Origins, Crisis and Transformation of the American World Order* (Princeton: Princeton University Press, 2011), 2.

22. Ikenberry, *Liberal Leviathan*, 3.

23. Held et al., *Global Transformations*, 3, 7–8.

24. S. Ganguly, "The Genesis of Non-Alignment," in *India's Foreign Policy: Retrospect and Prospect*, ed. S. Ganguly (New Delhi: Oxford University Press, 2010), 3–4.

25. J. Sengupta, *A Nation in Transition: Understanding India Economy* (New Delhi: Academic Foundation, 2007), 14.

26. Waltz, "Reductionist and Systemic Theories," 47.

27. A. Organski, *World Politics* (New York: Knopf, 1968), 363 and R. Gilpin, *War and Change in World Politics* (Cambridge: Cambridge University Press, 1981), 8.

28. B. Buzan and G. Lawson, *The Global Transformation: History, Modernity and Making of International Relations* (Cambridge: Cambridge University Press, 2015), 22.
29. See B. Moore, *Social Origins of Dictatorship and Democracy: Lord and Peasant in the Making of the Modern World* (Boston: Beacon Press, 1967).
30. See A. Kohli, *State Directed Development: Political Power and Industrialization in the Global Periphery* (Cambridge: Cambridge University Press, 2004).
31. See S. Huntington, *Political Order in Changing Societies* (Boston: Yale University Press, 1968).
32. See M. Lipset, *Political Man: The Social Basis of Politics* (Michigan: Doubleday Books, 1960).
33. Buzan and Lawson, *The Global Transformation*, 6.
34. Ibid., 150.
35. See S. Tharoor, *Nehru: The Invention of India* (New Delhi: Penguin Books, 2003).
36. D. Gupta, *Revolution from Above: India's Future and the Citizen Elite* (New Delhi: Rupa Publications, 2013), 196.
37. A. Varshney, *Battles Half Won: India's Improbable Democracy* (Gurgaon: Penguin Books, 2013), 3.
38. Gupta, *Revolution from Above*, 119.
39. See Kohli, *Poverty amid Plenty.*
40. D. Ollapally and R. Rajagopalan, "India: Foreign Policy Perspectives of an Ambiguous Power," in *Worldviews of Aspiring Powers: Domestic Foreign Policy Debates in China, India, Iran, Japan, and Russia*, ed. H. Nau and D. Ollapally (Oxford: Oxford University Press, 2012).
41. Kohli, *Poverty amid Plenty*, 12.
42. EIA: U.S. Energy Information Administration, "India." Last modified in June 26, 2014. http://www.eia.gov/countries/analysisbriefs/India/india.pdf
43. EIA, "India."
44. M. Park, "Top 20 Most Polluted Cities in the World, *CNN*, May 8, 2014.
45. J. Ablett et al., *The Bird of Gold: The Rise of Consumer Market* (São Paulo: McKinsey Global Institute, 2007), 13.
46. K. Vinayakam and S.P. Sekar, "Rural to Urban Migration in an Indian Metropolis: Case Study," in *Journal of Humanities and Social Sciences*, 6 (3), (2013): 32–35, 32.
47. R. Mohan et al, *Balance and Reform: An Agenda for the New Government*, (New Delhi: Observer Research Foundation, 2014), 27.
48. Mohan et al, *Balance and Reform*, 26.
49. Ibid.
50. H. V. Pant, *Contemporary Debates in Indian Foreign and Security Policy: India Negotiates its Rise in the International System* (London: Palgrave McMillan 2008), 154.
51. EIA, "India."
52. R. Sikri, *Challenge and Strategy: Rethinking India's Foreign Policy* (New Delhi: Sage 2009), 208–210.
53. Sikri, *Challenge and Strategy*, 199–200.
54. Buzan and Lawson, *The Global Transformation*, 159.

55. D. Mistry, "Domestic and International Influences of India's Energy Security," in *India's Foreign Policy: Retrospect and Prospect*, ed. S. Ganguly (New Delhi: Oxford University Press, 2010), 327.

56. C. Odgen, *Indian Foreign Policy* (London, Polity 2014), 27.

57. Mistry, Domestic and International Influences of India's Energy Security, 327.

58. S. Ganguly and R. Mukherji, *India Since 1980* (Cambridge: Cambridge University Press, 2015), 20.

59. Ganguly, "The Genesis of Non-Alignment," 4.

60. S. Raghavan, "Indira Gandhi: India and the World in Transformation," in *Makers of Modern India*, ed. R Guha (Boston: Harvard University Press, 2014), 236.

61. See Kohli, *Poverty amid Plenty*.

62. E. Luce, *In Spite of the Gods: The Rise of Modern India* (New York: Anchor Books, 2008), 290.

63. Khilnani et al, *Non-Alignment 2.0*, 141.

64. Ibid., 143.

65. EIA, "India."

66. I Saint-Mézard, *The International Dimensions of India Energy Security* (Paris: Asia Center, 2014), 6.

67. EIA, "India."

68. Khilnani et al., *Non-Alignment 2.0*, 144.

69. Mohan et al., *Balance and Reform*, 26.

70. Saint-Mézard, *The International Dimensions of India Energy Security*, 5–6.

71. Mohan et al, *Balance and Reform*, 26–27.

72. J Garver, *Protracted Contest: Sono-Indian Rivalry in the Twentieth Century*, (Seattle: University of Washington Press, 2001), 29.

73. Khilnani et al., *Non-Alignment 2.0*, 141, 20.

74. Pant, *Contemporary Debates in Indian Foreign and Security Policy*, 154.

75. Ibid., 159.

76. Ibid., 162, 166.

77. Ibid., 167.

78. Ibid., 154–155.

79. S. Patrick, "Impossible Stakeholders? The Difficulty of Integrating Rising Powers," in *Foreign Affairs*, 89 (6) (2010), 51.

80. B. Jones, *Still Ours To Lead: America, Rising Powers, and the Tension between Rivalry and Restraint* (Washington D.C.: The Brookings Institution, 2014), 66.

81. Pu X., "Socialization as a Two-way Process: Emerging Powers and the Diffusion of International Norms, in *The Chinese Journal of International Politics* (2012), 3.

82. I. Bremmer, *Every Nation for Itself: What Happens When No One Leads the World* (New York: Penguin Group, 2012), 10.

83. S. Goldenberg, "Obama's India Visit" Hopes for Clean Energy and Climate Deals," in *The Guardian*, January 23, 2015.

84. P. Baker and E. Barry, "As Visit Ends, Obama Presses India on Human Rights and Climate Change," in *The New York Times*, January 27, 2015.

85. The White House, "U.S.-China Joint Announcement on Climate Change," November 11, 2014. https://www.whitehouse.gov/the-press-office/2014/11/11/us-china-joint-announcement-climate-change.

BIBLIOGRAPHY

Ablett, Jonathan, et al. *The 'Bird of Gold': The Rise of the Consumer Market.* Sao Paulo: McKinsey Global Institute, 2007.

Baker, Peter, and Barry Ellen. "As Visit Ends, Obama Presses India on Human Rights and Climate Change." *New York Times*, January 27, 2015.

Bremmer, Ian. *Every Nation for Itself: What Happens When No One Leads the World.* New York: Penguin Group, 2012.

Buzan, Barry, and George Lawson. *The Global Transformation: History, Modernity and the Making of International Relations.* Cambridge: Cambridge University Press, 2015.

Buzan, Barry, and George Lawson. "The Global Transformation: The Nineteenth Century and the Making of modern International Relations." *International Studies Quarterly* 57 (2013): 620–634.

Buzan, Barry, Waever, Ole. *Regions and Powers: The Structure of International Security.* Cambridge: Cambridge University Press, 2003.

Carr, Edward Hallett. *The Twenty Years' Crisis: 1919–1939.* New York: Harper & Row, 1964.

Ganguly, Sumit. "The Genesis of Non-Alignment." In *India's Foreign Policy: Retrospect and Prospect*, by Sumit Ganguly, 1–10. New Delhi: Oxford University Press, 2010.

Ganguly, Sumit, and Rahul Mukherji. *India Since 1980.* Cambridge: Cambridge University Press, 2011.

Garver, John W. *Protracted Contest: Sino-Indian Rivalry in the Twentieth Century.* Seattle: University of Washington Press, 2001.

Gilpin, Robert. *War and Change in World Politics.* Cambridge: Cambridge University Press, 1981.

Goldenberg, Suzanne. "Obama's India visit: Hopes for Clean Energy and Climate Deals." *The Guardian*, January 23, 2015.

Guha, Ramachandra. *India after Gandhi: The History of the World's Largest Democracy.* New York: Harper Collins, 2007.

Gupta, Dipankar. *Revolution from Above: India's Future and the Citizen Elite.* New Delhi: Rupa Publications, 2013.

Held, Davis, Anthony McGrew, David Goldblatt, and Jonathan Perratom. *Global Transformations: Politics, Economics and Culture.* Stanford: Stanford University Press, 1999.

Huntington, Samuel P. *Political Order in Changing Societies.* New Haven: Yale University Press, 1968.

Ikenberry, G. John. *After Victory: Institutions, Strategic Restraint, and the Rebuilding of Order After Major Wars.* Princeton, Princeton University Press: Princeton University Press, 2001.

———. *Liberal Leviathan: The Origins, Crisis and Transformation of the American World Order.* Princeton: Princeton University Press, 2011.

Jones, Bruce. *Still Ours to Lead: America, Rising Powers, and the Tension between Rivelry and Restarint.* Washington D.C.: Brooking Institution Press, 2014.

Keohane, Robert O. *After Hegemony: Cooperation and Discord in the World Political Economy.* Princeton: Princeton University Press, 1984, 2005.

Keohane, Robert O. "Realism, Neorealism and the Study of World Politics." In *Neorealsim and Its Critics*, by Robert O. Keohane, 1–26. New York: Columbia University Press, 1986.

Khilnani, Sunil. *The Idea of India.* New York: Farrar, Straus & Giroux, 1999.

Khilnani, Sunil, et al. *Non-Alignment 2.0: A Foreign & Strategic Policy for India in the 21st Century.* Gurgaon: Penguin Books, 2013.

Kohli, Atul. *Poverty Amid Plenty in the New India.* Cambridge: Cambridge University Press, 2012.

———. *State Directed Development: Political Power and Industrialization in the Global Periphery.* Cambridge: Cambridge University Press, 2004.

Lake, David A. *Hierarchy in International Relations.* Ithaca: Cornell University Press, 2009.

Layne, Christopher. "The Unipolar Illusion: Why New Great Powers Will Rise." *International Security* 17, no. 4 (Spring 1993): 5–51.

Lipset, Martin Seymour. *Political Man: The Social Basis of Politics.* Michigan: Doubleday Books, 1960.

Luce, Edward. *In Spite of the Gods: The Rise of Modern India.* New York: Anchor Books, 2008.

Mearsheimer, John J. *The Tragedy of Great Power Politics.* New York: W. W. Norton & Company, 2001.

Mistry, Dinshaw. "Domestic and International Influences of India's Energy Security." In *India's Foreign Policy: Retrospect and Prospect*, by Sumit Ganguly, 323–342. New Delhi: Oxford University Press, 2010.

Mohan, C. Raja, Manoj Joshi, Ashok Malik, and Samir Saran. *Balance and Reform: An Agenda for the New Government.* New Delhi: Observer Research Foundation, 2014.

Moore, Barrigton. *Social Origins of Dictatorship and Democracy: Lord and Peasant in the Making of Modern World.* Boston: Beacon Press, 1967.

Ogden, Chris. *Indian Foreign Policy.* London: Polity, 2014.

Ollapally, Deepa M., and Rajesh Rajagopalan. "India: Foreign Policy Perspectives of an Ambiguous Power." In *Worldviews of Aspiring Powers: Domestic Foreign Policy Debates in China, India, Iran, Japan, and Russia*, by Henry R. Nau and Deepa M. and Ollapally, 73–113. Oxford: Oxford University Press, 2012.

Organski, A. F. K. *World Politics.* New York: Alfred A Knopf, 1958.

Pant, Harsh V. *Contemporary Debates in Indian Foreign and Security Policy: India Negotiates its Rise in the International System.* London: Palgrave MacMillan, 2008.

Pardasi, Manjeet. "Review Essay: Understanding the Rise of India." *India Review* 6, no. 3 (2007): 209–231.

Pardesi, Manjeet S., and Sumit Ganguly. "India and Energy Security: A Foreign Policy Priority." In *Indian Foreign Policy in a Unipolar World*, by Harsh V. Pant, 99–127. London: Routledge, 2009.

Park, Madison. "Top 20 Most Polluted Cities in the World." *CNN*, May 8, 2014.

Patrick, Stewart. "Irresponsible Stakeholders? The Difficulty of Integrating Rising Powers." *Foreign Affairs* 89, no. 6 (2010): 44–53.

Pu, Xiaiyo. "Socialization as a Two-way Process: Emerging Powers and the Diffusion of International Norms." *The Chinese Journal of International Politics*, 2012: 1–27.

Raghavan, Srinath. "Indira Gandhi: India and the World in Transformation." In *Makers of Modern India*, by Ramachandra Guha, 215–243. Boston: Harvard University Press, 2014.

Saint-Mézard, Isabelle. *The International Dimensions of India Energy Security.* Working Paper, Paris: Asia Center, February 2014.

Sengupta, Jayshree. *A Nation in Transition: Understanding Indian Economy.* New Delhi: Academic Foundation, 2007.

Sikri, Rajiv. *Challenge and Strategy: Retinking India's Froeign Policy.* New Delhi: Sage, 2009.

Tharoor, Shashi. *Nehru: The Invetion of India.* New Delhi: Penguin Books, 2003.

———. *Pax Indica: India and the World of the Twenty First Century.* London: Alan Lane, 2012.

Varshney, Ashutosh. *Battles Half Won: India's Improbable Democracy.* Gurgaon: Penguin Books, 2013.

Vinayakam, K, and S. P. Sekar. "Rural to Urban Migration in an Indian Metropolis: Case Study." *Journal of Humanities and Social Sciences* 6, no. 3 (2013): 32–35.

Walt, Stephen M. "International Relations: One World, Many Theories." *Foreign Policy*, no. 110 (Spring 1998): 29–32, 34–46.

Waltz, Kenneth N. "Reductionist and Systemic Theories." In *Neorealism and Its Critics*, by Robert O. Keohane, 47–69. New York: Columbia University Press, 1986.

———. *Theory of International Politics.* New York: Columbia University Press, 1979, 2010.

Chapter 8

Domestic Constraint's on Brazil's Role as Regional and Global Energy Power

Marcelo Rocha e Silva Zorovich

Brazil has not only become a powerful economy since the implementation of the Real Plan,[1] but also an important player in the widening struggle for regional and global energy security.[2] Brazil may strategically strengthen its geopolitical position as an important figure in this sector, despite its short term economic stagnation and country's state-run oil giant Petrobras' political scandals. Brazil's market size, along with its pre-salt discoveries and leadership role in usage of renewable energy sources may contribute to a more competitive scenario. Nevertheless, the country needs to overcome various domestic constraints, such as investments in infrastructure, bureaucracy, political and economic issues as well as work closer with neighboring countries and global partners.

The recent changes in the global political and economic realm as well as in the energy market—US shale gas developments, China's economic slowdown, Russia's sanctions, tensions in the Middle East and in the EU, the drop of the oil price—, just to mention a few, have contributed to put more pressure on energy security, while highlighting the interdependence character of the energy supply chain. According to Yergin, "the energy security system faces two major new challenges today."[3]

The first is to bring China, India and other major emerging markets into alignment with it. At the beginning of the 1970s, developed countries accounted for 80% of the world's oil consumption. Today, they are down to a little more than 50%; and their share will continue to decline, as virtually all the growth in oil demand will be in the developing world including Brazil as well as other countries in South America[4] and other parts of the world. The second challenge relates to the infrastructure and risks of energy security, which varies among developing nations. Energy security needs to be extended to the safety of the entire supply chain,[5] either in a global or a regional perspective. The expansion of Brazil's energy infrastructure, both for production and consumption, requires large investments, financial resources and political and economic will.[6]

Increasingly in a growing world interdependent political economy, energy trade traverses national borders and ties nations together and is no different in South America or elsewhere; nevertheless, despite the variety of energy sources, South American countries are unlikely to provide adequate energy security for their citizens.[7] Yet, energy security is not just about countering the wide variety of risks and threats; it is also about connecting nations, how they interact politically and economically with each other, and how energy impacts their overall national security. It entails policies and cooperation in the regional and global arena.[8] Therefore, assuring the security of markets requires coordination on both domestic and international fronts among national and multinational companies as well as governments.[9]

This is part of the gap Brazil needs to fulfill not only in South America, but in its potential role as a global energy power. On a global scale, particular attention will be given on Brazil's political and economic issues with China and the relationship with other BRICS, whether they occur in the energy sector. Hence, this paper aims to discuss the main domestic and international aspects associated with the political and economic challenges Brazil needs to overcome in order to become a more strategic actor in global energy integration and security. Meeting the challenges of the twenty-first century will not only require domestic efforts, but also inter-governmental alignment along with Brazil's energy policy coordination to mitigate environmental concerns as well as social costs.

Therefore, many questions arise when discussing Brazil's energy profile. What are the key sources? Where does the country face major challenges in terms of infrastructure? Have policy effectiveness and implementation been part of the challenges that the Brazilian government has to overcome regarding the lack of competitiveness (along with many other factors)? Is regional integration and interdependence a matter that only Brazil can resolve? What are the perspectives for Brazil as a regional (South America) and global energy supplier (China and the BRICS)? Who are Brazil's main domestic actors involved? Does politics matter in the Brazilian energy sector?

THE BRAZILIAN PROFILE OF ENERGY SUPPLY AND DEMAND

According to EPE,[10] despite the downward production of ethanol, the Brazilian government highlights the predicated future expansion of liquid biofuels consumption (ethanol and biodiesel) from 5.8% in 2012 (see Table 8.1) to

Table 8.1 Brazil's Final Energy Consumption by Source and Participation in 2012

Energy Sources	*10^3 tep*	*%*
Natural Gas	17,867	7.5
Mineral Coal	12,000	5.0
Firewood	12,961	5.4
Charcoal	5,733	2.4
Sugar Cane bagasse	29,032	12.2
Electricity	43,011	18.1
Ethanol	11,353	4.8
Biodiesel	2,339	1.0
Others	8,309	3.5
Sources (1)	*142,605*	*59.9*
Diesel Oil	44,435	18.7
Fuel Oil	5,211	2.2
Gasoline	22,512	9.4
Liquified Petroluem Gas	8,238	3.5
Kerosene	3,710	1.6
Others-oil	11,531	4.8
Petroleum sources (2)	95,637	40.1
Final Energy Consumption (1 + 2)	238,243	100

Source: Adapted from EPE, 2012.

9.6% by 2021. Biofuels, such as sugarcane, are expected to increase over the next decade from 29.3% to 33.3%. Table 8.1 indicates the composition of Brazil's energy matrix in 2012. Such forecasts are part of Brazil's Ten Year Plan for Energy Expansion,[11] including assumptions such as the country's continuous population growth (although slower) as well as an increase in the number of households.[12] In the last decade, experts have observed an increase in the usage of electrical appliances, driven in part by the increasing incomes of the lower and middle classes.[13]

Furthermore, this scenario takes into consideration the development of relevant economic sectors in 2012, focusing on sectors related to agriculture, commodities supply, and vehicles. All of them are deeply connected to China, and to a certain extent to some other BRICS. The population and income growth in Brazil associated with the increasing populations in cities may lead to greater demand for mobility, which translates into increased need for transportation. Regionally speaking, demographers show increasing population growth in both South America and Asia, which will result in augmented consumption patterns.[14] In addition to increasing consumption levels at the domestic level, demand is increasing at the international level, which has resulted in increases in commodity prices.[15].

The prospects for major events to be held in the country (World Cup 2014 and Olympic Games in 2016) have led experts to argue that the levels of investment must rise substantially. In addition to sporting events, which

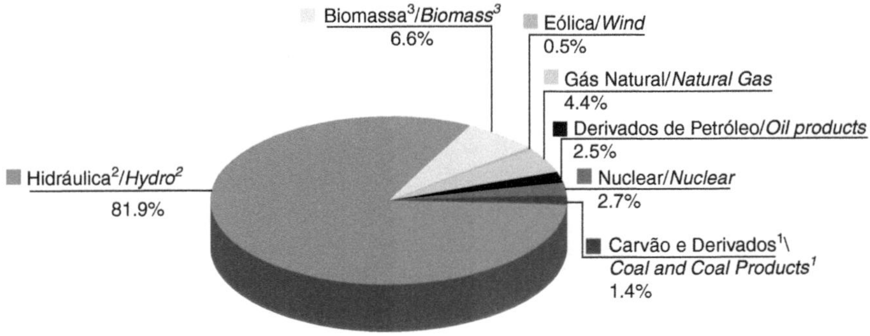

Graph 8.1 Brazil´s Structure of the Domestic Supply of Electricity in 2011–2012. *Source*: EPE, 2011–2012.

require increases in public and private investment, important advancements in the supply chain of petroleum and natural gas in pre-salt areas have been observed.[16] Another important consideration to understand Brazil's profile and natural resource potential can be viewed in Graph 8.1, which shows the structure of the domestic supply of electricity in 2011–2012. This is a critical subject due to recent price increment announced by the government in 2015.[17]

Brazil has a renewable electricity matrix, and the domestic hydraulic generation accounts for the higher percentage of the supply. By adding imports, which are also mainly from renewable sources, it can be inferred that more than 80% of electricity in Brazil comes from renewable sources. For the next decade (2012 to 2021),[18] the Brazilian government expects an overall increase of 54% in electricity consumption.

No less relevant is the domestic oil and shale oil production, which rose 2.4% from 2011 to 2012, reaching a record average of 2.21 million barrels per day produced by the end of the year. Offshore production accounted for 91% of the national total in 2011 and 2012. In this respect, Rio de Janeiro was responsible for the largest share of producing states: 74% of the annual amount. Part of the Chinese FDI is channeled into this area.[19] On the other hand, the biggest onshore producer was the state of Rio Grande do Norte (Northeast), with 28% of the total onshore gas share.[20]

Natural gas has always been a critical issue for Brazil and exemplifies part of the criticism related to the lack of infrastructure in the country. The natural gas share in the national energy matrix has been stifled at a level below 10%, compared to 20% of importance in the global market, suggesting a potential market for Brazil to explore as well its energy policy challenges.

As stated by Oliveira, in the 1990s, Brazilian energy policy made a significant turnaround as the import substitution policy was abandoned and the liberalization of the Brazilian economy began. Following these steps of change in its economic policy,[21] Brazil's energy sector was partially privatized. As a result, "Eletrobras, the national state-owned, holding company for the electricity sector, lost its coordinating role of the power system and Petrobras remained a state-owned company but lost its legal monopoly on the hydrocarbons market. A pipeline was built for natural gas imports from Bolivia and the state owned regional gas distribution companies were privatized. However, oil and hydropower remain the dominant primary sources of energy of the country," and Petrobras continues playing a major role, despite recent political and economic drawbacks. Management of these reservoirs remains a key aspect of the long-term planning decisions and economic development in Brazil.[22]

In order to understand the main issues in this specific sector, it is critical to emphasize that notwithstanding Brazil's potential as a regional supplier, the price of this fuel is US$ 16 per million of BTUs, which is significantly more than US$ 2 in the US[23] The lack of infrastructure, technology, and skilled labor imposes impacts on prices in the domestic market. The price difference between Brazil and the United States, for example, has been affecting the competitiveness of sectors in the Brazilian economy, such as the petrochemical industry, ceramics, and glass,[24] which depend on gas.

According to Pedrosa,[25] the commercial deficit of the industries that intensely use this fuel may continue to grow in the coming years as a result of developmental limitations related to natural gas. These industries, which recorded a surplus of US$ 15 billion in 2005, had a deficit of US$ 15 billion in 2011. Competitiveness is one factor that may help decrease prices, bringing more transparency and reducing the control of Petrobras. Brazil's natural gas price ranks as the top 10th most expensive. Comparatively, the self-sufficiency as a result of shale gas in the US led the price to drop to US$ 2 per million of BTUs, although it has recently increased to US$ 3.20.[26] As a result, part of business investments of the Brazilian petrochemical industry may migrate to the US market. Besides the US, the high prices in Brazil make natural gas less competitive than China, India, Mexico, and the United Arab Emirates, which are key international competitors.

The chemical composition of Brazilian gas is another limiting factor. In the pre-salt, gas is associated with oil and CO_2, which must be separated from the platform and re-injected, in a more costly process. Transporting this product up the coast, which can be a distance of 480 miles, also demands infrastructure investment. Furthermore, the stocks of gas are difficult and turning it into liquid for export makes the final product more expensive. Specialists disagree about the potential of natural gas in Brazil (see Table 8.2).

As Table 8.2 demonstrates, leading experts as well as companies specializing in energy production favor the implementation of a more strategic policy in order for Brazil to become more competitive in the gas sector. Basically, Brazil must increase gas production from 10% to 15%, which would also favor the regional supply and security.

Table 8.2 Opinion of Specialists on Natural Gas in Brazil

Name and Affiliation	Opinion of Specialists on Natural Gas
Ashley Brown, Harvard Professor of Energy Policies	There is a lack of open market access and transparency regarding what happens to the pipeline in Brazil. There are few products and not enough storage. There is no transparency in relation to prices. Petrobras' domination is negative to the Brazilian market. There are companies leaving Brazil due to the high price. Prices in Brazil are five times more expensive in comparison to the United States.
Marco Antonio de Almeida, Secretary of the Ministry of Mines and Energy, Brazil	The Brazilian market is new and one of the challenges to overcome is Petrobras' concentration.
Jose Anibal, Secretary of Energy of Sao Paulo	The fact that Petrobras is a dominant agent in this chain is a limiting factor. Prices are damaged.
Winston Fritsch, President of Petra Energia	It is fundamental to produce a cheaper gas. The challenge is the Brazilian cost.
Paulo Pedrosa, President of ABRACE	Having one of the highest prices in world, auctions would be an alternative to the loss of competitiveness.
Fernando Garcia, FIPE's Research Coordinator	A country that intends to achieve 20% of GDP in investments has to look at different mechanisms to attract resources.
Magda Chambriard, General Director of ANP1	There are evidences of gas in all parts of the country. It is not possible so limited. More efficient public policies are required to promote internal development.

Source: Adapted from interviews in the Forum "Futuro do Gás Natural," promoted by "Grupo Estado along with Projeto+Gas Brasil," 2012.

In a broader view, from a public safety perspective, according to ABRACE,[27] Brazil should recognize the importance of gas due to the fact that this fuel has been an element of global geopolitical change. Brazil must therefore develop a public policy more focused on the development on the gas sector. However, the debate should not only be a matter of price reduction, but a sustainable government policy to foster the strengthening of this sector in order to increase local and regional supplies. This policy should be based on the following pillars: (a) ample supply; (b) transportation; (c) efficient distribution; (d) competitive marketing (including costs); and (e) regional integration.

The aforementioned discussion implies that "energy security is all about interdependence." As oil markets in different countries are strongly linked and influence each other, a disruption or changes in the supply or demand impacts on the country's energy matrix. Whether producing or consuming has repercussions on other participants in the market, whether or not there are counterparts to the disrupted participant.[28]

Still according to the specialists mentioned above, despite the extensive priorities, gas transportation should receive special attention in order to foster the other pillars that compose the gas puzzle. According to OPEC,[29] transportation technology represents a major source of efficiency gains. The 2009 Gas Law allows for the possibility of pipeline development. Brazilian states must effectively work with all institutions as well as state governments to decrease the distribution costs. In this scenario, the balance between supply and demand requires revisions in energy policies since Brazil has proven reserves of natural gas (see Table 8.2). As a consequence, it implies the capacity to meet all needs without depending on third parties. Brazil currently imports part of its natural gas and consumption has been increasing since 2002 (Table 8.3).

Natural gas consumption as well as the flow and production of imported natural gas have increased from 2002 to 2011. In recent years, Brazil has identified large natural gas reserves in the Southeast (Table 8.2), largely due to the discovery of oil in Campos's (RJ) basin. Traditionally perceived as rich in hydropower, but poor as far as hydrocarbons are concerned, Brazil is celebrating its emerging oil self-sufficiency.[30] To ease its use, large investments must be made in transportation systems with joint collaboration of public and private sectors. In parallel to the national energy potential, Brazil has the

Table 8.3 Final Natural Gas in Brazil, in Volume

Energy Source	2002	2003	2004	2005	2006	2007	2008	2009	2010	2011
Natural Gas	5.6	6	6.4	6.8	7.1	7.2	7.3	6.9	7	7.6

Source: EPE, 2011–2012.

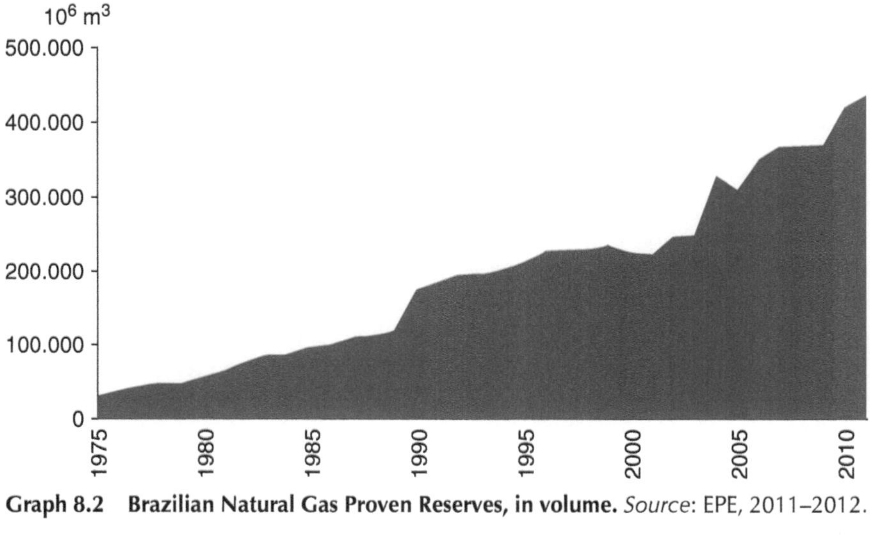

Graph 8.2 Brazilian Natural Gas Proven Reserves, in volume. *Source*: EPE, 2011–2012.

possibility to use and even expand the gas supply of Bolivia. Nevertheless, uncertainty exists about the usefulness of transporting Bolivian gas, considering past political problems with Petrobras as a result of Evo's initiatives toward nationalization of Bolivia's energy sector.[31]

Regionally speaking, around 5% of the world's natural gas reserves are located in Latin America, with Mexico and Venezuela standing out with the highest percentage of regional reserves. Bolivia is a fundamental player in this segment as well, as 6% of the country's regional reserves are from free natural gas.[32] Since the end of the 1990s, ANP[33] has been offering exploration blocks through annual licensing rounds. Some companies showed interest in the Brazilian sedimentary basins, located in the continental shelf.[34] The prices paid to acquire exploration rights suggest optimism concerning the future Brazilian hydrocarbons' production in order to become less dependent on Bolivia, for instance, or to become a more strategic regional supplier and increase its capacity.

Optimism has increased as a result of the Brazilian government's Decennial Plan for the Power Sector (2006–2015), as the government predicts power consumption to increase by an average of 5% along with GDP growth. Nevertheless, estimates concerning future consumption are largely dependent on the role envisaged for natural gas in the energy balance, besides Petrobras' recent corruption under investigation.[35]

In broad terms, Brazil needs to increase its energy availability to ensure greater economic progress, which will enable the government to improve the living conditions of all of the population.[36] In order to be viewed as a regional energy supplier, Brazil requires several sources of energy in addition to natural gas. For supply safety reasons, it is important to depend on several primary energy sources rather than only one or two (see Graph 8.3). However, it is relevant to highlight the political and economic priorities, which are often dictated by the abundance of certain natural resources that are more common in Brazilian territory. Thus, it is understandable to have large share of hydroelectricity—Brazil is one of the richest in water and energy resources—and the modest contribution of coal, as the country has few reserves and they are of low quality.[37]

Even with the abundance of water reservoirs, biggest city in South America, "São Paulo is struggling to overcome a water shortage. Experts are divided as to whether more investments by Sabesp (which is 50.3% owned by the state government of São Paulo) could have helped avert shortages. But São Paulo state environmental prosecutors in the nearby city of Piracicaba, supplied by the same reservoir system, have launched an inquiry and claim that the crisis could have been avoided. Benedito Braga, president of the World Water Council and a professor of civil and environmental engineering at the University of São Paulo, said the principal causes of the shortage were

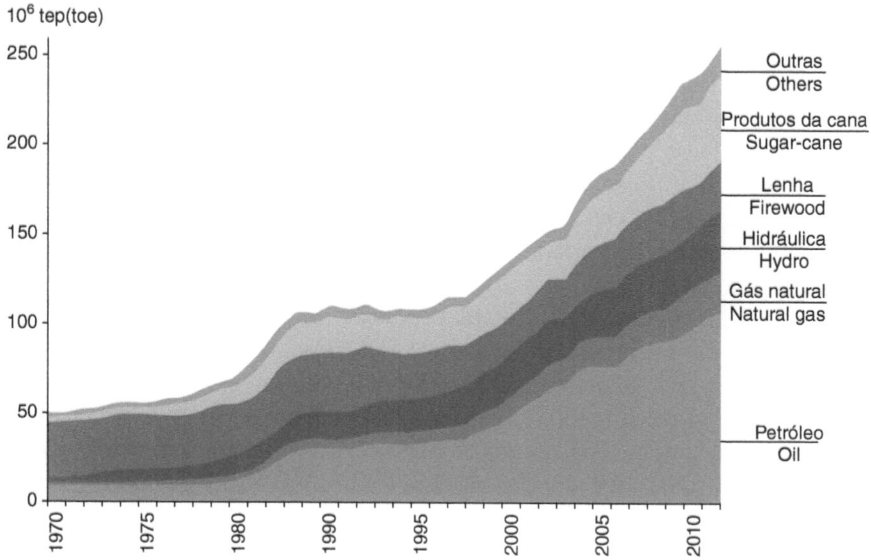

Graph 8.3 Brazilian Primary Energy Production in Volume. *Source*: EPE, 2011–2012.

the drought and high consumption by a population used to an abundance of water in Brazil's rainy south-east,"[38] with potential consequences to energy sector.

In spite of the importance of the hydro sector, historically, one of the main driver of Brazil's energy policy was to minimize the use of imported oil, a policy reinforced after the oil crisis in the 1970s. Brazil sought to rely on national energy inputs whenever possible due to the great difficulty in generating US dollars in the country and the huge expenses that imported oil and its derivatives have caused.[39] Chart 8.3 demonstrates the potential Brazil has in terms of increasing its capacity of primary energy production over the last 40 years.

Despite the variety of energy sources and abundance in some case, countries in South America are unlikely to provide adequate energy security for their citizens. Although some authors argue that the process of regional integration would benefit the energy supply, this situation triggers an alarming scenario with respect to national regulations that often impose barriers to investments that aim to promote regional energy integration. In addition, in the case of Brazil, the issues regarding its own infrastructure delay its internal development, thereby contributing to a more lengthy process at the regional level as well.[40]

Furthermore, in some cases, the cost of domestic energy has been increasing in specific segments during the last decades, not only in Brazil but also around the region, which has caused concerns among policymakers as the energy supply is important and regional energy cooperation remains essential in order to decrease risks.

Since the 1950s, regional integration has been part of Latin American countries' agendas.[41] In all cases, countries that struggle to achieve regional integration rely on many aspects, including political will, economic development as well as infrastructure investments. It is no different in the particular case of South American countries, as "the lack of infrastructure that can outweigh the unkind geography is a key hurdle that needs to be removed to push regional integration."[42]

Haghighi[43] suggests that "interdependence between energy producers, transporters and consumers provides strong functional need for co-operation." Padgett argues that "generating electricity or extracting gas, and transporting it to the point of consumption, requires intensive co-operation in large-scale, long-term investment. The central problem revolves around creating a stable framework to satisfy the security needs of consumers and producers and to handle the inherent investment risk."[44]

Haghighi also states that there is "disagreement over how to structure commercial energy relations between advocates of market and non-market forms of contracting. The former argue that the optimal way of satisfying the security needs of both consumers and producers is through the international market. The latter point to the inherent limitations of energy markets, particularly in natural gas."[45]

According to Oliveira,[46] there are many obstacles to overcome, but at least four are important to mention as more immediate measures to a policy framework and regional integration: (a) continuous coordination of the output of the region's power plants; (b) the environmental licensing of energy projects; (c) the legal deficit that adversely affects private investment in energy generation; and (d) the need for institutional mechanisms to resolve regional disputes.

In this sense, Brazil has increased its regional efforts in order to expand integration along with Venezuela, Peru, Bolivia, Colombia, Argentina, and Paraguay. However, various obstacles remain in each country. As a result, energy security in Brazil will play a significant role in the political and economic arena as current crises also force other countries to seek alternative energy sources as well. Brazil also occasionally faces energy shortages due to its poor infrastructure and efforts to conciliate public-private co-operation.

In summary, in the last decades, due to the oil crisis of the 1970s as well as Brazil's dependency on imported oil, the country has turned to alternative renewable energy sources. The overall situation suggests future investments in renewable biofuels. As part of its policies, Brazil has begun a transition to a low carbon economy, while it has sought to increase its energy supply through integration with its South American neighbors. Brazil has potential, particularly in terms of natural gas. Nonetheless, Brazil faces many challenges in terms of infrastructure and the need to implement policies to incorporate a better composition of the energy matrix.

POTENTIAL REGIONAL INTEGRATION WITH NEIGHBORING COUNTRIES

Oliveira[47] argues that "while other regions are under pressure to find reliable, environmentally clean, competitive sources to supply their energy consumption, South America can look optimistically at its future supply of energy," although the region has a low percentage of potential global energy supply compared to other regions.

Oliveira also suggests that "the region has large untapped energy resources, both renewable and non-renewable,"[48] as is the case of Brazil, particularly with regard to a successful biofuel program with ethanol and biodiesel. In line with this argument, in terms of renewable energy sources, sugar cane products have significantly improved their relevance in terms of Brazil's percentage of primary energy production. In this respect, as explained by Amatucci and Spers, the automotive sector—among other sectors—is one of the main competitive and internationalized markets. They argue that "notwithstanding the fact that the use of biofuels such as ethanol and biodiesel offers proven results in terms of emissions reductions, consumer and producer behaviors apparently remain focused on short-term economic optimization."[49]

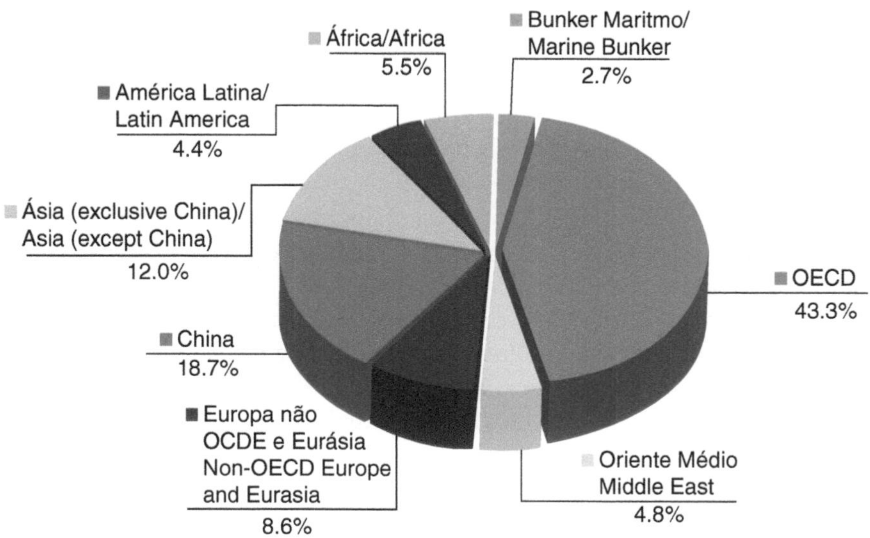

Graph 8.4 Energy Supply by Region in %. *Source*: OPEC[1] and EPE, 2009.
[1]OPEC Secretariat. World Oil Outlook, 2011.

The challenge in terms of energy remains not only in producing renewable sources, but also in meeting the ongoing market development that requires increases in the energy supply. OPEC forecasts a 2.4% car growth per year between 2008 and 2035 in Latin America (excluding Mexico in this case).[50] Furthermore, "when a citizen purchases any new "flex" car, the owner receives a printed table that lists ethanol and petrol price combinations in order for the consumer to be able to make a quick decision between the two fuels at any gas station. The decision is made purely in terms of pricing and efficiency considerations. Similarly, producers react to breakeven prices of crude oil and alternative feedstocks."[51]

Amatucci and Spers claim that the importance of institutional arrangements in order to make appropriate decisions as well as to address critical questions concerning alternatives and sustainable products such as biofuels. They argue that a characterization of the Brazilian biofuel institutional environment under formal—laws and rules, and informal—traditions, perceptions, and behaviors, or constraints, allows us to understand the incentives for adoption of biofuel technology.[52] "Formal and informal institutions provide incentives to individuals engaged in economic activities, in different levels."[53] They advocate for the use of alternatives source to fossil fuels which requires both appropriate technological development and the subsequent adoption by consumers. While electric or hybrid cars seem to be an international trend, the Brazilian automotive industry has instead focused on biofuels. Despite its ecological advantages, experts argue that there are some negative elements that may serve as a barrier to its widespread use, such as price, origin, and the accessibility of both the final product and raw materials. Their studies aim to identify factors that influence decision-making for the Brazilian economic agents.[54] Amatucci and Spers conclude that Brazilian experiences with ethanol and biodiesel differ significantly due to a different institutional economic framework. While ethanol has reached a mature level of sustainability, biodiesel still relies on institutional initiatives in order to be fully developed.[55]

According to OPEC, South American countries have developed plans to progressively adopt "clean" standards for both gasoline and diesel, which the Brazilian leadership might support (Table 8.5). The case of biofuels only enhances Brazil's potential as a regional energy supplier.

Oliveira states that "the benefits of the South American energy integration are already identified." The aforementioned example of biofuels is only part of various potential energy sources. Nevertheless, in many cases, domestic market protection is an impediment to increasing the efficiency of the regional integration. This is not an issue to be exclusively solved by Brazil; it suggests South American countries to bring together a regional framework to overcome the potential barriers, not denying or neglecting the leadership role of Brazil.

Table 8.4 Biofuel Production, 2006

	Country	*Ethanol (billion, liters)*	*Biodiesel*
1	USA	18.3	0.85
2	Brazil	17.5	0.07
3	Germany	0.5	2.8
4	China	1.00	0.07
5	France	0.25	0.63
6	Italy	0.13	0.57
7	Spain	0.4	0.14
8	India	0.3	0.03
9	Canada	0.2	0.05
10	Poland	0.12	0.13
11	Czech Republic	0.02	0.15
12	Colombia	0.2	0.06
13	Sweden	0.14	—
14	Malaysia	—	0.14
15	UK	—	0.11
World total		39.0	6.0

Source: REN21, 2008.

Since the 1950s, one of the main purposes of the Brazilian energy policies has been to develop "the energy islands of the country into a nationally interconnected energy system able to supply cheap energy to final consumers." Oliveira argues that in the 1970s, "the Itaipu project promoted the integration of the Brazilian South to its Southeast," During the 1980s, "the Tucurui project integrated the Northeast to the coastal area of the Amazonia." He emphasizes that "the North-Northeast interconnected grid was linked to the South-Southeast-Centerwest in the middle of 1990s with a long transmission line."[56]

However, Oliveira emphasizes that the regional ambition toward integration with the neighboring countries is a more recent movement, beginning in the 1990s in tandem with the liberalization of the energy system, and other regional efforts such as Mercosur.[57] In terms of energy imports, two major projects were developed in the 1990s: (a) a line of transmission from Argentina; (b) a gas pipeline from Bolivia. As for Mercosur, initial priorities were given to complementarity in natural gas and electricity.

More energy initiatives are envisioned[58] such as: (a) a transmission line could increase the supply of power from Venezuela to Porto Velho; (b) another line would link the power system of Guiana to Manaus; (c) *Garabi*, a hydropower plant along the Argentine border, would enhance the existing transmission connection between the power systems of Brazil and Argentina; (d) the gas pipeline linking the North of Argentina to the South of Brazil aims at creating a network of pipelines that would be able to provide natural gas around the Southern Cone.[59] In terms of regional benefits, these initiatives could save oil consumption, lead to increase in income per capita in other countries as well as employment. However, the potential benefits depend on regulation and political and economic efforts to increase stability and guarantee investments.

Furthermore, according to Pineau et al.,[60] three types of barriers exist with regard to regional integration and policy implementation: (a) material infrastructure; (b) the regulation of energy flows; and (c) rules for trading these flows as well as the prices. In the 1990s, there was significant progress in the construction of basic necessary infrastructure to make physical energy flows more feasible, especially in the Southern Cone. Nonetheless, this progress was not followed by improvements in the regulation of regional energy flows or in mechanisms for its trade.[61]

According to Oliveira, due to its geographical position as well as the size of its energy market, Brazil has a decisive role to play in the promotion of regional energy flows. Major investments have been made to develop infrastructures. Such policies were designed to promote the transportation of natural gas and electricity from neighboring countries to the Brazilian energy market.[62] Yet, bilateral agreements (see Table 8.5) adopted to utilize the infrastructure proved unable to resist the changes in the regional energy due to the political and economic context. As a consequence, this infrastructure remains largely idle because energy imports from neighboring countries are not perceived as safe and energy exports to these countries are viewed as a source of insecurity for the domestic market. Therefore, low-cost energy resources remain idle, while the cost of domestic energy supplies is increasing throughout the region.[63] In the 1990s, "South America adopted the institutions-market approach to promote regional energy integration."[64] As such, "the formulation of multilateral treaties to regulate the regional energy market was the object of various studies, but nothing was done about actually developing such treaties."[65] Based on Oliveira's as well as on Ramos' studies, the following framework demonstrates the international electricity operational connections in South America,[66] whose emphasis are on bilateral agreements (Table 8.5).

According to the aforementioned authors, although the fact that the potential economic benefits from regional energy integration are fruitful, it has not progressed as expected as energy has become more expensive for consumers in exporting countries, while energy producers in importing countries face competition from providers with lower production costs.[67] This situation tends to mobilize sociopolitical forces opposing the interdependence process. As a result, the appropriate identification of costs and benefits for the agents engaged in the process of integration is therefore a necessary step to make in order to adjust the various conflicting interests, as suggested by Keohane and Nye's approach.

Table 8.5 International Electricity Connections in South America—Bilateral Agreements

Southern Countries	*Operational-Location*
Argentina-Paraguay	Central de Partidas Vacyreta
	Central Salto Grande (Colonia Elia [Arg]-San Javier [Par])
Brazil-Paraguay (binational)	Saidas da central Itaipu
Brazil-Argentina	S.M. Ricon (Arg)-Garabi (Bra)
	Paso de los Libres (Arg)-Uruguaiana (Bra)
	Rincon (Arg)-Ita (Bra)
Argentina-Chile	Central Termo Andes (Arg)-Norte Grande (Chi)
Argentina-Paraguay	Clcrinda (Arg)-Guaramare (Par) Eldorado (Arg)-Carlos Antonio Lopez (Par)
Brazil-Paraguay	Fozdo Iguacu (Bra)-Acaray (Par)
Nothern Countries	*Operational-Location*
Brazil-Venezuela	Boa Vista (Bra)-EIGL.ri (Ven)
Ecuador-Peru	El Alamor (Ecu)-Piu-a (Per)
Colombia-Venezuela	Cuestecita (Col)-Cuatricertenario (Ven)
	Tibu (Col)-LaFria (Ven)
	San Mateo (Col)-Coro20 (Ven)
Colombia-Ecuador	Ipialcs (Col)-Tucan/Ibarra (Ecu)

Source: Adapted from Oliveira (2010), Ramos (2004).

It has been argued that "part of the costs and benefits of energy integration stems from the output of hydropower plants, as it determines the use of fossil fuels in thermal power plants,"[68] which plays a decisive role in the pricing of natural gas. Given these circumstances, a potential solution could be a formulation of supranational cooperative rules or multilateral agreements that lead to an agreed division of costs and benefits in terms of electricity output is a difficult task.[69]

Regardless the type of proposals and segments targeted, Brazil has a meaningful role in coordinating regional actions and supply of energy due to its capacity of production, renewable sources, and hydropower as well as market size. According to OPEC,[70] the combination of these factors along with rising crude supplies and ethanol production combined to an aggressive plan for refinery expansion for internal and external purposes would position the country in a more competitive situation and make Brazil a more influential regional actor.

BRAZIL—ENERGY POLICY FORMULATION, IMPORTANT ACTORS AND PETROBRAS' SCANDAL

Whether Brazil can act as a regional leader or a more impacting energy global player depends on many conjectural factors, including its domestic situation, and energy policy formulation within its political and economic realm. In this respect, the National Council of Energy Policy (*Conselho Nacional de Política Energética*) is responsible for formulating the energy policy and establishing guidelines aiming at national sustainable development. Under the leadership of the Ministry of Mines and Energy along with the *Empresa de Pesquisa Energética* (*EPE*), the "ten-year expansion plan 2022" is the result of an interactive process of energy planning in cooperation with many government agencies—Secretary of Energy Planning and Development, Center for Electric Power Research, Department of Petroleum, Natural Gas and Renewable Fuels, Special Advisor on Environmental Management, amongst others—also supported by public consultation, companies and agents of the energy sector. [71]

As part of the list of companies subject to the structure of the Ministry of Mines and Energy, in addition to *EPE*, other partners and regulatory agencies articulate a meaningful role within this sector, such as *CPRM* (*Companhia de Pesquisas de Recursos Minerais*), *PPSA* (*Empresa Brasileira de Administração de Petróleo e Gás Natural S.A.—Pré-Sal Petróleo S.A.*), Petrobras, Eletrobras (*Centrais Elétricas Brasileiras S.A.*), *Departamento Nacional de Produção Mineral* (*DNPM*), ANEEL (*Agência Nacional de Energia Elétrica)*, and ANP *(Agência Nacional do Petróleo, Gás Natural e Biocombustiveis).* [72]

Petrobras deserves a special attention due to its political and economic relevance, and more recently, because of the scandals and corruption the company is involved in. "According to revelations published in *Veja*, a leading weekly, and *Estado de São Paulo*, a newspaper, Mr. Costa, a former executive at Petrobras, who ran Petrobras' refining division from 2004 to 2012, has accused more than 40 politicians of involvement in a vast kickback scheme. The list reportedly includes a minister, three state governors, six senators and dozens of congressmen from President Dilma Rousseff's Workers' Party (PT) and several coalition allies. The beneficiaries are alleged to have pocketed 3% of the value of contracts signed with Petrobras in return for supporting the government in congressional votes."[73]

"Aécio Neves, leader of the opposition Party of Brazilian Social Democracy (PSDB) and a rival of Ms. Rousseff for the presidency, was quick to dub the scandal "*mensalão* 2." In the original *mensalão* ("big monthly stipend"), a scheme that operated from 2003 until it came to light in 2005, the PT paid a monthly bribe to congressional allies in exchange for their votes. The scandal hurt the government of Luiz Inácio Lula da Silva, Ms. Rousseff's predecessor and political patron, and landed some of the PT's leading lights behind bars."[74]

"Instead of surging, Petrobras' oil production has stagnated, heightening Brazil's reliance on imported oil. Petrobras finds itself mired in corruption investigations and claims of managerial incompetence. And its debt load is exploding: Petrobras now ranks as the world's most indebted company, dependent, more or less, on US mutual funds to finance its ambitious investment plans. The result is that Petrobras' losses in its refining, transportation and marketing operations reached $8 billion in 2013.[75] In addition, "nine of Brazil's biggest construction firms are now under investigation: Camargo Correa, OAS, UTC, Odebrecht, Mendes Junior, Engevix, Queiroz Galvao, Iesa and Galvao Engenharia. The group includes some of the most important donors to political party campaigns."[76]

Corruption seems endless. "As the next phase of the operation is set to focus on politicians favored by the scheme, local newspapers have reported that the raids have created an atmosphere of panic in the capital Brasilia. "In the short term, the scandal puts new pressure over the recently re-elected government of President Dilma Rousseff—already troubled by weak economic growth and inflation rates that are too close to the top of the Central Bank's target," says Adriano Pires, from the Brazilian Infrastructure Center (CBIE). "In the medium and long term, however, this scandal may impose a major overhaul of the relationship between private companies, politicians and state-run firms in Brazil," Pires adds, pointing to regulations in the area of campaign financing."[77]

What may sound striking is the fact that "Ms. Rousseff herself chaired the Petrobras board of directors from 2003 to 2010, although she denies knowledge of any wrongdoing, and stresses her government's full backing for the investigation, and determination that wrongdoers will be punished. According to public prosecutors, the corruption scheme has been in operation at Petrobras for at least 15 years—and if that is true it would have pre-dated the Workers' Party government. The impact of the case on the country's growth prospects is as yet unclear."[78]

"Economist Silvio Campos Neto, from the Consultancy firm Tendencias, is very pessimistic: "this is the most important company in the country and at times when we are struggling to recover growth, such a huge bribery scandal will not give a positive message to potential investors," he says. On the other hand, Sergio Lazzarini, a professor at the business school Insper, however, has a slightly more positive view. He says corruption is an endemic problem in emerging markets: "At least this scandal shows that Brazil has strong institutions, which are able to investigate and expose a case that involves powerful businessmen and politicians."[79]

Regardless of the position that one may take, the case underlines Brazil's vulnerability, and Petrobras' scandals may prevent the country to take over the leadership and become a more active regional and global energy player. Politics does matter in the Brazilian energy sector.

BRICS AND CHINA

Accelerating sustainable growth along with energy security is amongst the most important challenges in today's world. Energy security is a key element of BRICS' agenda for global recovery, sustainability and investment for future growth. BRICS' Fifth Summit declaration stressed on the fact that energy based on fossil fuels will continue to dominate the energy mix for the foreseeable future. These countries intend to expand sources of clean and renewable energy, the use of energy efficient, and alternative technologies in order to meet the increasing demand of their economies and people. In this context, they emphasize that international cooperation should prevail in the development of energy for peaceful purposes.[80]

Although these countries have different energy realities, what unites them in terms of energy is the fact that they will have an important role in the evolution of the global energy context, which accounts for 32% of global energy demand. Furthermore, "while foreign policy goals in energy differ amongst the BRICS, one important area of commonality is the depth of consumption subsidies used in each of the countries other than Brazil."[81]

Another aspect is geography. India, China and Russia are geographically more interconnected which doesn't happen to Brazil and South Africa[82]. For future perspectives in terms of consumption, Table 8.6 illustrates the energy demand by type, comparing the BRICS (except South Africa) with developed OECD countries in 2008 and forecasts for 2030.

Coal remains an important component of energy for the BRICS due to China's and India's energy matrix, as opposed to Brazil's and Russia's. China deserves more particular attention as the 2nd largest global economy, Brazil's major trading partner, the global largest polluter, whose energy efforts have been aiming at reducing the coal dependency.

Backed by its "renewable" Five-Year-Plan (2011–2015), China's guidelines continuously focus on searching for raw materials and commodities to

Table 8.6 Energy Demand by Type: 2008 and Forecasts for 2030 (Less South Africa)

Energy Demand by Type	*BRIC*		*OECD*	
	2008	*2030*	*2008*	*2030*
Coal	49.0%	49.4%	20.8%	18.6%
Oil	20.4%	20.2%	37.5%	31.5%
Gas	13.4%	13.8%	23.5%	25.2%
Nuclear	1.9%	4.4%	10.9%	11.6%
Hydro	2.9%	2.9%	2.1%	2.2%
Biomass and Waste	12.3%	8.0%	4.2%	7.2%
Other Renewables	0.2%	1.2%	1.0%	3.6%
Total	100%	100%	100%	100%

Source: Adapted from and IEA and The Norwegian Institute of International Affairs, 2012.

feed the booming domestic industry and consumption. PCR's geopolitical plans remain ambitious as a consequence of a domestic non-stopping infrastructure construction, urbanization process, and tightening regulations on natural resources utilization.[83] That is where Brazil fits in to supply oil and other commodities to the Chinese market.

To manage its policies, the Chinese government aims to have a balance between a highly driven exporter powerhouse and a heated and aging domestic consumer market. Ultimately, combining these initiatives is part of PRC's strategy to ensure an adequate level of energy security to sustain its one billion plus population and fierce international competition.

As a big producer and consumer of energy resources, China outstripped the United States in 2010 in terms of consumption. China's natural gas supply and demand prospects are a major change of Eurasia, the world gas trade patterns and geopolitical factors. Therefore, China's investment in the Western Hemisphere, and particularly in Latin America, cannot be seen as an isolated event. In this respect, energy takes an important percentage (74%) of the Chinese FDI (Figure 8.1) in the realm of its political and economic initiatives.

As mentioned in the beginning of this chapter, according to Yergin, part of the global political and economic challenges is to bring China and India to the global energy system. Thus, Asia accounts for over 30% of Brazilian crude oil exports with China and India being the most important,[84] as both countries are relatively scarce in natural resources as opposed to Brazil and Russia.

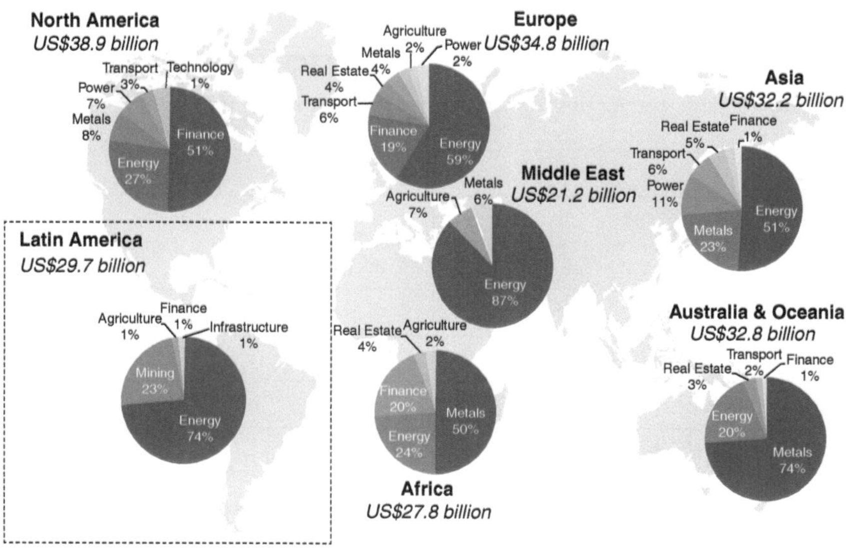

Figure 8.1

The case between Brazil and China deserves special attention. Thanks to the robust economic performance, accompanying a closer dialogue at state government level,[85] China (PRC) became Brazil's first trade partner in Asia in 2000. Furthermore, Brazilian companies, enthusiastic about deepening their bilateral ties, sought to establish a foothold in PRC, seeking to have opportunities in the biggest mass consumption market worldwide. Other political and economic initiatives underlined the aim of cooperation in 2003. Thus, a remarkable shift was observed from 2004 to 2006, thereby marking a political rapprochement.

Former President Lula visited the Central Empire in 2004, the same year that President Hu Jintao officially reciprocated a few months later, celebrating 30 years of diplomatic relations, and highlighting a Memorandum of Understanding on Trade and Investment. By this time, a High-Level Sino-Brazilian Coordinating Committee arose to foster their bilateral intentions. From 2007 on, other political initiatives have been undertaken by both sides. In 2009, former president Lula's visit to PRC endorsed the fact that China became Brazil's number one global trade partner. Graph 8.5 demonstrates the evolution of the bilateral ties between Brazil and China since 2000, reaching US$ 77.9 billion in 2014.

Ebinder and Avasarala (2012)[86] argue that "owing to their advantageous resource endowments and the nature of their energy consumption, Brazil and Russia are relatively energy-secure nations, whereas China and India are less self-sufficient, but superior infrastructure and a centralized government

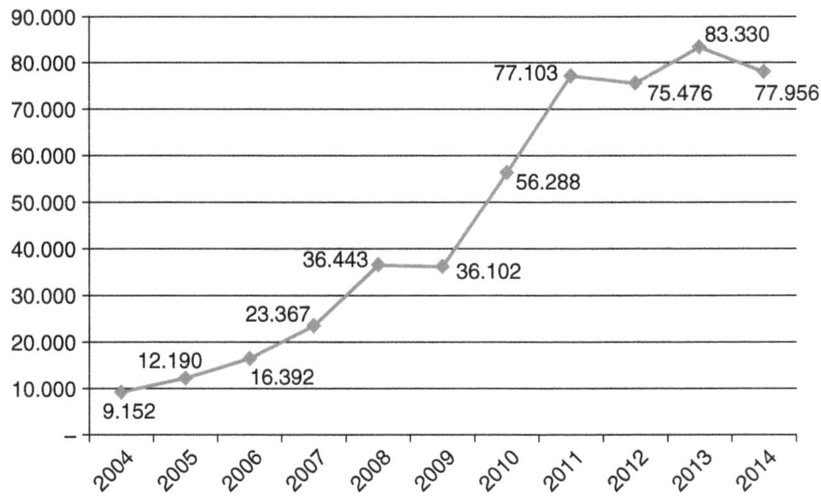

Graph 8.5 The Evolution of Bilateral Trade between Brazil and China—in US$ Billions. *Source*: Brazilian Ministry of Development, Industry, and Trade (MDIC), elaborated by CEBC 2015.

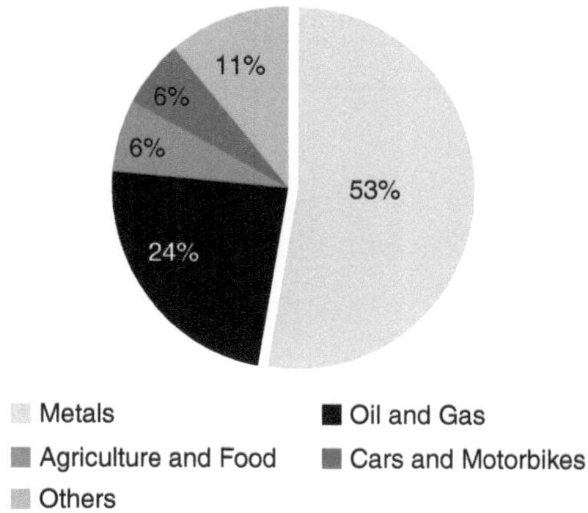

Graph 8.6 The Chinese Investment (FDI) by Sector in Brazil—in %. *Source*: CEBC, 2011.

position China in a better position to meet its rising energy demand. India has the most unstable energy condition in the BRICS, and possibly the most uncertain energy future of all."

CONCLUSION

These final remarks summarize the key points aforementioned and lead to policy options for Brazil, which has become an important energy security player in South America integration; yet with various political and economic limitations. This work aimed to underline some perspectives related to the energy production and consumption in Brazil. It emphasized the potential of natural gas, but casted light on the issues Brazil encounters in terms of infrastructure and the need to revise and implement a more accurate policy toward the country's capacity, production, supply as well as competitiveness to better compose the energy matrix. It also indicated the importance of use of diverse sources of energy for safety and economic reasons to become a more prepared regional and global energy supplier, including renewable sources. The latter will be part of Brazil's future investments (Table 8.7).

Over the next 10 years, according to the Brazilian government, the use of renewable energy in the country will grow at an average rate of 5.1% per year, from a total of 43.1% in the Brazilian energy mix to 45% in 2021.[87] Around R$ 1.2 trillion (US$ 500 billion) of investments will be focused not only on liquid biofuels, but also on electricity transmission, oil and natural gas, and infrastructure (2013–2022): 22.6% correspond to the supply of electricity, 72.5% to oil and natural gas, and 4.9% to the supply of liquid biofuels[88] (Table 8.7).

Table 8.7 Synthesis of Investment Estimates in R$ Billions for Future Energy Projects

Energy Investments in Brazil	*R$ Billions*	*%*
Electricity	269	24.5
Generation	213	19.4
Transmission	56	5.1
Petroleum and Natural Gas	753	68.4
Exploration and Production of Petroluem and Gas	547	49.9
Supply of Petroleum products	198	18.1
Refining Petroluem	174	15.8
Infra structure	24	2.2
Supply of Natural Gas	8	0.7
Supply of Liquid biofuels	79	7.2
Ethanol-production plans	71	6.5
Ethanol-infrastructure (pipelines and port)	7	0.6
Biodiesel-production plans	1	0.1
Total	1,101	100

Source: Adapted from the Brazilian Government, PDE 2021.

Undoubtedly, Brazil has a relatively "clean" energy system particularly when considering its renewable energy sources, such as ethanol and biodiesel. Amatucci and Spers positioned that alternatives to fossil fuels require both appropriate technological development and subsequent adoption by consumers. Despite its ecological advantages, certain negative elements may serve as a barrier to the use of biofuels, such as price, origin, and accessibility of both final product and raw materials. Brazil's experiences with ethanol and biodiesel differ significantly from one another. While ethanol has reached a mature level of sustainability, biodiesel still relies on institutional initiatives in order to be fully developed. Brazil can be an international export platform of ethanol. The government projects investment in infrastructure to supply it (Table 8.7).[89]

The social-environmental studies developed in Brazil's 10-year plan were guided by the concept of sustainability, marked by the reduction of environmental impacts on the energy supply expansion and discussions at national and international level on climate change. Important environmental issues have been identified with several sources for which such industry efforts should be geared to contribute to the minimization of risks and exploiting opportunities related to expansion. Indigenous peoples, protected areas, aquatic biodiversity and native vegetation have been considered priority areas for environmental management in this context.[90]

The climate issue has reiterated its relevance in Brazil by the enactment of Law 12.187/09 and Decree 7.390/10, which regulates this law. This new legal framework established the National Policy on Climate Change and set a goal of reducing greenhouse gas emissions by 36.1% to 38.9% for 2020.

The inclusion of biofuels in the energy matrix is considered strategic by many developed countries with the goal of obtaining energy security and compliance targets for reducing greenhouse gas greenhouse. However, due to the economic crisis in Europe and the US, the trend is to have more protectionist markets, thereby postponing targets of reducing greenhouse gases. The supply of ethanol points to an expansion of 154% of the production of biofuel over a 10-year horizon, from 28 billion in 2012 to 71 billion in 2021.[91]

Based on this argument, "a combination of policies could facilitate higher efficiency bagasse cogeneration as well as encourage use of leaves and tops for energy production, where appropriate. Some of these policies are similar to those needed to stimulate a combined heat and power systems (CHP) with natural gas:

(a) Require utilities to purchase excess power from sugar mills, and distribution costs via long-term contracts; (b) require utilities to interconnect CHP systems to the power grid without excessive delay or unreasonable technical requirements; (c) continue to develop and demonstrate more efficient technologies such as bagasse gasification and combined-cycle power generation in sugar mills; (d) provide long-term loans at attractive interest rates to sugar mills that adopt more efficient CHP technologies; and (e) finance and support the gradual adoption of mechanical harvesting systems.[92]

In line with the aforementioned paragraphs, the important question is whether such fuels are used to meet the regional needs of energy-intensive industries in the manufacturing sector (cars for example), highly intensive in the use of energy, increasing urbanization, and the rapid growth of transportation without causing excessive damage to the environment.

According to OPEC,[93] South American countries have developed plans to progressively adopt tighter "clean" standards for both gasoline and diesel, which could be potentially supported by Brazil's leadership and capacity of production in this sector, having Sao Paulo as a strategic cluster. Amatucci and Spers[94] advocate the presupposition that maybe the challenge to export ethanol depends on a combination of factors, such as positive sign from potential buyers (at global and South American levels) as well as trade barriers reduction to import it.

This chapter enhances the domestic movements promoted within Brazil to connect internal regions as well as South American attempts of integration in the electricity sector, along with bilateral agreements. One the one hand, these agreements are positive to initiate economic and political engagement; on the other hand, they could prevent deeper regional integration, and should be on a multilateral level. As per Oliveira's view,[95] "the next step in the integration process is to move from the current bilateral commercial contracts to a multilateral security of supply agreement." He advocates for a policy of "Multilateral Regional Security Reservoirs (MRSR) that could be depleted by any country participating in an agreement, under an economic regime in which depletion prices would be aligned to the long-term marginal cost of the regional supply of energy."

The "disappointing outcome of the regional energy integration process is largely due to the misplacement of the energy integration agenda on the convergence of national regulations." As a consequence, Oliveira argues that "the focus of the integration agenda should pragmatically move from trade to increase the security of supply and to lower the volatility of energy prices" and sustain a MRSR agreement.[96]

In order to do so, Brazil's policy should focus on large hydropower reservoirs that could be shared with neighbor countries to avoid risks of energy shortages, as has often occurred. Similarly, South American neighbors have large gas reservoirs that can be shared with Brazil with the same goal. In the search for the development of a common agenda, the Brazilian government has been trying to make agreements and Brazilian companies are participating in studies in countries in Central and South Americas.[97]

Brazil's government argues that regional energy integration will provide many benefits for both parties, such as the use of complementary hydraulic schemes for example. The association of these countries with Brazil mostly occurs with sources that generate renewable energy and low cost, as is the case of hydropower.[98]Another example of energy integration and co-operation could be with Peru, a country with hydroelectric potential of about 180 GW. There are plans for the construction of six hydroelectric plants totaling about 7 GW of installed capacity. The use of "Inambari" with 2 GW is that which is at the most advanced stage, but the start of construction of the plant has no prediction. Those studies that prioritize service to local demand indicate the possibility of exporting energy surplus to Brazil with the interconnection of power systems getting in the state of Rondonia (North).[99]

Geller et al. suggest "the expand use of combined heat and power systems fueled by natural gas, as there is potential for using cogeneration (CHP). The potential increase in natural gas supply in Brazil opens up new opportunities for CHP systems. Given the high efficiency of cogenerating electricity and useful thermal energy, the following policies could be implemented to eliminate the barriers to natural gas-based CHP systems:

(1) "Require utilities to purchase surplus power from CHP systems; (2) Require utilities to interconnect CHP systems to the power grid without excessive delay or overly burdensome requirements, as well as provide back-up power to owners of CHP systems at reasonable terms; (3) Give priority to CHP projects as new gas supplies become available and are allocated to commercial and industrial consumers; (4) Provide financial incentives such as long-term loans at attractive interest rates from the national development bank for CHP systems that meet certain conditions such as high overall

efficiency and low pollutant emissions; and (5) Reduce import duties on CHP equipment such as gas turbines but also promote the production of this equipment in Brazil."[100]

With Bolivia, there is an opportunity not only in natural gas but also in a hydroelectric project "Cachoeira Esperança," with 800 MW. If it occurs, the study will result in a Brazil-Bolivia binational undertaking, located close to Madeira River, also showing potential for export of generation surplus to Brazil. The hydroelectric potential of Bolivia corresponds to 20.3 GW.[101]

In addition Guyana has an approximate water potential of 8 GW. The energy surplus produced by hydroelectric plants that will be deployed can be imported by Brazil, through interconnections in the state of Roraima (North).[102] In sum, these potential projects listed above may be possible, among others, though mmultilateral agreements (MRSR) between Brazil and its neighbors.[103] As a consequence, it could reduce the investment needed to keep supply and demand balanced in the short-term, and it would improve the environmental performance of the energy system—both locally and regionally. It would save scarce oil resources for future generations and would diminish the volatility of energy prices of all countries in of the South Cone,"[104] as implied by the interdependence approach.

In summary, the relevance of these issues addresses a problem that has still been important in our contemporary world of international relations. In addition, energy security is associated with a number of complex factors, associated with international trade,[105] supply and demand and interdependence, as postulated by Keohane and Nye.[106] As stated by Yergin,[107] "the dependence on energy systems, and their growing complexity and reach, all underline the need to understand the risks and requirements of energy security in the 21st century."

Indeed, the perspectives for Brazil as a regional (South America) and global energy supplier (China and the BRICS) depend on a combination of political and economic factors. Part of the concerns remains associated with Brazil's domestic constrains, political scandals as the case of Petrobras, but also related to an issue not to be exclusively solved by Brazil. The discussion suggests South American countries to bring together a regional framework to overcome the potential barriers, not denying or neglecting the leadership role of Brazil. On China, its economic slowdown may impact on Brazil's future exports of commodities, although the bilateral trade remains shows long-term growth.

As for the BRICS, this topic also addresses a problem that has been important in our contemporary world of international political economy. For more than a century and since the vulnerability of the energy shocks of the 1970s as well as the asymmetries between the geographical distribution of resources, this subject allows us to explore relevant issues related to the security and oil geopolitics, the institutional approach, the new economics of energy, the technical systems, and the role of markets and investments.

In addition, the energy security *per se* is associated with a number of complex factors, such as international trade, supply, demand and pricing.[108] Aiming at the evolution of the field and comparing the 1970s or the previous decades, today's growing political and economic actors reshape a new dynamic in the BRICS energy security puzzle, reason why its scientific relevance contributes to a scholarly debate. Yergin (2012)[109] argues that "the dependence on energy systems, and their growing complexity and reach, all underline the need to understand the risks and requirements of energy security in the twenty-first century. Increasingly in a growing world economy, energy trade traverses national borders and ties nations together." "That is why energy security is not just about countering the wide variety of risks and threats." It is also about relations among nations, how they interact with each other, how energy impacts their overall national security, and about cooperation and collaboration in the international community.

Yergin[110] also explains several dimensions of energy security, from physical (protecting the assets; infrastructure; supply chains; trade routes; making provisions for quick replacements and substitutions) to commercial issues. In some cases, it occurs through Foreign Direct Investments (i.e., China-Brazil). It also reflects that energy security is a system, composed of the national policies and international institutions that are designed to respond in a coordinated way to disruptions, dislocations, and emergencies, as well as helping to maintain the steady flow of supplies. All of these issues apply to the BRICS` challenges ahead, but in different dimensions for each country.

Brazil may not want the BRICS to become a threat to global energy markets, and South Africa "has continued to underwrite Africa's openness to American and European interests. In some ways, Brazil and South Africa are both buffers between the Asian giants and the Western world."[111]

To conclude, the BRICS' unprecedented economic growth observed since the 1990s (although the current scenario in 2015 suggests economic stagnation in the case of Brasil and Russia) has aroused the relevance of energy security cooperation. In this sense, BRICS energy security requires policies and a business climate that promote investments and development to ensure that adequate supplies and infrastructure will be available in a timely way.

Some bilateral efforts have been made, but an effort to legitimize and support a liberal[112] engagement requires the BRICS to adopt a multilateral agenda to the challenges ahead. Due to their significant economic and political changes, whether successful bilateral and/or multilateral energy policies will be effectively implemented also depends on the quality of their governance,[113] and energy may either cement or break the BRICS apart.[114]

With regard to Petrobras, "perhaps its biggest challenge is that it is not just an energy company. It is also at the heart of a fierce debate here over the extent of the Brazilian government's use of its wealth to achieve political and economic goals." Furthermore, in an effort to keep inflation from accelerating during the electoral campaign, "Rousseff's government prevented Petrobras from raising fuel prices to meet the cost of importing refined gasoline and diesel. At the same time, domestic fuel consumption has surged since the Brazilian authorities offered incentives to car manufacturers to raise production."[115]

Politics may not have been the only influence on Petrobras' stock, as falling oil prices around the world, for instance, have also contributed to recent weakness. Notwithstanding, "the price of gasoline on the international market has slipped below the domestic price of the fuel in Brazil." As a consequence, the government has adjusted prices "to help Petrobras make up for losses that stem from government policies that had forced the company to sell fuel to local consumers at prices below international ones."[116] Plus, if oil prices remain at historically low levels for a long period, there might be pros and cons for each country.

For "China, which is set to become the largest net importer of oil, should gain from falling prices. However, lower oil prices won't fully offset the far wider effects of a slowing economy."[117] This may have a direct impact on Brazil's oil exports to the Central Empire.[118] The projects expected for the pre-salt in Brazil may be affected. "The lower the oil price, the higher the pressure Petrobras will have to not increase it, which will further hinder investments," says Rodolfo Oliveira, economist Trends Integrated Consulting.[119]

Moreover, other measures have been taken by the Brazilian government. As argued by some specialists, by cutting the transfer of R$ 9 billion (US$ 3.5 billion) of Treasury provided for the electricity sector in 2015, the government aimed a bird and might have hit two in an attempt to rebalance the fiscal situation of the country. In addition to saving resources to pay the interest payments, the so-called primary surplus, the measure will eventually reinforce the states' revenue and the Union itself. With the end of the Treasury subsidies on electricity bills, the National Electric Energy Agency (ANEEL) prepares higher tariffs on electricity. The national average bills of consumers should be about 30% more expensive, which means a similar increase in the collection of State taxes.[120] This also means more concerns on "energy insecurity" along with the droughts that have affected Brazil and the shortage of water in some reservoirs, such as the case of Sabesp in São Paulo. Brazil urges to have more energy security to become a relevant regional and global player.

NOTES

1. L.C Bresser-Pereira,. Macroeconomia Pós- Plano Real: as relações básicas. Em Novo-desenvolvimentismo—Um projeto nacional de crescimento com eqüidade social Sicsú, J.; de Paula, L. F.; Michel, R., 2005 World Bank and the IMF, 2013.

2. Adilson de Oliveira. “South Cone Energy Integration: A Look From Brazil.” *International Journal of Energy Sector Management*. Vol. 1. No. 2. pp. 122–140. 2007.

3. Daniel Yergin, *The Quest. Energy, Security, and the Remaking of Modern World*, 2012.

4. IEA, 2013.

5. Ibid.

6. Also see Table 8.7.

7. Adilson de Oliveira. “South Cone Energy Integration: A Look From Brazil.” *International Journal of Energy Sector Management*. Vol. 1. No. 2. pp. 122–140. 2007.

8. Daniel Yergin. *The Quest. Energy, Security, and the Remaking of Modern World*. 2012.

9. Daniel Moran. “Energy Security” in *Handbook of Oil Politics*, ed. Robert E. Looney. London and New York: Routledge, 2012.

10. EPE. Empresa de Pesquisa Energética. Brazilian Energy Balance. *Ministry of Mines and Energy*, Brasilia, 2012.

11. The document “PDE 2021 Public Consultation” is available at http://bit.ly/eNQBmm.

12. IBGE, 2008.

13. Marcelo Zorovich and Manoel Guedes. “A emergência de quase 4 milhões de consumidores.” *Revista da ESPM*, 2011.

14. OPEC Secretariat. *World Oil Outlook*, 2011.

15. Ibid.

16. Ibid.

17. O Estado de SP . “Corte de subsídios na energia elevará receita nos Estados,” 01/15/2015. Available at http://economia.estadao.com.br/noticias/geral,corte-de-subsidios-na-energia-elevara-receita-nos-estados,1619991 Accesed January 10, 2015.

18. The document “PDE 2021 Public Consultation” is available at http://bit.ly/eNQBmm.

19. CECB, 2015.

20. Ibid.

21. Giambiagi, F. and Villela, A. (2005), Economia Brasileira Contempora^ nea (1945–2004), Eslevier, Rio de Janeiro.

22. Adilson de Oliveira. “South Cone Energy Integration: A Look From Brazil.” *International Journal of Energy Sector Management*. Vol 1. No. 2. pp. 122–140. 2007

23. “Fóruns Estadão. Brasil Competitivo. Economia e Negócios.” *O Estado de* Sao Paulo, 2012.

24. Ibid.

25. Paulo Pedrosa. In “O Futuro do Gás Natural,” *Grupo Estado and Projeto+Gas Brasil*, O Estado de Sao Paulo, 2012.

26. OPEC Secretariat. *World Oil Outlook*, 2011–2012.

27. ABRACE: *Associação Brasileira de Grandes Comsumidores Industriais de Energia e de Consumidores Livres*, 2012.

28. Ibid.

29. OPEC Secretariat. *World Oil Outlook*, 2011–2012.

30. Ibid.

31. "Petrobras scraps Bolivia project." BBC, 05/04/2006.Available at http://news.bbc.co.uk/2/hi/business/4971832.stm.

32. SIEE OLADE. *Energy Economic Information System*, 2009.

33. Agência Nacional de Petróleo (ANP), the national hydrocarbons regulator.

34. Adilson de Oliveira. "South Cone Energy Integration: A Look From Brazil." *International Journal of Energy Sector Management*. Vol. 1. No. 2. pp. 122–140, 2007.

35. BBC. "Petrobras scandal: Brazil's energy giant under pressure," 11/21/2014. Available at: http://www.bbc.com/news/business-30129184 Accessed January 10, 2015.

36. José Goldemberg and José Roberto Moreira. "Política energética no Brasil." *Estudos Avançados* (55), 2005.

37. Ibid.

38. The Guardian. "São Paulo faces a critical water shortage as the World Cup prepares to kick off," 05/21/2014 , http://www.theguardian.com/sustainable-business/sao-paulo-water-shortage-world-cup Accessed January 10, 2015.

39. José Goldemberg and José Roberto Moreira. "Política energética no Brasil." *Estudos Avançados* (55), 2005.

40 Adilson de Oliveira. "Series on Trade and Energy Security. Energy Security in South America: The Role of Brazil." *International Institute for Sustainable Development*, 2010.

41. Araminta de Azevedo Mercadante; Umberto Celli Junior; Leandro Rocha de Araújo (Coords.). "Blocos Econômicos e integração na América Latina, África e Ásia." Curitiba: Juruá, 2006.

42. Adilson de Oliveira. "South Cone Energy Integration: A Look From Brazil." *International Journal of Energy Sector Management*. Vol. 1. No. 2. pp. 122–140. 2007.

43. Haghighi, S.S. (2007) Energy Security: The External Legal Relations of the European Union with Major Oil and Gas-Supplying Countries (Oxford and Portland, OR: Hart Publishing).

44. Stephen Padget. "Energy Co-operation in the Wider Europe: Institutionalizing Interdependence." *JCMS*. Vol 49. No. 5. pp. 1065–1087. 2011

45. Haghighi, S.S. (2007) *Energy Security: The External Legal Relations of the European Union with Major Oil and Gas-Supplying Countries*. Oxford and Portland, OR: Hart Publishing.

46. Adilson de Oliveira. Series on Trade and Energy Security. Energy Security in South America: The Role of Brazil. International Institute for Sustainable Development, 2010.

47. Adilson de Oliveira. "South Cone Energy Integration: A Look From Brazil." *International Journal of Energy Sector Management*. Vol. 1. No. 2. pp. 122–140, 2007.

48. Ibid.

49. M. Amatucci and E. E. Spers. "The Brazilian Biofuel Alternative." *Int. J. Automotive Technology and Management*, Vol. 10, No. 1, 2010.

50. OPEC Secretariat. *World Oil Outlook*, 2011.

51. Ibid.

52. North, D.C. *Institutions, Institutional Change and Economic Performance*, Cambridge University Press, p. 152, 1990.

53. Alston, L.J., Eggertsson, T. and North, D.C. *Empirical Studies in Institutional Change*. Cambridge University Press, United Kingdom, 1998.

54. Marcos Amatucci and Eduardo E. Spers. "The Brazilian Biofuel Alternative." *Int. J. Automotive Technology and Management*, Vol. 10, No. 1, 2010.

55. Ibid.

56. Adilson de Oliveira, "Series on Trade and Energy Security. Energy Security in South America: The Role of Brazil." *International Institute for Sustainable Development*, 2010.

57. Araminta de Azevedo Mercadante; Umberto Celli Junior; Leandro Rocha de Araújo (Coords.). "Blocos Econômicos e integração na América Latina, África e Ásia." Curitiba: Juruá, 2006.

58. IIRSA. "Initiative for the Integration of Regional Infrastructure in South America." Available at www.iirsa.org . Accessed January 05, 2012.

59. Ibid.

60. Pierre-Oliver Pineau, Anil Hira and Karl Froschauer. "Measuring International Electricity Integration: A Comparative Study of Power Systems Under the Nordic Council, Mercosur and Nafta." *Energy Policy* 32: 1457–1475, 2004.

61. Ibid.

62. Adilson de Oliveira, "Series on Trade and Energy Security. Energy Security in South America: The Role of Brazil." *International Institute for Sustainable Development*, 2010.

63. Ibid.

64. Ibid.

65. Ibid.

66. A. Moniz Ramos. "Fundamentos para la constitución de un Mercado común de electricidad." *CEPAL serie recursos naturales e infraestructura no. 73*. Santiago, 2004.

67. Adilson de Oliveira. "Series on Trade and Energy Security. Energy Security in South America: The Role of Brazil." *International Institute for Sustainable Development*, 2010.

68. Ibid.

69. Ibid.

70. OPEC, 2013.

71. EPE. Plano Decenal de Expansão de Energia—PDE 2021. "Derivados da cana e eólica sustentarão aumento de participação das fontes renováveis ao longo dos próximos 10 anos." Rio de Janeiro, 09/26/2012.

72. Brazilian Ministry of Mines and Energy. Available at http://www.mme.gov.br/web/guest/acesso-a-informacao/institucional/estrutura-organizacional Accessed in January 10, 2015.

73. "The Petrobras affair." The Economist, 08/08/2014. Available at http://www.economist.com/blogs/americasview/2014/09/scandal-brazil Accessed January 10, 2015.

74. Ibid.

75. "Brazil's Star, Petrobras, Is Hobbled by Scandal and Stagnation." *The New York Times*. 04/15/2014. Available at http://www.nytimes.com/2014/04/16/business/international/brazils-star-petrobras-is-hobbled-by-scandal-and-stagnation.html?_r=0

76. BBC. "Petrobras scandal: Brazil's energy giant under pressure," 11/21/2014. Available at http://www.bbc.com/news/business-30129184.

77. Ibid.

78. Ibid.

79. Ibid.

80. BRICS Fifth Summit. http://www.brics5.co.za/joint-statement-of-the-brics-business-forum-2013/ Accessed August 14, 2013.

81. Eriksen, S.S.; Lodgaard, S.; Melchior, A; Rich, K.; Rowe, E.W.;Sending, O.J. "BRICS, Energy and The New World Order." *Norwegian Institute of International Affairs*, 2012.

82. IEA. *Renewable Energy. Markets and Prospects by Region.* 2011.

83. SinoLatin Capital. *Why China and Latin America*? May 2011.

84. Ibid.

85. O. Biato Jr, "A Parceria Estratégica Sino-Brasileira: Origens, Evolução, Perspectivas (1993–2006)." *Fundação Alexandre de Gusmão, Ministério das Relações Exteriores*, 2010.

86. Ebinger, C.K., Avasarala, G. Five Major Energy Problems the Next President Has to Face.2012.

87. EPE. Plano Decenal de Expansão de Energia—PDE 2021. "Derivados da cana e eólica sustentarão aumento de participação das fontes renováveis ao longo dos próximos 10 anos." Rio de Janeiro, 09/26/2012.

88. Ibid.

89. Ibid. The document "PDE 2021 Public Consultation" is available at http://bit.ly/eNQBmm.

90. Ibid.

91. Plano Decenal de Expansão de Energia—PDE 2021. "Derivados da cana e eólica sustentarão aumento de participação das fontes renováveis ao longo dos próximos 10 anos." Rio de Janeiro, 09/26/2012.

92. Howard Geller, Roberto Schaeffer, Alexandre Szklo, Mauricio Tolmasquim. "Policies for Advancing Energy Efficiency and Renewable Energy Use in Brazil." *Energy Policy.* 32, 1437–1450, 2004.

93. OPEC, 2013.

94. M. Amatucci and E. E. Spers. "The Brazilian Biofuel Alternative." *Int. J. Automotive Technology and Management*, Vol. 10, No. 1, 2010.

95. Adilson de Oliveira. "Choosing Energies in Brazil: Sugar or Oil?" *Appropriate Technology,* Vol. 15, No. 3, 1988.

96. Ibid.

97. EPE. The document "PDE 2021 Public Consultation" is available at http://bit.ly/eNQBmm.

98. Ibid.

99. Ibid.

100. Howard Geller, Roberto Schaeffer, Alexandre Szklo, Mauricio Tolmasquim. "Policies for Advancing Energy Efficiency and Renewable Energy Use in Brazil." *Energy Policy.* Vol. 32, 1437–1450, 2004.

101. EPE. The document "PDE 2021 Public Consultation" is available at http://bit.ly/eNQBmm.

102. Ibid.

103. Adilson de Oliveira. "Series on Trade and Energy Security. Energy Security in South America: The Role of Brazil." *International Institute for Sustainable Development*, 2010.

104. Ibid.

105. United Nations. *Trade Agreements, Petroleum and Energy Policies*. 2000.

106. Robert O. Keohane and Joseph Nye. *Power and Interdependence*. New York: Harper Collins, 1989.

107. Daniel Yergin. "The Quest." *Energy, Security, and the Remaking of Modern World*, 2012.

108. Polsky, M. *Energy Security and the New Institutional Economics*. 2007

109. Yergin, D. "The Quest." *Energy, Security, and the Remaking of Modern World.* 2012.

110. Ibid.

111. Eriksen, S.S.; Lodgaard, S.; Melchior, A; Rich, K.; Rowe, E.W.; Sending, O.J. "BRICS, Energy and The New World Order." *Norwegian Institute of International Affairs*, 2012.

112. Russett, B. "Liberalism," In "International Relations Theories." *Discipline and Diversity*. 2nd ed. Eds. Tim Dunne, Milja Jurki, and Steve Smith.

113. SGI: Sustainable Governance Indicators—Governance Capacities in the BRICS. Bertlesmann Stiftung, 2012.

114. Hulbert, Matthew and Christian Brutsch. 2012. "Energy: Charging the BRICS to Power the World." *Global Policy Journal*, 25 January.

115. The New York Times. "Brazil's Star, Petrobras, Is Hobbled by Scandal and Stagnation," 04/15/2014. Available at http://www.nytimes.com/2014/04/16/business/international/brazils-star-petrobras-is-hobbled-by-scandal-and-stagnation.html?_r=0 Accessed January 10, 2015.

116. MercoPress. South Atlantic New Agency. "Brazil will Rise Fuel Prices to Help Government Managed Debt-laden Petrobras," 10/15/2014. Available at http://en.mercopress.com/2014/10/15/brazil-will-rise-fuel-prices-to-help-government-managed-debt-laden-petrobras Accessed January 10, 2015.

117. BBC. "Falling Oil Prices: Who are the Winners and Losers?," http://www.bbc.com/news/business-29643612 Accessed Accessed January 10, 2015.

118. Ibid.

119. O Estado de SP. "Corte de subsídios na energia elevará receita nos Estados," 01/15/2015.Available at http://economia.estadao.com.br/noticias/geral,corte-de-subsidios-na-energia-elevara-receita-nos-estados,1619991 Accessed January 10, 2015.

120. Ibid.

Part III

ENERGY PRODUCTION, TECHNOLOGICAL DEVELOPMENT AND ENVIRONMENTAL SUSTAINABILITY IN THE 21st CENTURY

Chapter 9

Shale Gas Development in Latin America

Christian Gomez and Eric Farnsworth

Shale gas is revolutionizing the global energy landscape and the impressive size of proven reserves puts the Western Hemisphere in an enviable position to capitalize on new realities. In addition to the United States and Canada, Argentina, Brazil, and Mexico all rank within the top 10 countries worldwide with technically recoverable shale gas resources; together, they make up approximately 40% of the world's total supply.[1] Additionally, Colombia also has significant potential.

Nonetheless, while the existence and recoverability of these resources is apparent, the economic viability of extracting them is unclear, given production costs, volumes, and wellhead prices. Global prices including the price of substitutes are also critically important factors in determining economic viability. Beyond purely economic considerations, a variety of factors including the regulatory framework, infrastructure, access to capital, and environmental and other political considerations could potentially retard development of the sector.

The primary point of comparison within the Western Hemisphere is the United States where, in 2014, there were over 20,000 active horizontal wells. This compares with 250 unconventional wells in Argentina and just 10 in Colombia during the same time period.[2] Investors spent $90 billion in the United States on developing shale gas in 2012 alone; in contrast, foreign direct investment in every sector in Latin America in 2013 totaled $180 billion.[3]

Driving much of the US sector is the participation of smaller and medium sized companies, which are more nimble and willing to take risks that other companies might not consider. Such companies face a more difficult value proposition in Latin America given the challenge of accessing capital and working within complicated political environments. Improvements in capital market efficiency and easier access to funding would assist small and medium sized companies, both foreign and domestic, to play a similar role in developing Latin America's own revolution if given the opportunity to do so.

Policymakers possess several tools to aid in the process of shale gas development. In addition to support for increasing access to finance, governments can prioritize gas demand over more expensive, dirtier fuels to incentivize further investment in the natural gas sector. They can develop standards and best practices which will allow for the transparent, efficient development of the sector in a manner that limits environmental consequences. They should consider common regulations that will create larger internal markets and economies of scale. Additionally, given the need to develop human resources within the energy sector in order to reach peak effectiveness, governments should consider supporting workforce development as appropriate.

SHALE GAS DEVELOPMENT IS A PATH TO PARTNERSHIP WITH THE UNITED STATES

The United States possess a comparative advantage in the sector and therefore it is appropriate for the United States to engage in the sector regionally at the public and private sector levels. The presence of natural gas in shale deposits has long been known, but technology was not advanced enough to make exploitation commercially viable. The development of techniques such as hydraulic fracturing and horizontal drilling, among others, provided the catalyst for rapid development in the United States and elsewhere. Once technology became available, the shale revolution in United States then began to thrive due to inexpensive access to basins, subsoil rights given to property owners rather than to the state, and sound and enforceable contract law. In addition, environmental regulations are adequate and generally give confidence for appropriate sectorial development.

As a result, potential for partnership with Latin America and the Caribbean in the development of the shale gas sector is significant. The United States enjoys technology, capital, and managerial expertise which is critically important, especially on environmental issues. The United States has experience developing guidelines and regulations for the sector that promote the full development of shale gas resources. Furthermore, the United States government and industry can assist in the development of regional shale gas resources through technology transfer and training, infrastructure, and workforce development. These are common issues facing all nations in the region that seek to participate in the shale gas revolution, and can form the basis of regional partnership based on mutual self-interest and concrete results, even when other avenues for US cooperation with the region may be temporarily inconvenient.

THE WESTERN HEMISPHERE SHALE GAS COUNCIL

One idea to develop this strategy would be the formation of a Western Hemisphere Shale Gas Council, for the purpose of coordinating best practices and exchanging sectoral views at the government level. Initially, membership in the Council, headed by the Secretary of Energy or his designee, would be drawn from those nations in North America, Latin America, and the Caribbean Basin with the largest regional shale gas resources. Over time, non-hemispheric nations which enjoy significant proven shale gas reserves, including Australia, China, Russia, and South Africa, could be included as observers, as could nations including India and Japan which have a vested interest, as consumers, in the success of Western Hemisphere shale gas development. In the meantime, significant attention must be paid to the efficient and effective development of shale gas in Latin America specifically if the region is to participate fully in the global shale revolution.

Argentina

According to some estimates, Argentina is second only to China in technically recoverable shale gas reserves, at 802 trillion cubic feet (Tcf).[4] Argentina is the country with the most advanced shale industry in Latin America, with hundreds of wells drilled and numerous projects in production. This activity is primarily located within the Neuquen Basin. The Nequen Basin contains the Los Molles and Vaca Muerta shale formations, with Vaca Muerta considered one of the premier shale gas development areas in South America if not the world. It is estimated that the Vaca Muerta formation has technically recoverable shale gas and shale oil resources of 308 Tcf of gas,

Table 9.1 Top 10 Countries with Technically Recoverable Shale Gas Resources

Rank	*Country*	*Shale Gas (trillion cubic feet)*	
1	China	1,115	
2	Argentina	802	
3	Algeria	707	
4	US*	665	(1,161)
5	Canada	573	
6	Mexico	545	
7	Australia	437	
8	South Africa	390	
9	Russia	285	
10	Brazil	245	
	World Total	7,299	(7,795)

*EIA estimates used for ranking order. ARI estimates in parentheses.

approximately 38% of Argentina's total technically recoverable shale gas reserves.[5] The country also has three other untested sedimentary basins which could prove to contain shale gas. These include the Parana, San Jorge, and Austral Magallanes basins.

Argentina has great potential to attract investment. In 2013, state energy company YPF reported holding 3 million acres in the Neuquen Basin, making it the top acreage holder in the country. Together with older Repsol-operated wells, YPF had drilled 37 Vaca Muerta wells through 2012. This constitutes over 70% of the wells drilled in Vaca Muerta, indicating the significant involvement of YPF and the Argentine government in shale discovery and exploitation. YPF has made a number of deals with foreign corporations which signal openness to oil and gas partnerships. According to YPF, Chevron aims to invest $1.6 billion in shale development in Vaca Muerta, and plans to drill 170 wells in a 96,000 acre area.[6] YPF has also consummated deals with Bridas, Shell, and Chinese corporation CNOOC; President Xi Jinping's 2014 visit and President Cristina Fernandez de Kirchner's return visit in 2015 showcased China's interest in the sector.

Despite current market conditions, Argentina is taking the long view and state-owned YPF stated that it will not reduce its $6 billion investment plan for 2015. Indeed, long-term investment in the Vaca Muerta is not likely to be affected significantly by the price drop.

Mexico

Historic energy sector reform signed by President Enrique Pena Nieto in August 2014 will provide a needed boost for Mexico's shale gas production. By opening the country to foreign investment, among other steps, reforms will usher in capital and technology in shale gas and oil. Mexico's technically recoverable natural gas is estimated at 545 Tcf, potentially larger than the country's proven conventional reserves.[7] Mexico holds the 6th largest shale gas resources in the world. The best documented play is the Eagle Ford shale in the Burgos Basin, which extends from South Texas. This play has an estimated 343 Tcf of risked, technically recoverable shale gas potential. Based on similarities with Eagle Ford shale in Texas, this is considered to be Mexico's top-ranked prospect for shale exploitation. Other basins in Mexico in which drilling has not yet occurred, and where potential is far less certain than the Burgos Basin, include the Sabinas, Tampico, Tuxpan, and Veracruz basins.

As of 2013, Pemex had drilled less than 10 shale gas and oil exploration wells in the Eagle Ford, and Pemex's initial shale exploration wells have been unusually costly ($20–25 million per well) while only providing modest results. There are also security concerns, including coastal zones were drug-trafficking organizations have engaged in large-scale violence to gain control of routes and branching out into other illegal activities such as kidnapping and oil theft. Additionally, shale gas development requires significant water resources, and much of Mexico's shale potential lies in arid areas, requiring thought about ways to satisfy water resource requirements. Nonetheless, Pemex plans to increase shale activity in the next few years, budgeting 6.8 billion pesos ($575 million) in 2014, highlighting some 200 shale gas opportunities in five geologic provinces in eastern Mexico.[8]

Emilio Lozoya, Pemex's CEO, has invited the world's largest energy companies to join the exploration opportunity. On March 4, 2014, he said that Pemex aims to attract as much as $1 trillion in energy investment to exploit Mexico's shale oil and gas reserves. He also said that, "Capital from all over the world is welcome in Mexico . . . We hope to have hundreds of companies operating in any type of rock formation, be it shale, or shallow water, or mature fields, or deep water projects."[9] With regard to the drop in the price of oil in late 2014 and early 2015, companies in Mexico are taking a long view and a temporary drop in oil prices will not alter investment plans in the medium to long-term.

Brazil

Access to capital is the most significant challenge facing shale production in Brazil. Most investors are focused on offshore pre-salt oil fields. While much of Brazil's oil and gas potential lies offshore, it has 18 mostly undeveloped basins onshore. Three of these, the Paraná in the south and the Solimões and Amazonas in the north, already produce conventional oil and gas and have significant potential for shale. There are also six other basins which may have shale potential, but they are unproven. In total, Brazil's technically accessible shale gas resources are around 245 Tcf.

The Paraná extends to Argentina, Uruguay and Paraguay. The Brazil-Bolivia, and Uruguaiana-Porto Pipelines already exist in this region. No shale gas activity has occurred in the Brazil portion of the Basin, although Amerisur Energy has discussed the Paraguay portion. Risked, technically recoverable shale gas is estimated at 81 Tcf in the Parana Basin. The Solimões are the country's most productive onshore basin for conventional oil and gas. However, it has not been explored for shale. There are estimates that this basin contains 65 Tcf of risked, technically accessible shale gas. The Amazonas are extremely remote and conventional petroleum was only commercialized in 2009, and this basin is estimated to contain 100 Tcf of risked, technically accessible shale gas.[10]

Given the large potential of shale gas development in Brazil and the country's growing clean energy needs, particularly as a long-term drought continues to impact Brazil's hydroelectric capacity, the government has decided to

Map 9.1 Brazil Shale Gas/Oil Assessment.

stimulate shale gas exploration and put onshore shale gas blocks for auction in 2013. In early 2013, Petrobras drilled its first shale oil well in Argentina, and during the November 2013 auction of exploration concessions, Petrobras dominated the bidding. When exploration concessions were first auctioned off, many top companies sat out the sale since many blocks were not considered prime territory.[11] Only Shell and Total were registered to participate. Environmental issues may also complicate investor interest, which could have implications for other Latin American countries.

Colombia

Despite not having abundant shale gas resources compared to the other three countries discussed, Colombia has the most competitive energy framework and most energy-friendly regime of any country in the region. Colombia has only 55 Tcf of shale gas.[12] These are still sizeable reserves, and the relatively stable regulatory environment combined with the country's openness to investment is encouraging. Three main basins exist in Colombia that could be lucrative for shale production, although security remains a risk in some locations. Initial shale drilling by Ecopetrol as well as the entrance of

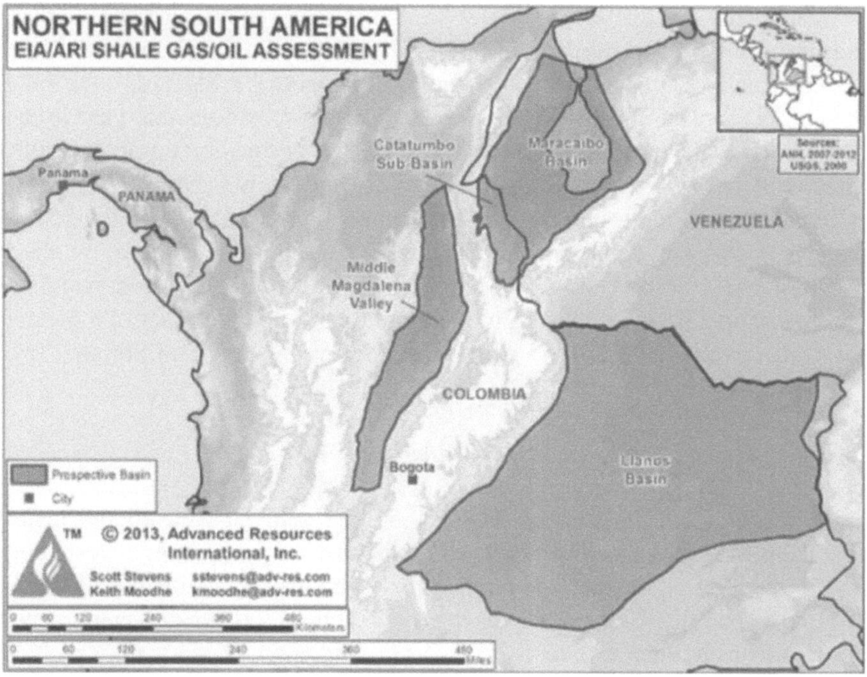

Map 9.2 Northern South America Shale Gas/Oil Assessment.

private companies such ConocoPhillips, Exxon, Shell and others has been positive.

The Middle Magdalena Valley Basin has been the focus so far, both for shale leasing and drilling and also conventional onshore drilling. There are estimates of 18 Tcf and 4.6 billion barrels, out of risked shale gas and shale oil in-place of 135 Tcf and 79 billion barrels. In comparison, Ecopetrol has estimated the basin to have 29 Tcf of shale gas potential. The Llanos Basin is another large basin in eastern Colombia. The Maracaibo/Catatumbo Basin is shared with Venezuela and is one of South America's richest petroleum basins. The Putamayo may contain shale but has not yet been assessed.[13]

Colombia's Agencia Nacional de Hidrocarburos (ANH) regulates oil and gas exploration and development. The model for contracts established by the ANH includes eight year exploration and 24 year production terms, with preferential terms for shale gas including a 40% reduction in royalties and higher oil prices. ANH conducted Colombia's first shale block auction in 2012. ANH had high hopes for the second unconventional auction through Ronda Colombia 2014. However, only 26 of the 95 available concessions secured offers. The auction raised about $1.4 billion, according to ANH, significantly less than the $2.6 billion originally forecast.

Despite this, Ecopetrol continues to control the majority of the market, managing about one third of the oil and gas licenses in Colombia. Shale exploration has become a critical facet of Ecopetrol's business plan, and the company plans to drill for unconventional hydrocarbons this year in the Middle Magdalena and Catatumbo areas, including nine stratigraphic wells, three exploratory wells and three productivity and study pilots.[14]

Because of the shifting economic environment, some international oil companies are delaying deployments of capital, including shale exploration plans in Colombia.[15] Some Colombian energy companies have lowered spending on exploration as they focus on production. State-owned Ecopetrol's budget for shale development in 2015 was $40 million versus $240 million in 2014.

CONCLUSION

Shale resources in the Western Hemisphere are abundant, clean, and lower cost than alternatives, and can fuel economic growth by attracting investment, boosting power generation and employment, and making national economies more cost competitive, among other things. But success is not guaranteed, even with an abundance of resources. Gas pricing regimes, regulations including the rule of law, access to technology and capital, and infrastructure are fundamental variables that investors consider before committing to move ahead with project development. When it comes to energy resources including shale gas in the Western Hemisphere, conditions above ground are just as important as conditions below ground. Establishment of a cooperative regional mechanism could be one way to encourage development of these vital resources in a cost effective, efficient, and environmentally effective manner.

Since mid-2014, market conditions have evolved in the Western Hemisphere, which have negatively impacted the scenario for investment in unconventional wells. In June 2014, West Texas Intermediate sold at $106 per barrel. The price closed at just under $50 per barrel in early 2015.

The fall in prices may have a significant impact on spending and investment plans for both energy companies and governments. During planning, companies budget for a certain oil price in order to meet a break-even price where exploration and production would be profitable. For example, Venezuela's break-even in January 2015 was $117.50 per barrel, making it extremely challenging to turn a profit.[16] The drop in oil prices impacts gas investment because national and international oil companies often prospect the two at the same time and thus must make appropriate spending decisions based on their relative prices. As such, the two markets are closely intertwined.

As a result the oil price environment should be leveraged to encourage integration efforts in the region as investment levels are cut back, in order to improve conditions that will make investment more attractive even in a price constrained environment. For example, the Pacific Alliance, which includes Mexico and Colombia, along with Chile and Peru, should seek to create larger internal markets and stronger investment conditions to draw investment.[17] Shale gas development is a priority for the Alliance, and the creation of a development bank to financing infrastructure projects is one recommendation for further development.

Still, despite the downturn in 2014 and early 2015, shale gas development will continue in Latin America due to the sector's attractiveness and the ample reserves. While oil prices may stay at depressed levels for the medium-term, investment is a long-term value proposition. Steps taken now may very well be rewarded in the future.

NOTES

1. "Technically Recoverable Shale Oil and Shale Gas Resources: An Assessment of 137 Shale Formations in 41 Countries Outside the United States." *U.S. Energy Information Administration*. June 2013.
2. Arthur, Alexis. "Shale Oil and Gas the Latest Energy Frontier for South America," *Oilprice.com*. June 5, 2014.
3. "Shale Gas Development in Latin America." *Americas Society/Council of the Americas*. October 22, 2013.
4. EIA.
5. Ibid.
6. Gonzalez, Pablo. "Argentina Prepares China Shale Deal to Boost Gas Reserves." *Bloomberg*. August 21, 2013.
7. EIA.
8. Ibid.
9. Williams, Adam. "Pemex CEO Invites World to Tap Mexico's Energy Shale Reserves." *Bloomberg*. March 4, 2014.
10. Ibid.
11. Fick, Jeff. "Brazil Auctions Shale Oil, Natural Gas Blocks." *The Wall Street Journal*. November 28, 2013.
12. EIA.
13. Ibid.
14. Willis, Andrew and Dan Murtaugh. "World Oil Stock Seeks Shale Lift as Colombian Boom Fades." *Bloomberg*. March 12, 2014.
15. Willis, Andrew. "Exxon Said to Apply Brakes on Colombia Shale in Oil Rout." February 10, 2015.
16. Bentley, Elliot, Pat Minczeski, and Jovi Juan. "Which Oil Producers are Breaking Even?" *Wall Street Journal*. January 5, 2015.
17. "An Energy Agenda for the Pacific Alliance." *AS/COA*. February 12, 2015.

Chapter 10

The Pitfalls of Lithium Mining in Bolivia[1]

Bruce Bagley and Olga Nazario

Lithium's potential as a key ingredient in the new generation of electric car batteries has raised interest in the mineral around the globe. It is the lightest metal on the planet and used in most electronic equipment and advanced batteries. Lithium is also used in ceramics and glass production, bipolar medications, air conditioners, lubricants, nuclear weaponry, and other products.

The world's largest lithium mining company predicted that in the near future 10% of all new cars—close to 5 million vehicles per year—would be powered by lithium-ion batteries. It also predicted that this rate would rise to 20% by 2020. Given existing, known supplies, this means that the world could run out of lithium, a non-renewable resource, in 2020.

Powerful multinational corporations plan to invest billions in future lithium production. Some industry predictions estimate that lithium car battery sales could jump tenfold, from $100 million to $103 billion per year over the next two decades. If such investments take place, countries that possess lithium, like Bolivia, and are willing to export it could become important suppliers in the global economy.

The largest lithium deposits in the Western Hemisphere are located in Argentina, Bolivia and Chile. Bolivia's lithium reserves are estimated at between 5.5 and 8.9 million tons and rank among the largest in the world. They are found in the high Andean Salar de Uyuni (Uyuni Saltpan), a 3,860 square mile high-desert (Altiplano) area of salt-embedded minerals in the Bolivian Andes situated in the country's southwestern Potosí Department, bordering Chile's Atacama Desert.

The concentration of lithium varies widely in different parts of the Salar de Uyuni. Production may focus on small areas. Much of the lithium could remain inaccessible or take decades to extract. Considering the real grade and distribution of lithium in the Uyuni, the deposits may not be a particularly attractive resource and the real exploitable reserve could be only approximately 300,000 tons rather than the estimated millions. In 2009, a pilot operation was initiated in the Salar de Uyuni to determine the technical and economic feasibility of mining and refining the lithium-rich brine.

Bolivia will have to invest heavily over the next few years in order to produce lithium carbonate for export. It is virtually inevitable that Bolivia will have to seek partnerships with foreign companies in coming years to obtain badly needed capital and technology. It remains to be seen if the current administration of President Evo Morales (2006–present) will be able to take advantage of growing market demands for lithium or the opportunity will be left for future Bolivian governments.[2]

President Morales is attempting to change the past patterns of foreign exploitation by asserting national control of Bolivia's huge natural gas reserves and its massive lithium deposits. To date, his government has maintained strict state control of the country's lithium resources and limited foreign access and investment. Also, several of Bolivia's indigenous organizations, active in the Salar de Uyuni region, have demanded a "fair" share in the eventual bounty to be reaped from future lithium exploitation. As a result, mining of Bolivian lithium reserves will depend for now on the capacity of the Morales administration to negotiate and balance the benefits of all the stockholders versus his own political agenda.

Since 2006, foreign corporations and governments have lobbied the Bolivian government for access to the lithium riches. Among them, Japanese automakers, Brazil, Canada, French electric car maker, Bolloré, and others. In 2009, COMIBOL signed a Memorandum of Understanding with Korean Resources Corporation and with the French Bolloré Group. Also in 2009, during the visit of Iran's former president Mahmoud Ahmadinejad to La Paz, an agreement reached committed Iran to assist Bolivia on conducting lithium research. Iranian experts joined Brazil's Ministry of Science and researchers from companies such as France's Eramet SA and Bolloré SA, to hold a joint study on the lithium reserves. Bolivia's ongoing tensions with Washington have effectively left American companies on the sidelines as other foreign enterprises continue to actively negotiate lithium deals in Bolivia.

Serious potential environmental problems stemming from lithium mining in Bolivia cannot be ignored. The ecologically fragile Salar de Uyuni could become an environmental disaster, if sufficient precautions are not taken. Also, lithium development could seriously damage two key industries in the region—agriculture and tourism. The Salar de Uyuni, the brightest spot on earth as seen from space is an area of outstanding natural beauty. More than 60,000 tourists visit the area every year. The ecosystem as well as the tourist industry in the area could be seriously compromised in the search for lithium.

Finally, the chronic institutional deficiencies and lack of stability of the Bolivian state pose serious questions regarding its capacity to manage an ambitious lithium development program. Problems of inadequate administrative capacity within the Bolivian government clearly antedate the current Morales administration.

As the world seeks energy alternatives and the efficient ways to use and store it, Bolivia has been thrown into the limelight as the country with the largest estimated lithium reserve base. Foreign corporations and governments alike have been lobbying the Bolivian government for access to its lithium riches. Yet, Bolivia faces daunting challenges in developing the lithium from its huge reserve base. The most significant of these challenges are as follows:

1. Need to invest heavily over the next few years to produce lithium for export.
2. Lack of technical and managerial capabilities to develop the lithium on its own.
3. A pattern of nationalization of natural resources which scares away investors with the required know how and capacity to develop the lithium.
4. The Morales government hesitation in choosing partners.
5. Geological assessment of the grade and distribution of lithium that indicate that the concentration of lithium varies widely and production would concentrate in small areas. Most of the lithium could remain inaccessible, not be economical, or would take decades to extract.
6. The risks of environmental damage to the Salar de Uyuni ecosystem and to surrounding communities could surpass benefits of lithium extraction.

Largely because of internal political tensions and nationalism, Bolivia continues in 2015 to vacillate regarding how best to tap its lithium reserves. This delay has permitted nations with smaller reserves, like Chile and China, to step up to meet growing world demands for lithium. Yet, geologists and economists are unsure whether world lithium reserves outside of Bolivia are enough to meet exploding global demand, although some are convinced that accessible lithium resources are significantly larger than estimated by the United States Geological Survey. If true, Bolivia's current window of opportunity to exploit its lithium could eventually be eclipsed by lithium finds in other regions or by the emergence of new, alternative technological breakthroughs that do not rely on lithium at all. As Bolivian economist Juan Carlos Zuleta has observed: "We have the most magnificent lithium reserves on the planet, but if we don't step into the race now, we will lose this chance. The market will find other solutions."[3]

This paper looks at the uses of lithium, the geological assessments that have been made of the Bolivian lithium reserves and the implications for both Bolivia and for the United States in terms of the decisions that Bolivia will make on how best to exploit its potential richness. It also offers some conclusions and recommendations in the context of US security interests. An appendix provides a brief synopsis of the Morales government.

THE DEMAND FOR LITHIUM

Lithium's use as a key ingredient in the new generation of electric car batteries has raised international interest. It is the lightest metal on the planet and used in most electronic equipment and advance batteries. As of 2007, 60% of the world's cell phones and some 90% of laptops contained lithium batteries and the percentages continue to rise each year. Lithium is also used in ceramics and glass production, bi-polar medication, air conditioners, lubricants, nuclear weaponry, and other products.[4]

Powerful multinational corporations plan to invest billions in future lithium production. Some industry predictions estimate that lithium car battery sales could jump ten-fold from $100 million to $103 billion per year over the next two decades. If such investments take place, countries that possess lithium, like Bolivia, and are willing to export it could become important suppliers in the global economy.[5]

Despite the enthusiasm, there are also concerns about lithium batteries on both the economic, technical, and environmental fronts. The process for transforming lithium into its commercially valuable form, lithium carbonate, remains complex and expensive. The electrical vehicle batteries currently being developed with lithium are still too large and heavy and charge too slowly. The batteries are so expensive that they push the cost of electricity-driven cars beyond the reach of most consumers, although lithium accounts for less than 5% of the estimated cost of battery today.[6] Lithium batteries also have a record of catching fire. So while lithium car batteries might become a massive global market in the future, they could also fizzle.[7]

Environmental studies claim that mass production of lithium carbonate is not environmentally sound. The Meridian International Research, for example, has concluded that, "it could cause irreparable ecological damage to ecosystems that should be protected and that LiIon propulsion is incompatible with the notion of the "Green Car."[8]

Yet, in 2007, SQM, the world's largest lithium mining company predicted that in the near future 10% of all new cars—close to 5 million vehicles per year—would be powered by lithium-ion batteries. It also predicted that this rate would rise to 20% by 2020. Based on such predictions, the demand for battery-grade lithium carbonate would rise from 85,000 tons per year in 2007 to 160,000 tons per year in 2015. Given existing known supplies, this means that the world would run out of lithium, a non-renewable resource, in 2020. In that context, some experts belief that depletion rates of lithium could exceed current oil depletion rates and switch dependency from one diminishing resources to another.[9] These forecasts, however, depend on unreliable forecasts of increasing production of electric vehicles and do not take into consideration technological changes that would affect content of lithium in batteries and suspected lithium reserves in unsurveyed regions, such as Africa. At the moment, lithium seems to be abundant and inexpensive.[10]

The United States is the main importer of lithium materials and compounds and the leading producer of value-added lithium materials. In 2009, the US Department of Energy funded US $2.4 billion in grants to accelerate the development of US manufacturing capacity for batteries and electric-drive components and for the deployment of electric-drive vehicles. The grants are designed to launch an advanced battery industry in the United States.[11]

LITHIUM RESERVES AND PRODUCTION

Lithium reserves are contained in five raw natural mineral forms: Continental Brines, Pragmatites, Hetolites, Oil Filed Brines and Geothermal Brines, having the first one the richest contain of lithium. Experts disagree on availability of lithium reserves. The US Geological Survey, for example, indicates that lithium resources total 2.5 million tons in the United States and approximately 23 million tons in other countries, mainly Bolivia, Chile, Argentina and China.[12] It also predicts that that there are still huge supplies of lithium that remain untapped world-wide.[13]

The largest lithium reserve bases are found in Argentina, Bolivia and Chile.[14] Nearly 80% of the known global lithium reserve base is located in the "The Lithium Triangle," an area bordered by the three large South American salt flats: The Salar de Atacama in Chile, the Salar de Uyuni in Bolivia, and the Salar del Hombre Muerto in Argentina. In 2005–2008, the United States imported 63% of its lithium needs from Chile.[15] In 2008–2011, the US imported 52% of lithium from Argentina and 44% from Chile.[16]

The largest known lithium reserves in North America are found in the near surface lithium clay deposit located in the State of Nevada. This is the main domestic supplier for the United States.[17] The deposits are being mined by Canada's Western Lithium Corporation. An estimated 11 million tons of lithium carbonate equivalent are located within five mineralized near surface zones. Just one of the five zones could produce 20,000 tons a year, or approximately one-fifth of the world production.[18]

Table 10.1 Global Lithium Production and Reserves

Country	Production 2012 Meter Tons	Reserves*
Chile	13,000	7,500,000
Australia	13,000	1,000,000
China	6,000	3,500,000
Argentina	2,700	850,000
United States	1,000**	38,000
Portugal	820	10,000
Zimbabwe	500	23,000
Brazil	490	46,000
Bolivia	—	—

*The US Geological Survey discontinued reserve base estimates in 2009.
**The figure corresponds to 2008 as US companies currently withhold this information.
Source: US Geological Survey, Mineral Commodity Summaries, January 2010.

Chile is the world's largest producer and exporter of lithium carbonate. Chilean lithium is being mined by Chemetall, a subsidiary of the New Jersey-based Rockwood Holdings Inc., and Soquimich (SQM), a Chilean state-owned company. Together they produce 70% of all lithium carbonate from Chile's lucrative Salar de Atacama. With an output of 40,000 metric tons of lithium carbonate per year, Chile currently supplies a third of the global market. Estimates of Chile's remaining lithium reserves range from 3 million to 7.5 million tons. Chile's lithium reserves are also among the purest in the world and Chile's easy access to sea transport helps keep its total production costs low.[19]

Australia, China, Argentina and the United States follow Chile as the largest producers of lithium. Serbia and Finland also produce lithium. Large deposits exist in several other countries but production is low due to the high cost of extraction. These limitations mean that the future of the world's lithium supplies remains uncertain and may be insufficient to meet potential future demand.

BOLIVIA'S LITHIUM

With governments around the globe (including the United States) demanding increased fuel efficiency to reduce their dependence on imported oil, private foreign corporations are keenly pursuing access to Bolivia's lithium. Bolivia's lithium reserves are estimated at 5.5 million tons.[20] Some analysts believe that the reserves could be as high as 8.9 million tons, the largest in the world. The reserve base is found principally in the Salar de Uyuni (Uyuni Saltpan), a 3,860 square mile high-desert (Altiplano) area of salt-embedded minerals in the Bolivian Andes located in the country's southwestern Potosí Department, bordering Chile's Atacama Dessert.[21] The beautiful Salar de Uyuni is the

Map 10.1 Map of Bolivia.

largest salt flat in the world and the brightest object on the Earth's surface visible from space. Many experts believe that efficient exploitation of Bolivia's lithium deposits could substantially increase present world supplies.

There is no definite data on Bolivia's lithium reserves. TRU Group, a lithium consulting firm, indicates that harvesting and processing lithium from the large Salar de Uyuni is a difficult and tricky task. TRU argues that further strategic deep drilling exploration is required before the real potential of Bolivia's to produce lithium carbonate for batteries is determined.[22]

The French Meridian Research group also believes that estimates of the concentration of lithium vary widely in different parts of the Salar. Production would concentrate in small areas. Most of the lithium could remain inaccessible or would take decades to extract.[23] The Meridian Research Reports that:

> The structure of the Salar de Uyuni is very different to the Salar de Atacama (Chile) . . . . Whereas the brine containing halite layer in the Salar de Atacama is 35 meters thick, the Uyuni halite deposit is very thin . . . The halite is . . . porous all the way through, with a much higher porosity of 35% and is filled with interstitial brine. This means that the quality of lithium available per unit surface is much lower and a correspondingly greater area of the Salar will have to be exploited for an equivalent lithium production.[24]

The report concludes that, considering the real grade and distribution of lithium in the Uyuni, its lithium might not be a particularly attractive resource and that the real exploitable reserve could be only approximately 300,000 tons rather than the estimated millions. Also, the available methods for mining could be highly environmentally damaging.[25]

The current Bolivian government of President Evo Morales (2006–present) has put forward a general plan for the development of its lithium deposits, but the specific details of how Bolivia's lithium will be mined, by whom and under what terms have not yet been clarified. In March 2008, the Morales government issued a decree investing $5.7 million to set up a state owned "pilot extraction plant" at the edge of the Salar de Uyuni. The plant is operated by a "General Directorate of Evaporative Resources of the Salar de Uyuni," under the state mining company COMIBOL (*Corporación Mineral de Bolivia*). Drilling started in August 2009.[26]

The pilot plant was intended to determine the technical and economic feasibility of getting the lithium-rich brine out from under the Salar's crust and separating it into its distinct, marketable products. In 2014, in its strive for independent lithium production, the Bolivian government opened an industrial-scale pilot plant in the Yocalla province (southwest of the Salar) for commercial production of batteries. By March 2015, the plant, built by a Chinese company, produced more than 4,000 cellular batteries. Besides, two rural communities in Cochabamba region have access to electricity using lithium batteries produced at the plant.[27] The plant's production goal is to produce up to 30,000 to 40,000 metric tons of lithium carbonate per year. A third phase, in which marketable lithium compounds will be produced in association or partnership with foreign investors, will then follow.[28]

The cost for making the Bolivian lithium industry operational is very high. Current estimates suggest that it could cost at least $200 million to make the main plant operational. Substantial additional investments to support needed chemical industries and substantial infrastructure development in the Salar de Uyuni—still a remote and largely undeveloped region of the country—would also be required at a price of at least US $1 billion or more. In light of both the costs and the technological problems involved, Bolivia is exploring partnerships with foreign investors, a strategy that some local community groups and many Bolivian nationalists do not support.[29]

CHOOSING A PARTNER

Since 2006, foreign corporations and governments have lobbied the Bolivian government for access to the lithium riches. Bolivia has a long, often bitter history of foreign mining and mineral exports that stretches back to the exploitation of its legendary Potosí silver mines by the Spanish during the colonial era. The administration of President Morales is attempting to change the past patterns of foreign exploitation by asserting national control of Bolivia's huge natural gas reserves and its massive lithium deposits.[30] To date, the Morales government has maintained strict state control of the country's lithium resources and limited foreign access and investment. Also, several of Bolivia's indigenous organizations, active in the Salar de Uyuni region, have demanded a "fair" share in the eventual bounty to be reaped from future lithium exploitation.[31]

Among the major international investors interested in Bolivia's lithium are Brazil, Iran, the Japanese companies Mitsubishi and Sumitomo, South Korea and the French electric vehicle manufacturer, Bolloré.[32] In 2009, COMIBOL signed a Memorandum of Understanding with Korean Resources Corporation.[33] Also in 2009, during the visit of Iran's former President Mahmoud Ahmadinejad to La Paz, an agreement reached committed Iran to assist Bolivia on conducting lithium research. Iranian experts will join Brazil's Ministry of Science and researchers from companies such as France's Eramet SA and Bolloré SA, to hold a joint study on the lithium reserves.[34] The agreement with Iran also calls for exchange of scientific and technical information, Iranian training of Bolivian professionals on the subject and promoting faculty exchanges. The Iranian-Bolivian technical commission will also analyze potential for marketing lithium by-products in Iran and other countries.[35] During a previous visit to Bolivia, President Ahmadinejad also showed interest in Bolivia's uranium, although reserves are considered minimal.

Since September 2009, the French electric car consortium, Bolloré-Eramet, has been pressing the Morales government for an agreement. They have satisfied demands placed by the Morales government proposing an industrialization plan that is nature-friendly. The plan, "A French-Bolivian Project for the Well-Being of Bolivians in Harmony with *Pachamana* (Mother Earth)," includes establishing an electric manufacturing plant, a prerequisite of the Morales administration for investors interested in the Salar de Uyumi.[36] Yet, the Bolivian government has yet to respond to Bolloré or any other investor. "We want partners, not patrons," is President Morales' motto for refusing to accept offers from the interested investors. As such, mining of Bolivian lithium reserves will depend for now on the capacity of the Morales administration to negotiate and balance the benefits of all the stockholders vs. his own political agenda.

Overall, COMIBOL has signed Memorandums of Understanding in the area of industrialization planning, lithium research and professional training with Brazil, Iran, Japan, South Korea and China.[37]

INSTITUTIONAL AND POLITICAL OBSTACLES FOR DEVELOPING BOLIVIAN LITHIUM

It remains to be seen if the Morales administration will be able to take advantage of growing market demands for lithium or the opportunity will be left for future Bolivian governments. After two decades of foreign interest in Bolivia's reserves, the country's natural resource policies, regional posture and poor infrastructure have led foreign investment in lithium mining to increasingly favor Chile and Argentina in recent years.[38] Bolivia will have to overcome daunting obstacles before it will be able to export significant amounts of lithium to international markets. Key developmental challenges are outlined below.

Lacking a Road Map

Bolivia might not have the technical capacity to achieve the goals stated by the Morales administration regarding its lithium reserves. President Morales alleges that Bolivia itself will produce electric cars. There is no clear indication when the electric car will be fully developed—since the industrial capacity for this type of complex industrial process is inexistent in the country—how big that market will be, and when it will peak. Bolivia could aim at more traditional lithium markets, such as glass and ceramics, but demand in these markets is not likely to be as profitable. A middle course would be for Bolivia to supply already proven types of lithium batteries for products such as watches, cell phones, iPods, laptops and other electronic devices. No matter which option is finally selected, Bolivia will have to come up with a clear plan or road map to maximize the development potential of its lithium reserves. It does not yet have one.[39]

Institutional Weaknesses

The chronic institutional deficiencies and lack of stability of the Bolivian state pose serious questions regarding its capacity to manage an ambitious lithium development program. Problems of inadequate administrative capacity within the Bolivian government clearly antedate the current Morales administration. To undertake lithium development successfully, the Bolivian state will need trained and qualified experts in the technical and scientific aspects of lithium mining, environmental controls, in business and financial management, as well as in international legal contracts, social and environmental impact assessments. At present, those capacities required for effective, professional development of its lithium resources are not in place.[40]

Environmental Impacts

Serious potential environmental problems stemming from lithium mining in Bolivia cannot be ignored. Bolivia's ecologically fragile Salar de Uyuni could become an environmental disaster, if sufficient precautions are not taken. The adequacy of Bolivia's environmental strategy for lithium development in Southwest Potosí is questioned by many Bolivian and international environmental organizations. Lithium mining could cause a major water crisis in the Uyuni region, which already suffers from serious water shortages that negatively impact the farming, llama herders, tourism industry, and drinking water supplies. While Bolivian officials claim that the national lithium project's water requirements will be minimal, their estimates are based on only limited and incomplete information.[41]

Contamination of the air, water and soil is also a potential concern. Large quantities of toxic chemicals will be used to process the predicted 30,000 to 40,000 tons of lithium per year that the project expects to mine. The escape of such chemicals via spills or air emissions could threaten the communities in the area and the ecosystem as a whole. Reports from Chile's Salar de Atacama describe a landscape marred by toxic piles of discarded salt and extensive canals filled with chemically contaminated water.

Bolivian officials have often minimized such risks and the institutional systems in place to protect the environment are inadequate at best. Public institutions, such as Bolivia's Ministry of the Environment and Water, that are responsible for enforcing compliance with national environmental regulations and standards, do not have the professional expertise, technical capabilities or legal authority to intervene effectively.[42]

THREATS TO BOLIVIA'S RURAL COMMUNITIES AND TOURISM

Some groups and communities in the region openly endorse lithium development as an opportunity for increased income and development. But there are also groups with serious objections that oppose such development. Quinoa producers and tourism operators have repeatedly expressed concerns about the benefits that the Bolivian government has promised from lithium on the grounds that the supposed benefits would not meet local needs. Moreover, lithium development could seriously damage two key industries in the region—agriculture and tourism. Bolivia's laws guaranteeing community participation are as weak as its environmental protections.[43]

The Salar de Uyuni often classified as a natural wonder of the world remains an area of outstanding natural beauty. Flamingos breed in the lagoon created over the Salar during the summer rains. More than 60,000 tourists visit the area every year. The ecosystem as well as the tourist industry in the area could be seriously compromised in the search for lithium.[44]

BOLIVIA–US RELATIONS

US relations with Bolivia under the administration of President Morales have been tense and antagonistic. President Morales has consistently opposed free-trade policies advocated by the United States and adamantly refused to negotiate with the US-sponsored regional Andean free trade agreement, as did Ecuador and Venezuela.

The US has been concerned about rising Bolivian coca cultivation. Bolivia is the world's third largest producer of coca leaf, the raw material used to manufacture cocaine. Crop-eradication programs in the 1990s and early 2000s, funded by the US, incensed many of Bolivia's poorest farmers—mainly Morales supporters—for whom coca is often their only source of income. A former coca farmer himself, Morales tolerant policies toward coca cultivation for traditional and medicinal uses has progressively worsened US-Bolivian bilateral relations since 2006.[45] The close alliance between the Morales administration with Venezuela's president Hugo Chavez has also made the Bolivian government antagonistic to US policies in the region.

In 2008, US suspended most US drug control funding for Bolivia and recalled Ambassador Philip Goldberg for "consultations." President Morales also took actions against the US He expelled the US Agency for International Development (USAID) alternative development workers from the Chapare coca growing region.[46] He formally expelled Ambassador Goldberg, after accusing him of meddling in Bolivia's internal affairs. Soon after, the US government temporarily suspended Peace Corps operations in Bolivia. Washington followed by declaring that Bolivia had "failed" to uphold its international counterdrug obligations, although President George W. Bush did grant Bolivia a "national security waiver" so that US assistance would not automatically be cut off. President Bush also suspended Bolivia's US trade preferences under the Andean Trade Promotion and Drug Eradication Act (ATPDEA), again citing Bolivia's failure to adhere to anti-narcotics commitments. In November 2008, Morales ordered the US Drug Enforcement Agency (DEA) out of Bolivia.[47]

In June 2010, the Barack Obama administration indicated that a framework agreement had been signed with the Bolivian government to improve relations. Three days later, however, President Morales threatened kicking USAID out of the country if continued supporting individuals who do not support Morales' policies.

BOLIVIA'S LITHIUM GAMBLE

The strains in US-Bolivian bilateral relations have not dampened efforts by non-US foreign corporations to court the Bolivian government. US companies, however, do seem to have been disadvantaged by the recurrent tensions in bilateral relations. But even non-US-based foreign companies seeking to tap Bolivia's lithium reserves have had to navigate carefully around Morales' resource nationalism. Indeed, President Morales has clashed repeatedly with American, European and even South American investors. In 2006 Morales shocked Brazil by nationalizing Petrobras' (Brazil's state-owned petroleum company) natural gas projects in Bolivia and then demanding a sharp rise in the prices paid by the Brazilians for Bolivia's natural gas. He carried out further nationalizations just before the 2009 vote on the Constitution by sending Bolivian soldiers to occupy the operations of the British oil giant BP.[48]

Bolivia will have to invest heavily over the next few years in order to produce lithium carbonate for export. With world economic growth slowed as a result of the 2008–2009 global recession, it remains unclear how Bolivia will be able to make such investments on its own.[49] Given domestic realities and international circumstances, it is virtually inevitable that Bolivia will have to seek partnerships with foreign companies in 2010 and beyond to obtain badly needed capital and technology. Bolivia's current tensions with Washington have effectively left American companies on the sidelines as other foreign enterprises continue to actively negotiate lithium deals in Bolivia.[50]

Yet, largely because of internal political tensions and nationalism, Bolivia continued in 2010 to vacillate regarding how best to tap its lithium. This hesitation has permitted nations with smaller reserves to step up to meet growing world demand for lithium. Chile leads the pack. China has also emerged as a top lithium producer, tapping reserves found in a Tibetan salt flat. But geologists and economists are unsure whether world lithium reserves outside of Bolivia are enough to meet exploding global demand, although some are convinced that accessible lithium resources are significantly larger than estimated by the United States Geological Survey. If true, Bolivia's current window of opportunity to exploit its lithium riches in the national interest under advantageous international conditions would appear to be only a limited one that could eventually be eclipsed by lithium finds in other world regions or by the emergence of new, alternative technological breakthroughs that do not rely on lithium at all.

CONCLUSIONS

Bolivian lithium does not appear critical in satisfying US demand. Nor is it clear if Bolivia's real reserves could produce as much lithium as the reserve base indicates or of desirable quality. Moreover, lithium-carbonate batteries might not provide the environmentally sound and sustainable alternative the US seeks to reduce its demand on fossil oil. Yet, considering the unpredictability of the technological developments, having access to potential resources is generally a sound policy.

The perspectives for the Morales administration capacity to develop its lithium potential, under current conditions, are not encouraging. Bolivia lacks the technical and managerial capacity to do so on its own. Also, most experts believe that Bolivia's lithium potential does not match that of its oil and gas reserves. The Morales administration's reluctance to accept the offers being presented by foreign investors may result in Bolivia losing an opportunity to profit from the growing demand for lithium. Also, investors with the needed know-how may be reluctant to investing under an environment charged by recent nationalizations and lack of assurance of negotiating in a leveled field and under sustainable conditions. Yet, countries like Iran may be approaching Bolivia hoping to have access to its resources both for industrial or nuclear purposes and thus, the lithium may become increasingly politicized.

The progressive worsening of US-Bolivian relationship in recent years, characterized by mutual suspicion and distrust, presents formidable obstacles to any rapid improvements. Yet, the Obama administration has indicated a willingness to strengthen relations with Bolivia. Considering that natural resources often last longer than governments, despite current antagonistic relationship, US would be well-served to have a long term policy strategy in dealing with Bolivia and its resources.

Finally, the United States and the international community should raise awareness of the environmental risks involved in tapping lithium and other resources in the unique Salar de Uyuni.

Recommendations for US Policy

1. Develop a long term-policy for Bolivia. The lithium and the coca growing issues will be there long after the Morales government. It will be important economically and strategically for the region. Lithium exploitation should also be of concern for its potential for environmental damage to the communities near the Salar de Uyuni.

2. Seek improved relations with the current Bolivian administration. The Assistant Secretary for Western Hemisphere Affairs, Arturo Valenzuela's trip to Bolivia in early June 2010 suggests a willingness on the part of the Obama administration to mend relations with Bolivia. Mr. Valenzuela indicated that both countries are working on a framework agreement to restore ambassadors and normalize relationships. While Secretary Clinton skipped Bolivia during her Andean tour in June 2010, efforts should be increased at different levels to restore relations with Bolivia on the basis of mutual respect. This would provide opportunities for frank discussions on key issues, particularly the US concern over coca cultivation in Bolivia and opportunities for restoring USAID development assistance.
3. Increase US-Bolivian academic and professional exchanges.
4. Improving people to people relations contributes in diminishing distrust. Programs such as the Fulbright and Humphrey Scholarships and the State Department's International Visitor Program could allow professional exchanges among US and Bolivian experts on lithium mining, and other scientific and technical fields of mutual interest.
5. Promote exchanges with Bolivian institutions of higher learning and US universities.
6. USSOUTHCOM could take advantage of the valuable experiences in implementing clean technology pilot projects with seven Latin American countries in close collaboration with civil and military organizations, local civil society organizations and universities. This experience has proven effective in building bridges with Latin American governments and institutions that have a high level of distrust for any initiative coming from the US.

APPENDIX I

Bolivia under the Morales Government

A nation of statistical extremes, landlocked Bolivia is the highest and most isolated South American country. Approximately 60% of its 9.9 million (2009) inhabitants are indigenous, the largest proportion in the hemisphere. Although rich in mineral and energy resources, Bolivia ranks among South America's poorest countries. Wealthy urban elites, mostly of Spanish ancestry, traditionally dominated economic, social and political life of the country while the great majority of Bolivians, mostly low-income subsistence indigenous peasant farmers, miners, small traders and artisans, had little or no voice in national affairs.[51]

Victorious in the presidential campaign of 2005, Evo Morales is the first full-blooded, Indian president ever elected in Bolivia. Morales belongs to an Aymara impoverished family. He rose to prominence in Bolivian politics in the 1990 as a champion of the marginalized rural and urban poor, first as the leader of Bolivia's militant coca farmers' (cocaleros) association and, subsequently, of the leftist Movimiento al Socialismo (MAS) party. The movements he led were increasingly imbued with nationalism and heightened ethnic awareness.

In May 2006, President Morales put Bolivia's energy industry under state control. Foreign energy firms were given six months to sell at least 51% of their holding to the state and negotiate new contracts or leave the country. Although vehemently opposed by Bolivian business elites and foreign oil executives as a radical move, Morales nationalization was in practice more a renegotiation of terms with the foreign energy companies than an outright expropriation. As world energy prices rose in 2006 and beyond, the country's finances improved dramatically. At the same time, President Morales allied his government closely with the leftist, populist and anti-American Venezuelan President, Hugo Chavez, and adopted an increasingly critical, public stance toward the United States.[52]

In June 2006, Morales' MAS party won a decisive majority in special elections for a new constituent assembly that was tasked with rewriting the nation's constitution. The resulting draft constitution promised more rights to Bolivia's indigenous majority and more autonomy to the departments. It also contained provisions authorizing an incumbent president to stand for re-election for a second five-year term. Violent demonstrations by opponents, mainly from the wealthy elites, ensued and four of the country's richest regions defiantly declared autonomy in protest. The new charter was finally approved by more than 60% of Bolivian voters in a referendum held in January 2009. The passage of the new Constitution shifted political power toward the highland provinces, where poor, indigenous voters predominate, and away from the lowlands, where most of Bolivia's food is grown and where its petroleum and natural gas reserves are located.[53]

Talks between Morales's congressional backers and the fragmented opposition ultimately produced a compromise that modified earlier versions of the proposed new charter. One of the most polemical articles in the final draft reversed a plan to allow Morales to run for re-election indefinitely, limiting him (and future Bolivian presidents) instead to a single five-year term without the possibility of re-election. Other, often vaguely worded items among the new Constitution's 411 articles broadened definitions of property to include communal ownership; allowed Indians to mete out corporal punishment under their own legal systems; extended limited autonomy to regional prefects (governors); and reaffirmed state control over Bolivia's ample natural gas reserves and other resources.[54]

Opposition leaders, especially in the eastern lowlands, campaigned vigorously against approval of the proposed new charter. The Constitution was finally approved in a referendum on January 25, 2009, with close to 60% of the country in favor. The four departments in Bolivia's rebellious eastern lowlands rejected the charter by wide margins. Despite the intensity of his opposition in the eastern lowlands, President Morales scored a comfortable win over his conservative opponents when he stood for re-election in December 2009, improving on his 2005 majority and becoming the first incumbent Bolivian president since 1964 to win a second term. His party also won control of both chambers of congress, though in the lower house it fell just short of the two-thirds majority needed for constitutional changes.[55]

On April 4, 2010, Bolivians went to the polls to elect 500 officials in local and provincial elections, including the governors of nine departments. The election was seen as a crucial test of the spreading strength and influence of President Morales and his MAS party. They won handily. On May 1, 2010, the Morales government took control of four privately owned companies that generate electricity. The companies accounted for more than half of Bolivia's electricity market.[56]

The key to Morales's success has been his appeal to the 65% of Bolivians who identify as indigenous and who see him as "one of their own." Under Morales, they have benefited from increased levels social spending, boosted by high international prices for hydrocarbons and more taxes on foreign oil and gas companies. Cash payments have been made to poor families to encourage school attendance. Extra pension payments have been provided to the elderly and pre-natal and post-natal care has been extended to mothers without access to health care. Some estimates suggest that the payments reached a quarter of Bolivia's almost 10 million people in 2010.[57]

NOTES

1. An earlier version of this chapter, co-authored with Olga Nazario, was published by the Western Hemisphere Security Analysis Center (WHEMSEC) in June 2010.

2. For a History of Evo Morales Three Presidential Terms, see Linda C. Farthing and Benjamin H. Kohl, *Evo's Bolivia: Continuity and Change*. Austin: University of Texas Press, 2014, passim.

3. Jason Mick, "Tensions Soar among Industrialized Nations Over Bolivia's Lithium Supplies," *Daily Tech* (http://www.dailytech.com/Tensions+Soar+Among+Industrial+Nations+Over+Bolivias+Lithium+Supply/article14173.htm), February 6, 2009.

4. Lithium use world wide by volume in 2007 broke down as follows: Polymer 4%; Chemical Process 3%; Battery 25%; Air-conditioning 6%; Lubricant 12%; Glass

8%; Ceramics 10%; Aluminum 4%; Continuous Casting 3%; Pharmaceutical 3%; Others 22%, in Brian W. Jaskula, "Lithium," Washington D.C.: US Geological Survey (USGS), 2007.

5. Rory Carroll and Andres Schipani, "Multinationals Eye Lithium Reserves Beneath Bolivia's Salt Flats," *The Guardian*, June 17, 2009; Stephen Markley, "Will Lithium-Air Battery Be the Silver Bullet?" *The Miami Herald*, May 07, 2010. Demand for lithium has skyrocketed over the last five years as manufacturers of batteries for BlackBerrys, iPods and other electronic devices have increasingly relied on the mineral. It is, however, the automobile industry that looms as the largest potential consumer of lithium over the next decade and beyond, because lithium batteries would allow electric cars to store more energy and be driven longer distances than the heavier nickel batteries.

6. Interview with Jay Chmelauskas, President, Western Lithium Canada Corporation, in Jon Knox, "All Set to Power New Generation Lithium Batteries," *Automotive Industries*, March 22, 2009.

7. Simon Romero, "In Bolivia, Untapped Bounty Meets Nationalism, "*The New York Times*, February 3, 2009; Christine Negroni, "Lithium Battery Risks Draw Scrutiny," *The New York Times*, October 27, 2009, Saqib Rhaim, "Will Lithium-Air Battery Rescue Electric Car Drivers from 'Range Anxiety?'" *The New York Times*, May 7, 2010.

8. Meridian International Research, "The Trouble with Lithium 2: Under the Microscope." (Le Legers, France: 29 May 2008, pp. 1).

9. William Tahil, "The Trouble with Lithium," *Meridian International Research*, December 2006, p. 1.

10. Toby Procter, "How long will Lithium last?," BLOG, 20 April 2009 (http://www.climatechangecorp./content.asp?

11. Brian W. Jaskula, "Lithium," In U.S. Geological Survey, *Mineral Commodity Summaries* (January 2010), pp. 92–93.

12. Ibid. p. 93.

13. Jaskula (2009), "Mineral Commodities Summaries."; Keith Johnson, "Peak Lithium: Will Supply Fears Drive Alternative Batteries?," *Environmental Capital*, February 3, 2009.

14. The Reserve Base is the part of the identified resource that meets specified minimum physical and chemical criteria related to current mining and production practices. It is a rather nebulous figure. Only "reserves" should be considered available at the moment, of which Bolivia is yet to be determined, in Tahil, W. op cit. "The Trouble with Lithium," *Meridian International Research*, December 2006, p. 4.

15. Brian W. Jaskula, op. cit. January 2010.

16. Brian W. Jaskula, "Lithium," in U.S. Geological Survey, Mineral Commodity Summaries (January 2013), pp. 94–95.

17. In 2013, a study conducted by the University of Wyoming found an extensive lithium deposit in southwest Wyoming that exceed those found in Nevada (http://oilprice.com/Energy/Energy-General/New-Wyoming-Lithium-Deposit-could-Meet-all-U.S.-Demand.html).

18. Interview with Jay Chmelauskas, President, Western Lithium Canada Corporation (2009) op. cit.

19. Sean Goforth, "Argentina, Bolivia and Chile: The ABCs of Lithium," *World Politics Review*, 01 Apr 2010.

20. Brian W. Jaskula (2007), op. cit.

21. Juan Carlos Zuleta, "El enorme potencial de litio de Bolivia,"(La Paz, Bolivia: Hora 25, No. 69, November 18–December 2, 2009.

22. TRU Group Inc., "Development of the World's Largest Lithium Brine," Uyuni Lithium Brine Salar Resource Bolivia Development, (http://trugroup.com/Uyuni-lithium.html).

23. Meridian International Research, p. 13.

24. Ibid.

25. Ibid., p. 14.

26. Ibid.

27. GNRE, "Baterías de litio aportan con electricidad a 2 comunidades." evaporiticos.gob.bo, March 23, 2015, http://www.evaporiticos.gob.bo/?p=1704#more-1704.

28. John M. Broder, "U.S. Issues Limits on Greenhouse Gas Emissions from Cars," The New York Times, April 1, 2010; Rebecca Hollender and Jim Schultz, "Bolivia and its Lithium: Can the "Gold of the 21st Century" Help Lift a Nation out of Poverty?" (Cochabamba, Bolivia: A Democracy Center Special Report) *The Democracy Center*, May 2010, pp. 24–25.

29. Ibid, pp. 34–36; Romero, 2009.

30. Hollender and Schultz (2010), pp. 3–4; Barjak, Frank and Carlos Valdez, "Bolivia Pins Hopes on Lithium, Electric Vehicles," *Associated Press*, February 28, 2009.

31. Romero (2009) op. cit.; Damian Kahya, "Bolivia holds key to electric car future," *BBC News*, 9 November 2009.

32. Carroll and Schpani, 2009; Demand for lithium carbonate doubled from 2003 to 2007, and the market for lithium-ion batteries may expand to 14 times its 2009 size by 2030. In the shorter term, lithium's potential has led Credit Suisse to expect a 13% rise in sales of lithium-ion (Li-Ion) batteries and a 10% rise in sales of liquid crystal displays (LCD) within a 3–4 year horizon due to growing demand for portable electronic instruments. Credit Suisse, Research Flash, "Investment Ideas Bright outlook for lithium-ion batteries and liquid crystal." *Zurich*, 14 October 2009.

33. Simon Moores, "Bolivia/Korea Strike Lithium Deal," Industrial Minerals (August 17, 2009); "French/Bolivia Lithium," *Industrial Minerals*, 10 July 2009.

34. "Iran to Study Bolivia's Lithium Reserves," *Press TV*, 25 November 2009.

35. Anf Agencia, "Irán también apoyará el desarrollo de la industria de litio en Bolivia," *Los Tiempo.Com*, Noviembre 24, 2010.

36. "Francesa Bolloré reitera interés en litio boliviano y pide acelerar proyecto," *Bolivia: Lo Mejor que Tenemos* (29 Abril 2010), (http://boliviateamo.blogspot.com/2010/04/francesa-bollore-reitera-interes-en.html).

37. GNRE, "Acuerdos firmados fortalecen proyecto de industrialización del litio," Septiembre 18, 2011, http://www.evaporiticos.gob.bo/wp-content/uploads/notasdeprensa/2011/180911hidr.pdf.

38. Goforth, 2010; Matthew Smith and Matthew Craze, "Lithium for 4.8 Billion Electric Cars Lets Bolivia Upset Market," *Bloomberg*, December 7, 2009.

39. Hollender and Schultz, 2010, pp. 29–34.

40. Ibid., p. 48.

41. Hollender and Schultz, pp. 39–41; Phil Plait, "Want a Planet? You Might Want to Avoid Lithium," *Discover Magazine*, 11 November, 2009.

42. Hollender and Schultz, 2010, 42–43.

43. Ibid., 45–46.

44. Meridian International Research, p. 16.

45. Andres Schipani, "Coca casts shadow on Bolivian election," BBC NEWS, 2009/11/22 13:49:08 GMT. http://news.bbc.co.uk/go/pr/fr/-/2/hi/americas/8364909.stm

46. Peter DeShazo, "Testimony before the House Committee on Foreign Affairs, Subcommittee on the Western Hemisphere Hearing on U.S.-Bolivia Relations: Looking Ahead." Washington D.C., March 3, 2009, p. 2; Viola Gienger and Bill Faries, "U.S. Ousts Bolivian Envoy in Tit-for-Tat With Morales (Update1), *Bloomberg*, September 11, 2008.

47. DeShazo, 2009, "Déjà vu all Over Again: 'Decertification' Memo of Justification Inaccurate and Misleading," *Andean Information Network*, p. 3, October 15, 2009.

48. El Economista, "Lula y Morales firman acuerdos para construir una carretera sobre el litio y la formación profesional," *El Economista*, 22 August 2009; Jonathan Levin, "Mitsubishi, Sumitomo Propose Lithium Joint Venture in Bolivia," *Bloomberg*, May 15, 2009.

49. Susan Kaufman Purcell, "Why Greece is Important to Latin America," *AméricaEconomía*, June 2010.

50. Romero, 2009; on business conditions in Bolivia, see *The World Bank, Doing Business in Bolivia*. Washington D.C.: The Word Bank, 2010. http://espanol.doingbusiness.org/ExploreTopics/DealingLicenses/.

51. BBC, "Country Profile: Bolivia," *BBC NEWS*, 2009/12/19 14:39:39 GMT. http://news.bbc.co.uk/go/pr/fr/-/2/hi/americas/country_profiles/1210487.stm. *Library of Congress, A Country Study, Bolivia*. Washington D.C., American University, 1991.

52. BBC, Country Profile, 2009; Painter, 2009.

53. Ibid.

54. BBC, "Bolivian Indians in Historic Step," *BBC News*, 2009/08/03 08:47:01 GMT. http://news.bbc.co.uk/go/pr/fr/-/2/hi/americas/8180790.stm.

55. Miguel Centellas, "Bolivia's Regional Elections: A Setback for Evo Morales," In *Americas Quarterly*, April 8, 2010. (http://www.americasquarterly.org)

56. Centellas, 2010; Diego Ore and Eduardo Garcia, "Bolivia Nationalizes Four Power Companies," *The Washington Post*, 2 May, 2010; A14

57. Ibid.

BIBLIOGRAPHY

Anf-Agencia, "Iran también apoyará el desarrollo de la industria de litio en Bolivia," LosTiempo.Com, November 24, 2010.

Andean Information Network, "Déjà vu all Over Again: "Decertification" Memo of Justification Inaccurate and Misleading," October 15, 2009.

Barjak, Frank and Carlos Valdez, "Bolivia Pins Hopes on Lithium, Electric Vehicles," Associated Press, February 28, 2009.

BBC, "Bolivian Indians in historic step," BBC News, August 3, 2009 http://news.bbc.co.uk/go/pr/fr/-/2/hi/americas/8180790.stm.

———, "Country Profile: Bolivia, BBC NEWS, December 19, 2009 http://news.bbc.co.uk/go/pr/fr/-/2/hi/americas/country_profiles/1210487.stm.

Broder, John M, "U.S. Issues Limits on Greenhouse Gas Emissions from Cars," The New York Times, April 1, 2010.

Carroll, Rory and Andres Schipani, "Multinationals eye lithium reserves beneath Bolivia's salt flats," The Guardian, June 17, 2009.

Centellas, Miguel, "Bolivia's Regional Elections: A Setback for Evo Morales," in Americas Quarterly, April 8, 2010. (http://www.americasquarterly.org).

Credit Suisse, Research Flash, "Investment Ideas Bright outlook for lithium-ion batteries and liquid crystal." Zurich, 14 October 2009.

DeShazo, Peter, "Testimony before the House Committee on Foreign Affairs, Subcommittee on the Western Hemisphere Hearing on U.S.-Bolivia Relations: Looking Ahead." Washington DC, March 3, 2009.

El Economista, "Lula y Morales firman acuerdos para construir una carretera sobre el litio y la formación profesional," El Economista, 22 August 2009.

"Francesa Bolloré reitera interés en litio boliviano y pide acelerar proyecto," Bolivia: Lo Mejor que Tenemos (29 April 2010), (http://boliviateamo.blogspot.com/2010/04/francesa-bollore-reitera-interes-en.html).

Gienger, Viola and Bill Faries, "U.S. Ousts Bolivian Envoy in Tit-for-Tat With Morales (Update1), Bloomberg, September 11, 2008.

Goforth, Sean, "Argentina, Bolivia and Chile: The ABCs of Lithium," World Politics Review, 01 Apr 2010.

Hollender, Rebecca and Jim Schultz, "Bolivia and its Lithium: Can the "Gold of the 21st Century" Help Lift a Nation out of Poverty?" (Cochabamba, Bolivia: A Democracy Center Special Report), The Democracy Center, May, 2010.

"Iran to Study Bolivia's Lithium Reserves," Press TV, 25 November 2009.

Jaskula, Brian W, "Lithium," Washington DC: U.S. Geological Survey (USGS), 2007.

———, "Mineral Commodities Summaries." Washington DC: U.S. Geological Survey, January 2009.

———, "Lithium," in U.S. Geological Survey, Mineral Commodity Summaries (January 2010), pp. 92–93.

Johnson, Keith, "Peak Lithium: Will Supply Fears Drive Alternative Batteries?" Environmental Capital, February 3, 2009.

Kahya, Damian, "Bolivia holds key to electric car future," BBC News, 9 November 2009.

Kaufman-Purcell, Susan, "Why Greece is Important to Latin America," AméricaEconomía, June 2010.

Knox, Jon, "All Set to Power New Generation Lithium Batteries," Automotive Industries, March 22, 2009 (www.allbusiness.com/automotive/automotive-industry environmente/12290847-1.html).

Levin, Jonathan, "Mitsubishi, Sumitomo Propose Lithium Joint Venture in Bolivia," Bloomberg, May 15, 2009.

Library of Congress, A Country Study, Bolivia. Washington DC, American University, 1991.

Markley, Stephen, "Will Lithium-Air Battery Be the Silver Bullet?" The Miami Herald, May 07, 2010.

Meridian International Research, "The Trouble with Lithium 2: Under the Microscope." (Le Legers, France: 29 May 2008).

Mick, Jason, "Tensions Soar Among Industrialized Nations Over Bolivia's Lithium Supplies," Daily Tech, Feb. 6, 2009. (http://www.dailytech.com/Tensions+Soar+Among+Industrial+Nations+Over+Bolivias+Lithium+Supply/article14173.htm).

Moores, Simon, "Bolivia/Korea Strike Lithium Deal," Industrial Minerals (August 17, 2009); "French/Bolivia Lithium," Industrial Minerals (10 July 2009).

Negroni, Christina, "Lithium Battery Risks Draw Scrutiny," The New York Times, October 27, 2009.

Ore, Diego and Eduardo Garcia, "Bolivia Nationalizes Four Power Companies," The Washington Post, May 2, 2010.

Painter, James, "Bolivian president to deepen social revolution," in BBC News, December 7, 2009 http://news.bbc.co.uk/go/pr/fr/-2/hi/americas/8400639.stm.

Plait, Phil, "Want a Planet? You Might Want to Avoid Lithium," Discover Magazine, November 11, 2009.

Procter, Toby, "How long will Lithium last?" BLOG, 20 April 2009 (http://www.climatechangecorp./content.asp?

Rhaim, Saqib, "Will Lithium-Air Battery Rescue Electric Car Drivers from 'Range Anxiety'?" The New York Times, May 7, 2010.

Romero, Simon, "In Bolivia, Untapped Bounty Meets Nationalism," The New York Times, February 3, 2009.

Schipani, Andres, "Coca casts shadow on Bolivian election," BBC NEWS, November 22, 2009, http://news.bbc.co.uk/go/pr/fr/-/2/hi/americas/8364909.stm.

Smith, Matthew and Matthew Craze, "Lithium for 4.8 Billion Electric Cars Lets Bolivia Upset Market," Bloomberg, December 7, 2009.

Tahil, William, "The Trouble with Lithium," Meridian International Research, December 2006.

US Geological Survey, Mineral Commodity Summaries, January 2010.

World Bank, Doing Business in Bolivia. Washington DC: The Word Bank, 2010. http://espanol.doingbusiness.org/ExploreTopics/DealingLicenses/.

Zuleta, Juan Carlos, "El enorme potencial de litio de Bolivia," Hora 25, No. 69, (La Paz, Bolivia: 18 November-2 December 2009).

Chapter 11

25 Years of Contemporary History of PDVSA

Killing the Goose That Lays the Golden Eggs

Gonzalo E. Vázquez

During an interview in 2006, Hugo Chávez was asked:

> "What happens when the oil money runs out, what happens when the price of oil falls as it always does? Will the Bolivarian revolution of Hugo Chávez simply collapse because there's no money to pay for the big free ride?"

Chávez replied:

> "I don't think it will collapse, in the unlikely case of oil running out today. The revolution will survive. It does not rely solely on oil for its survival. There is a national will, there is a national idea, a national project. However, we are today implementing a strategic program called the Oil Sowing Plan: using oil wealth so Venezuela can become an agricultural country, a tourist destination, an industrialized country with a diversified economy. We are investing billions of dollars in the infrastructure: power generators using thermal energy, a large railway, roads, highways, new towns, new universities, new schools, recuperating land, building tractors, and giving loans to farmers. One day we won't have any more oil, but that will be in the twenty-second century. Venezuela has oil for another 200 years."[1]

Chávez was also unwavering in his conviction to his chosen method:

> "It's impossible for capitalism to achieve our goals, nor is it possible to search for an intermediate way [therefore] I invite all Venezuelans to march together on the path of socialism of the new century."[2]

Venezuela depends on oil revenues for approximately 96% of export earnings and 45% of budget revenues[3]and it has the world's largest known oil reserves, therefore the national oil company, Petróleos de Venezuela, S.A. (PDVSA), is frequently referred to as the government's "golden goose." The country has great potential to invest in the manner that Chávez described, but development projects were impeded as he executed a neo-authoritarian strategy that included increasing the government's control over Venezuela's oil industry. Although progress was made in health care and education during his regime, many infrastructure projects were abandoned and diversification of Venezuelan industry failed.

When Chávez was elected president of Venezuela in 1998, the price of oil was at historic lows (around $7/barrel). He became actively involved with the Organization of the Petroleum Exporting Countries (OPEC) and advocated maintaining limits on oil production. For this and for other reasons, by 2003 the price of oil was rising sharply. Oil prices exceeded $100/barrel in 2008 and generally maintained high levels during the rest of Chávez's presidency, but the unprecedented revenues did not free Venezuela from the resource curse, the phenomenon that countries with abundant natural resources tend to have lower long term economic growth. It has also been noted that the "curse" (a.k.a. the paradox of plenty) is often accompanied by political conflict and weak institutions, and by corruption and an unequal distribution of revenues.

Venezuela has been no exception to these tendencies. Since 1998 Venezuela's oil exports have grown by 410% and imports have grown by 130% on a nominal basis, but in real terms exports have fallen by 40% while imports have more than doubled.[4] Although revenue windfalls from oil led to several years of high growth in GDP, the average annual GDP growth for Venezuela from 1999–2011 was 2.8% while for other large countries in South America it was between 3.3% and 5.2%.[5] Under Chávez, Venezuela spent over $517 billion of its oil income and increased total debt (debt for the central government and PDVSA) from $31.3 billion in 1998 to an estimated $96 billion by the end of 2010.[6] By 2013, total debt was estimated at $219.1 billion and more than 70% of GDP.[7]

The policies and actions of the Chávez regime, especially the policy changes that explicitly bound PDVSA to the state, that is, the "renationalization" of the oil industry through the 2001 Hydrocarbons Law and tax reforms, and heavy use of oil moneys (which comprise about 25% of GDP)[8] for social programs and development projects helped to bolster Chávez's popularity, but they repelled foreign investors and created tensions between Chávez and PDVSA that helped to precipitate management's participation in the general strike of 2002–2003.[9] "Socialism of the 21st Century" policies such as petro-diplomacy and price controls strained Venezuela financially, increasing the country's dependence on China and squeezing PDVSA between its need for funds to enlarge production and financial demands from the government.

In March 2013, the death of Hugo Chávez triggered the constitutional requirement for a presidential election to be held within 30 days. Vice President Nicolás Maduro assumed the role of interim president following Chávez's death until April 14th, when Maduro claimed the presidency in an election that was widely questioned. The government has continued to alienate large segments of society as a result of its exchange controls, the limits placed on free press, and expropriations of private companies and lands. The present slow economic growth in Venezuela, a homicide rate that is among the world's highest, shortages of many basic goods and inflation of over 50% catalyzed student protests that first erupted into street demonstrations in early 2014. The government has jailed opposition leaders and violently suppressed the demonstrations, including the killing of numerous Venezuelan citizens.

Politics before, during and after the Chávez administration have affected the operational and financial performance of PDVSA and the forecasts for the company's future. Relative to its regional counterparts, most scholars have viewed Venezuelan democracy favorably but it did not continue to evolve to operate within the strict parameters of truly democratic governance. The country's oil industry has had to contend with the collapse of elected governments and political parties, and with experiments in governance—notably, the pact of "Puntofijismo."

THE LAST TEN YEARS OF THE FOURTH REPUBLIC UNDER PUNTOFIJISMO

In Venezuela, the contribution made by oil is so dominant that it is impossible to understand much about the country without appreciating the ways in which its politics are shaped by—and shape—Venezuela's oil industry. It is oil wealth and the fact that the state "*presents itself as the sole owner of the petroleum resource*"[10] that underlie the largesse of the government and give it the power to encroach on other sectors, to purchase the cooperation of constituents such as unions and the armed forces, and to marginalize its critics.

Venezuela held direct elections in 1948, but that attempt at democracy was truncated by a coup d'état that ushered in the military dictatorship of Marcos Pérez Jiménez. In 1958, Pérez Jiménez left office and a governing board that had military and civilian members was in place for several months while an electoral procedure was reestablished. The three main political parties at the time were Accion Democrática (AD), Comité de Organización Política Electoral Independiente (COPEI) and Unión Republicana Democrática (URD). All of them were oriented to center-left, but none were considered radical.

In October 1958, an agreement between the three parties was reached that, no matter who won the election, they would form a coalition government.

This power sharing alliance was called the Pacto de Punto Fijo ("Punto Fijo") came from the name of Rafael Caldera's residence in Caracas—Caldera was the head of COPEI at this time and he would later serve two separate terms as president) and it purposely excluded groups (e.g., the Communist party) that these leaders felt might destabilize democratic practices, which were not well established.[11] During this period, which is also referred to as the Fourth Republic,[12] the military did not interfere with politics and the two main parties (AD and COPEI) governed within a democratic framework that allowed for more civil liberties, for example, a free press. The coalition survived until Chávez came to power in 1998.

The efforts to institutionalize democracy even reached into foreign policy with the Betancourt Doctrine (Rómulo Betancourt of AD was the first president elected under the pact), which declared that any government in South America that had come into being through a coup d'état would not be recognized by Venezuela. Betancourt is also credited with successfully holding elections in 1963, which were won by Raúl Leoni (also of AD) and were noted for representing the first time in Venezuela's history that one elected president succeeded another. In 1968, Rafael Caldera (COPEI) was elected, which was the first time that one party in Venezuela would peacefully take over from another. During Caldera's term of 1969–1974 he gave amnesty to the revolutionaries on the left and he renounced the Betancourt Doctrine, reopening ties with Latin American dictatorships and establishing ties with Cuba and the Soviet Union.

In 1974, Carlos Andrés Pérez (AD) took office during an oil boom and he nationalized the oil industry in 1976. He splurged on public works projects and held lavish dinners, and between year-end 1970 and year-end 1978 Venezuela's total outstanding debt grew 42% annually.[13] From 1979 to 1984, Luis Herrera Campins (COPEI) was president and because he continued to spend and to provide large subsidies even as oil prices fell precipitously, Venezuela had to rely on debt even more. Low interest rates and a strong worldwide economy had supported generous international lending, but as interest rates were raised to curb inflation in industrialized countries the "*unsustainable economic activity—based on high domestic consumption, heavy borrowing from abroad, unsustainable currency levels, and excessive intervention by government into the economy*" created the Debt Crisis of the 1980s and the "lost decade" in Latin America.[14] A drop in oil prices coincided with the Black Friday devaluation and crisis (February 18, 1983), causing an economic collapse in Venezuela.

Jaime Lusinchi (AD) served as president from 1984 to 1989 and restructured Venezuela's debt, eliminated the value-added tax and put exchange controls in place, but his actions were limited by an escalating banking crisis. The redistribution of oil wealth was restricted and the legitimacy of the Puntofijismo and its institutions was declining so Lusinchi created the Comisión para la Reforma del Estado (COPRE), a commission to recommend reforms to the political system, which called for more direct government representation. This desire was an indication of the forces that would later help Hugo Chávez sweep to power.

In the 1988 elections, Carlos Andrés Pérez heeded the trend and ran on a populist platform that included denouncing neo-liberalism and Venezuelans reelected him. However Pérez abandoned his campaign promises and, at least partly in response to loan conditions of the IMF, began to implement an economic adjustment program. An announcement that fuel subsidies would be cut, in addition to other austerity measures, triggered massive street protests in Caracas and the surrounding area that began on February 27, 1989. Although the protests were severely repressed, the rioting and looting on one side and wanton shootings by security forces and the military on the other continued for a week. By official figures, over three hundred people were killed and unofficial death counts were in the thousands. This event, which came to be known as the Caracazo, further discredited and destabilized the Puntofijismo.

Pérez achieved some short-term macroeconomic success with the austerity program, but by 1992 the country had returned to fiscal deficits, a negative balance of payments, high inflation and increased unemployment. In February 1992, Lieutenant Colonel Hugo Chávez Frías led a coup attempt and was arrested. Nine months later, a second coup was also successfully suppressed, but discontent remained and in 1993 Pérez was removed from office for corruption.

In 1994, Rafael Caldera was reelected president and he returned to moderate neoliberal policies, but the actions that were taken to stabilize the economy and to make structural adjustments have been critiqued as "haphazard."[15] Venezuela's experience with economic orthodoxy was brief and foreign investment was not restored nor did efforts to privatize state companies prove to be effective.[16] Despite the vast oil revenues, figures for 1996 show that wages for most Venezuelans had stagnated while prices had doubled. In February 1997, a 77% pay raise for government workers was announced and plans to restructure the public sector's workforce of 1.3 million people were put on hold.[17]

It has been argued that Chávez and Bolivarianism became popular because they offered an alternative to Puntofijismo and not because they provided a response to neoliberalism.[18] The failures of democracy under the Punto Fijo pact, which caused the government to increasingly be viewed as insular and unresponsive, are considered by many scholars to have contributed to the rise of a political outsider. Regardless, the coup attempt had made Chávez a national figure and had helped to crystalize a popular movement.[19]

Sensitive to this, President Caldera pardoned Chávez after he had served only two years in prison.

Chávez began his formal political career by gathering politicians on the left into the Movimiento *Quinta* República (MVR) party. He made his first visit to Fidel Castro and, using radical language, campaigned on an anti-corruption and anti-poverty platform. Many observers argue that Chávez also received essential assistance from business elites such as Gustavo Cisneros Rendiles, who offered substantial financial backing and owned the television channel Venevisión, and Chávez had support from the Movimiento al Socialismo (MAS) party, particularly their organizational expertise.[20] In calling his agenda a Bolivarian Revolution, Chávez appropriated Venezuela's ancestry and exploited the legacy of the popular statesman Simón Bolívar as "*a concrete political instrument, in the name of which was launched a fight against the immense corruption of Venezuelan institutions, in the name of which was justified the frontal attack on traditional parties* [Puntofijismo] *and the threat to close institutions such as Congress and the Judiciary*."[21] While the cause can be debated, it is undeniable that on December 6, 1998 Hugo Chávez secured the Venezuelan presidency—his first political office—in a landslide (over 56% of the vote).

CHÁVEZ AND HIS PROJECT OF "SOCIALISM OF THE TWENTY-FIRST CENTURY"

The evolution of revolutionary movements in Venezuela has been described as a Bolivarian process in which different leftist movements converged, but did not fuse.[22] In 1958, the Partido Comunista de Venezuela (PCV), the AD party and labor unions had joined in overthrowing the dictatorship of Pérez Jiménez, but the Pacto de Punto Fijo had excluded the PCV. Also, although the first president elected under the pact (Betancourt) was from AD, the center-left youth within AD quickly became disenfranchised and in 1960 they formed the Movimiento de Izquierda Revolucionaria (MIR), which considered itself Marxist-Leninist and was greatly influenced by the Cuban Revolution. In 1963, radical members of the URD party and of the military formally aligned with the PCV and MIR to form an insurgent organization, Las Fuerzas Armadas de Liberación Nacional (FALN). But most Venezuelans did not support the guerrilla movement[23] and by 1965 intense anti-guerrilla efforts had reduced the numbers and the effectiveness of the FALN and its associates, which responded by directing more effort into social movements. A split within the PCV guerrillas in the mid-1960s led to the creation of the Partido de la Revolución Venezolana (PRV). The PRV provided the essential precedent for Chávez of associating their Marxist, revolutionary ideals with the "great liberator," Simón Bolívar.[24] In 1971, another split within the PCV led to the formation of the Movimiento al Socialismo (MAS), which played an important role in the 1993 presidential elections.[25]

It may have been more difficult for the guerrilla movement to reach into the general population because by 1958 Venezuela was experiencing growth rates nearly five times the average GDP growth of the other large countries in South America. "*By 1970, Venezuela had become the richest country in Latin America and one of the twenty richest countries in the world, with a per capita GDP* [. . .] *only 13% lower than that of the United Kingdom*" then, from 1978 to 2001, Venezuela experienced a cumulative decline of 18.64% in GDP.[26] This macro-volatility and the dependence on oil revenues helped to create a situation in which income is viewed to be determined more by luck (proximity to the revenue source) than by effort (participating in production).[27] In *The Paradox of Plenty*: *Oil Booms and Petro-States* (1997), Terry Lynn Karl suggests that the influx of oil revenues to government coffers also produces cultural resistance to taxation and blindness to the need for political systems that provide fiscal accountability.

By the late 1980s, Venezuela was in economic crisis. Accompanied by speculation and food shortages, inflation roughly tripled and it exceeded 100% in June 1989. The effects of the loan conditions imposed by the IMF heightened the perception that neoliberal policies were the cause of Venezuela's woes, and the corruption that appeared to keep the elites and the politicians insulated from the troubles helped to erode the credibility of the existing parties.

The process of drafting a constitution through a representative body, which has consistently been referred to as a constituent assembly in Venezuela, was used in 1811 (after independence from Spain was declared) and in 1830, when Gran Colombia was separated into the Provinces of Venezuela and the Republic of Colombia. From 1858–1953, there were seven other constituent assemblies that were convened to create a new constitution as a result of "*a de facto rejection of the existing constitution, a coup d'* état, *a revolution, or a civil war.*"[28] During the Puntofijismo, the idea of holding another constituent assembly had been discussed by members of AD and COPEI well before the coup attempts of 1992.[29] However, Venezuela's constitution of 1961 required ratification by both houses of Congress and by a majority of the states' assemblies or by a popular referendum before constitutional amendments could be made.[30] All of the presidential candidates therefore considered the possibility of a constituent assembly to be invalid or unrealistic—all except for Chávez, who used it to demonstrate that he was "for the people." In February 1999, the first decree that Chávez made after taking the oath of office was to establish a referendum on creating a constituent assembly, to start a process for drafting a new constitution.

Fewer than 40% of the voters participated in the referendum of April 1999. Chávez had described the constituent assembly as having "*the purpose to transform the state and create a new juridical order allowing the effective functioning of a social and participative democracy*" and over 80% of those who did vote approved the formation of such an assembly.[31] But the referendum also gave Chávez the power to determine the mission, procedure and duration of the assembly as well as its form and the election resulted in candidates from Polo Patriótico—the alliance of political parties that had backed Chávez in the 1998 election—receiving 94% of the 131 seats.[32] Also, while the work was distributed across twenty commissions and numerous constituencies submitted draft constitutions, only two weeks were allowed for integrating the drafts.[33] The new constitution, which was approved in December 1999, eliminated one house of the legislature, abolished municipal elections, extended the president's term to six years from five and gave the president the authority to dissolve the National Assembly.

During the 1990s, PDVSA operated independently from the government and from OPEC[34]and, although PDVSA was then recognized as the world's best-managed national oil company,[35] Chávez decried this situation as a state within a state. He began to make policy changes to increase control of Venezuela's oil industry by the state and he became active with OPEC. The September 11, 2001, attacks on the US and rising demand from China and India helped oil prices surge above OPEC's targets, but the general perception was that Chávez had reenergized OPEC. The increased oil prices brought enormous financial benefits to Venezuela.

In November 2001, shortly before the executive powers granted to him under an enabling law that had been authorized by the National Assembly were due to expire, Chávez established a new hydrocarbons law that included a provision that even international disputes would be arbitrated in Venezuela. Royalties for heavy crude oil production were raised from 20% to 30% and for other activities the rates increased from 16.67% to 30% of the sales price/barrel, although the income tax levied on the extraction of oil was reduced from 67.6% to 50%.[36] Chávez also dictated that 10% of PDVSA's annual investment budget was to be spent on social programs. These interventions were not welcomed by PDVSA, which was focused on maximizing oil production profitably.[37]

Resistance was also building outside of PDVSA. There were mass demonstrations in Caracas and in April 2002 a general strike was called to pressure Chávez into resigning. PDVSA workers joined the strike in December, causing a near-complete halt in Venezuelan oil production. Chávez left office, but briefly, so a few months later a combination of labor unions and business groups called for an "indefinite national strike" that in many places turned into a lockout. Chávez retaliated by naming a new board of directors for

PDVSA and the government ultimately fired about 18,000–20,000 PDVSA employees and replaced them with workers loyal to the Chávez regime, many of which were unseasoned. Fired employees were blacklisted and the brain drain from PDVSA was significant.[38] Most figures show that PDVSA has never regained its pre-strike production volumes.

In the early 1990s, PDVSA had begun a policy of courting outside investments that was known as the *apertura petrolera*. This policy had offset many costs and had given PDVSA access to the large amounts of capital required to support its production and to invest in its operations, which had been an ongoing challenge. For example, during the first phase of the projects that were intended to attract foreign investment (1992–1993), fourteen contracts were awarded for reopening inactive or abandoned fields. The estimated production for these fields was 125,000 barrels per day (bbl/d), but by 1997 they were yielding 260,000 bbl/d.[39] The four strategic associations, thirty-three operating agreements and eight exploration projects created during the apertura petrolera brought approximately $40,000 million to PDVSA and to Venezuela, with production that peaked at 600,000 bbl/d from fields under operating agreements plus 650,000 bbl/d from projects in the Orinoco Belt, which holds most of Venezuela's vast deposits of heavy crude.[40]

By naming his minister of energy and petroleum as PDVSA's president in 2002, Chávez codified his view that the company was a political vehicle and he began to aggressively "renationalize" after the December 2002 to February 2003 oil strike. Under a second hydrocarbons law, in April 2006 existing operating agreements in the oil industry were terminated and new contracts with private parties, except as a minority investor (40%), were prohibited. Although the affected oil companies filed suit, Venezuela took over the fields and facilities. However, PDVSA had neither the personnel nor the funds to maintain the properties and their production dropped from about 500,000 bbl/d to an average of 360,000 bbl/d in 2012.[41]

The 1999 constitution contained a provision for a recall referendum that could prematurely end a president's term of office and the required signatures were gathered, but such a referendum (held in August 2004) failed. The opposition decried the elections as being skewed by governmental off-balance spending, but the claims were not taken seriously. Opposition leaders then decided to boycott elections for the National Assembly in December 2005, which only gave more seats to pro-Chávez candidates. Chávez's bid for reelection in 2006 was supported by a coalition and in 2007 he solicited the party leaders with a proposal to formally consolidate the parties, which became the Partido Socialista Unido de Venezuela (PSUV).

In January 2007, the National Assembly granted Chávez another enabling law that permitted him to rule by decree and this time the duration was 18 months. Thus strengthened, in August 2007 Chávez brought forward proposals for constitutional reforms intended to create economic and political socialism and the National Assembly approved the measures in November. They were defeated in the popular referendum, but the institutions and stable rules to guide constitution-making that had been in place had by now been dismantled, aiding Chávez in ignoring the constituent power of the people—ultimately, all of the measures were legislated into being.[42] In 2008, Chávez used his decree powers to counter gains that the opposition had made in municipal and state elections by appointing politicians at the regional level and giving them discrete budgets.[43] In 2009, voters passed a referendum that amended the constitution to abolish term limits for the president and for other elected officials. Scholars continue to analyze the factors and mechanisms that enabled Chávez to move the country from democracy toward authoritarianism, but few doubt that his use of Venezuela's oil moneys was an essential factor.

Chávez had terminated Venezuela's existing social programs shortly after assuming the presidency and had initially created a fund that was under the supervision of the military.[44] It was under the specter of the 2004 referendum that sought to remove him from office that Chávez first drew funds from PDVSA to create the *misiones*, which bring health care and educational programs to impoverished areas. In 2005, the National Development Fund (FONDEN)—one of several notoriously opaque funds that have received tens of billions in revenue from PDVSA but, contrary to constitutional provisions, were personally administered by Chávez—was established. In February 2008, Chávez created several new subsidiaries of PDVSA through which to channel the funds that were to be diverted to social and development programs. Chávez thus transformed PDVSA from an important corporate player in the international oil industry into a vehicle that he controlled and used to fund and support his own political agenda.

PETRO-DIPLOMACY

In geographic size, Venezuela ranks 33rd in the world and it holds 46th place in population rankings,[45] but its oil reserves of 298.35 billion barrels are the world's largest, representing almost 25% of the reserves among OPEC countries[46] and 18% of the world total.[47] Oil and politics have always gone together and Venezuela's leaders had historically used this to their advantage, but the high oil prices of the late 1990s increased the opportunity for Chávez to use oil as a foreign policy tool.

Under Chávez, foreign policy was focused on increasing Venezuela's international authority rather than disseminating Bolivarianism per se.[48] Venezuela's international alliances were intended to obtain financing, military equipment and technological expertise and to create new axes of power.[49] Chávez adopted Bolívar's dream of a Pan-American alliance and later expanded it to include the concept of a "south-south cooperation" among developing countries around the world that would establish a "multi-polar world order" and would counter US influence in world affairs.[50] An image of declining US hegemony (reinforced by the September 11, 2001 attacks) and the unpopularity of US policies, including those of organizations closely affiliated with the US, such as the IMF, helped increase receptivity to Chávez's efforts.

Chávez actively participated in existing organizations and he built organizations to manage and share oil, but the idea of a cooperative approach based on energy resources was not new. Fidel Castro had sought to establish such a relationship with Venezuela as far back as 1959, although the first such pact between Venezuela and Cuba was not signed until 2000.[51] In 1980, Venezuela and Mexico had committed to sell oil at favorable prices to countries in the Caribbean and Central America under the San José Agreement. Building on that accord, Belize, Costa Rica, El Salvador, Guatemala, Honduras, Haiti, Jamaica, Nicaragua, Panama and the Dominican Republic joined Venezuela in signing the Caracas Agreement on Energy Cooperation in 2001.

Venezuela had historically considered the Caribbean as part of its zone of influence and the idea of opposing US power from there had made Cuba an essential partner for the Soviet Union during the Cold War. In 2004, the existing agreement with Cuba was extended to become La Alianza Bolivariana para los Pueblos de Nuestra América, a.k.a., ALBA. ALBA is both ideological (e.g., it continuously advocated for an end to the US blockade of Cuba) and focused on regional development.[52] Although it was intended to compete with alliances such as the Organization of American States (OAS) and Chávez was lobbying countries to join ALBA while George W. Bush was promoting the North American Free Trade Agreement (NAFTA), ALBA members are not prohibited from joining other groups that focus on regional integration.[53]

Petrocaribe was established in 2005 to further the offering of preferential terms for the sale or barter of oil to countries in the Caribbean, Central America, and South America and it is implemented by a subsidiary of PDVSA. Typically, members must pay 40% of their bill within 90 days and they may defer the balance at a 1–2% rate for up to 25 years. Part of the payment may also be made in goods and services.[54] For example, the original agreement committed Cuba to send 30,000 medical professionals to work in Venezuela; to train thousands of Venezuelan doctors and healthcare workers; to provide more than a thousand health clinics, diagnostic centers and other facilities; and to perform thousands of surgeries, in exchange for 53,000 bbl/d of oil.[55] Petrocaribe has helped many countries to sustain their vulnerable economies, but one criticism is the indebtedness that other member countries have to Venezuela.

Although Venezuela has received some payment for oil deliveries made under Petrocaribe, including barter for staples such as beans and rice, the various aid and subsidy programs that Venezuela sponsors are estimated to have cost the country $100 billion from 1999 to 2013.[56] Between programs such as Petrocaribe and the oil that goes to repay loans from China, observers estimate that Venezuela does not receive direct payment for more than 40% of its oil exports. The government's critics and some academics claim that these moneys should have gone toward improving Venezuela's infrastructure and point toward unfinished projects such as the plan to build sixteen high-speed railway lines, most of which have not been started.

While Maduro professes commitment to maintaining these programs, Venezuela's financial needs, increased domestic fuel requirements, and an inability to significantly grow oil production have led to cutbacks in shipments and less favorable terms and/or less oil for 2014 contracts. Deliveries to Cuba, which had been consistently monitored at over 100,000 bbl/d, were observed at 70,000–80,000 bbl/d in 2013.[57] Participating countries in Central America are expected to be able to secure affordable supplies of energy from alternate sources relatively easily, should that be needed, but in July 2014 the US quietly announced the Caribbean Energy Security Initiative to help develop a diverse energy matrix in that region and to preclude the destabilizing effects of any cutbacks in Petrocaribe.

MADURO: IS CHINA MY ONLY OPTION?

From the time that Chávez assumed the presidency until he died on March 5th 2013, Venezuela took in revenues of nearly one trillion petro-dollars[58] yet today the country is in a financial crisis. Under Chávez, more than $300 billion went to remote regions for health care and education and, according to the Finance Ministry's 2013 annual report, $112 billion has been transferred to FONDEN since it was established in 2005. The funds that PDVSA has been required to contribute for these and for various other purposes more than tripled from 2004 ($16.5 billion) to 2011 ($58.6 billion).[59] According to some reports the amount has declined in recent years, but the financial relationship between PDVSA and the government is complex and it has been changing as the country's financial situation has worsened. By 2007, all but $700 million of the $6 billion from oil revenues that had been reserved for years when oil prices were low had been spent.[60] Also, during Chávez's presidency the number of government employees swelled to 2.4 million from 1.3 million: an expansion that Chávez announced during his 2012 campaign that he intended to continue.[61] But maintaining price controls and a fixed exchange rate may have taken a greater toll on Venezuela than overspending has.

The fixed price for gasoline was raised to the equivalent of $.05–$.06 per gallon in 1996 and was kept at that level throughout the Chávez regime even though the market price for oil had more than quadrupled. Most Venezuelans realize that gas prices are unsustainably low, but view them as a guaranteed way that the country's oil wealth will reach them. Subsidized food prices benefited the poor in the short term, however most of the efforts to improve agriculture and the food supply undertaken during the Chávez regime appear to have backfired or fizzled.

Economists and academics point out that by maintaining an overvalued currency the Venezuelan government has exacerbated a situation known as Dutch disease, in which the increased revenue that accrues to a country from exporting its natural resources drives the real value of its currency up, making exports from other sectors more expensive in relative terms. This situation tends to favor imports and to discourage the diversification of the exporting country's industry. Venezuela is, in fact, heavily dependent on imported goods but as of September 2014 the government was in arrears more than $2.4 billion for imported food items and owed over $3.5 billion to pharmaceutical importers.[62] After paying on some of its bonds, the country's cash reserves declined to $1.3 billion at the end of October 2014[63] and international reserves dropped to $19.64 billion on November 7, 2014, their lowest level since 2003.[64] Venezuela's fiscal deficits for 2013 were estimated to equal 11.5% of GDP.[65] Inflation was 56.3% in 2013[66] and the official figure for inflation over the prior twelve-month period was 63.6% as of November 2014.[67] Venezuela's central bank stopped reporting its scarcity index late in 2013 and the consumer-price index for April and May 2014 was not published until June.

When Maduro first assumed the presidency, under Venezuela's exchange rate controls the official exchange rate was 10 bolívares to the dollar and 6.3 bolívares per dollar for medicines and food. In February 2014 the government announced Sicad II, which exchanges dollars at a rate of about 50 bolívares. The availability of dollars has become more limited and the demand for dollars remains high, causing black market rates to rise above 100 bolívares to the dollar in September 2014 and to exceed 175 in early January 2015, according to web sites that track it. Hampered by the inability to withdraw money from the country, many foreign companies have restricted their offering of products and services in Venezuela.

There is one investor that appears especially interested in taking advantage of Venezuela's situation, however: China. Venezuela has received $50.6 billion in loans from the China Development Bank (CDB) from 2007–2014, which was 52.4% of China's lending in the region.[68] Moneys have been put into joint ventures with PDVSA and into the Venezuelan government's China Fund, with much of the repayment to be made in oil shipments. Of the $36 billion loaned by CDB to the [Venezuelan government's] Banco de Desarrollo Económico y Social de Venezuela (BANDES) through 2013, $15.4 billion remained outstanding as of December 2013.[69] In response to the fall in oil prices that began in June 2014, in October 2014 China removed the requirement that a minimum of 330,000 bbl/d be shipped as loan repayment and in November 2014, $4 billion of the credit from China that was originally to be allocated to infrastructure projects was used to supplement Venezuela's foreign reserve account.[70] In January 2015, Maduro returned to China to ask for more funds (which were granted, but allocated to various projects) and he visited OPEC members to lobby for supply constraints.

Some moneys from China have been invested directly into ventures in the Orinoco, to develop new oil fields and to gain further production from old ones. China's participation in Venezuela's oil industry is not new: in 1997 (during the apertura petrolera) the Chinese National Petroleum Company (CNPC) received a contract to manage two mature oil fields and in 2004 a third field was added. The increased production from those fields was directed to China.[71] In 2001, PDVSA and CNPC formed Sinovensa, a joint venture that produced heating fuel. In 2008, CNPC paid $900 million to help develop the Junin 4 field in the Orinoco oil belt and committed to contribute 40% of the money needed (estimated at $16.4 billion), however PDVSA could not fund its share and progress was slow.[72] During 2013, Sinovensa received an infusion of $4 billion, while CNPC and Sinopec signed contracts to develop other Orinoco fields.[73]

While China imported 19,000 bbl/d of oil from Venezuela in 2005, by 2011 the amount had risen to 230,000 bbl/d[74] and on November 27, 2013, Rafael Ramírez (at the time, president of PDVSA and Venezuela's minister of energy and petroleum and vice president of the economy) told reporters that China was receiving 640,000 bbl/d, of which 310,000 bbl/d went toward servicing the debt.[75] PDVSA and CNPC have agreed to build a refinery with a capacity of 400,000 bbl/d in Guangdong and Petrochina is picking up the tab for eight tankers that will go to PDV Marine. PDVSA will pay for its portion of both through oil deliveries.[76]

Such arrangements are part of an increasingly close relationship between the countries. Relations between China and Venezuela were formally established in 1974. In 2000, an agreement for economic and technical cooperation was signed, followed by a structured "strategic partnership" in 2001. One condition of the $20 billion that Venezuela received from China in 2010 was that Venezuela had to spend at least 75% of the money on "cooperation projects."[77] There have been roughly 200 agreements reached between China and Venezuela over the past decade, with more than 400 projects in 15 different sectors.[78] In July 2014, President Xi Jinping made his first visit to Venezuela and formalized 38 agreements to allocate China's cash infusion of $5 million. Details for 22 of the projects were not made public, although they include refurbishing, expanding and building factories; work on an airport and a harbor; and building more than 4,000 apartments.[79,80] Recent announcements included the construction of a factory for household appliances and one to further the countries' cooperation in the automotive industry.[81] Military and technological assistance included Venezuela's first satellite launch, which took place in 2008 due to China, and Venezuelan scientists will now be participating in the design and test of a satellite project for the first time.[82] The relationship may have been established initially as part of Chávez's move away from the US and its ideology, but Chinese investment is essential to developing the potential of the Orinoco, China's cash has helped to sustain Venezuela's social programs, and China may be Venezuela's lender of last resort as other avenues of borrowing become increasingly prohibitive or unavailable.

Russian oil companies entered the Orinoco after the companies that were active during the apertura petrolera had been expelled. In 2010, PDVSA joined with a consortium of five Russian oil firms and the first extraction of that oil took place in October 2012. Gazprom, the largest of the Russian energy firms, has production contracts for oil fields and, through its bank, has committed to lend $1 billion to PDVSA to help it ramp up production.[83] Rosneft entered into a joint venture with PDVSA in 2013 in which it receives a 40% share in a 132-square-mile section of the Orinoco belt with reserves estimated at 40 billion barrels and a production target of 400,000 bbl/d. Rosneft will pay PDVSA $1.1 billion and will provide a loan of up to $1.5 billion, with the provision that PDVSA may borrow up to $300 million annually.[84]

India has become more active in seeking out sources of energy and has signed contracts to purchase Venezuelan oil. In 2008, ONGC Videsh became part of a venture in the Orinoco and was joined by India Oil in 2010 and further agreements were reached in 2013. Other partners have provided PDVSA with an estimated $10 billion in loans (this figure includes unpaid accounts receivable that have been converted to debt). Of this, $1.2 billion is owed to Repsol, $1.5 billion to Schlumberger and $2 billion to Chevron.[85]

The amount of oil that Venezuela sends to China and to India (estimates for 2013 exports to India were 400,000 bbl/d) has been growing rapidly, but in 2013 the United States still received the largest share of Venezuela's oil exports (755,000 barrels).[86] In 2013, Venezuela was the fourth-largest supplier of imported crude oil and petroleum products to the United States and this trade has kept the countries connected despite their shaky political relationship.[87] The United States, which exported 7,000 bbl/d of petroleum products (primarily those used for blending into heavy oil) to Venezuela a decade ago, shipped 84,000 bbl/d in 2013, as PDVSA's ability to produce lighter grades of oil has declined.[88]

Within and outside of the country, critics have been calling on Maduro to reverse the declines at PDVSA and in Venezuela's economy and, starting in the spring of 2014, he had hinted that a shake-up of government officials was coming. On September 2, 2014, Maduro removed Rafael Ramírez from his long-standing positions as oil minister and president of PDVSA and from his roles as vice president of the economy and as Venezuela's finance minister. Moving aside Ramírez, who supported reducing the gasoline subsidies and modifying the foreign exchange regime, was seen as a victory by the hard-left elements within the government and as a sign that Maduro was not willing to make the changes that are needed to restore Venezuela's economy.

THE FUTURE OF PDVSA

PDVSA's technical expertise was depleted by the removal of PDVSA's experienced management team and by the wholesale firing of workers after the strike. Canada, which shares Venezuela's dependence on producing petroleum products from heavy crude, has welcomed many former PDVSA employees. Of the Venezuelans that found jobs at the University of Calgary in Alberta, about two-thirds came due to the mass sacking of 2003 and all of them went on to jobs in Canada's oil industry.[89] The workers who remain at PDVSA receive compensation that is among the lowest in the industry. Oil industry professionals receive an average salary of $50,000 in Venezuela and $100,000 in Colombia and the low wages may be why many young professionals are working at PDVSA for five or six years and then leaving.[90]

During the apertura petrolera, major international oil companies put $16 billion into creating a network of refineries and other facilities.[91] After 2001, most of them left Venezuela or were forced out, receiving only book value for their assets and properties. When LUKoil (a Russian oil company) withdrew from a multi-billion dollar project in the Orinoco in 2013, its vice president stated that production from the Orinoco could be rapidly increased if Venezuela offered stability in its contracts and a good investment environment.[92] Critics charge that without such participation PDVSA has been investing only 50% of what it needs to maintain its operations and about 25% of what it needs to increase production.[93] In 2012, PDVSA borrowed a total of $7.5 billion from CNPC, Chevron, and Schlumberger for such expenditures.[94]

Despite the depletion of funds, in 2011 crude oil production returned to 2009 levels of 100,000 bbl/d, but overall production was approximately 75% of what it was in 2001[95] and, according to the figures that PDVSA provides to OPEC, in 1998 the company produced about 3.6 million bbl/d while for 2013 the average was 2.78 million bbl/d.[96] Prior to its "renationalization," PDVSA's strategic plan had called for 4.6 million bbl/d to be produced from the Orinoco fields by 2019, however, performance at the end of 2013 was 1.2 million bbl/d against the year's goal of 1.5 million bbl/d.[97]

In 2011, the company had a production cost of $7.53 per barrel of oil, but in 2012 costs soared to $11.09 per barrel.[98] Revenues from crude oil and petroleum byproducts for 2013 dropped 8% from the previous year to $113.97 billion while net profits for the year rose to $15.83 billion, which was attributed to the sale of part of PDVSA's stock portfolio to Venezuela's central bank for $12 million and a $4 billion reduction in PDVSA's transfers to the government for its social programs.[99]

PDVSA's funds are also limited by Venezuela's domestic gasoline subsidy, which keeps growing as the heavily discounted price fuels ever-greater consumption. As of December 2013, Venezuela's cost to break even on a gallon of premium gasoline was $1.62 while the average market price for the year was $2.70 per gallon and although oil prices have declined a price spread remains. As part of its production processes, Venezuela imports tens of thousands of gallons of petroleum products, e.g., methyl tertiary butyl ether (MTBE), for blending into its own oil to produce gasoline and other end products. The amount has been increasing due to demand and to refinery problems[100] and Venezuela made headlines in October 2014 when, for the first time in history, it imported light crude (from Algeria and Russia). Venezuela's oil minister has stated that the gasoline subsidy costs the country $12 billion annually (at the high prices of early 2014).[101] If opportunity costs are included, analysts' estimates are as high as $50 billion.[102]

PDVSA receives no money for the domestic subsidies, for the oil that goes to repay the loans from China, and for barter programs such as Petrocaribe, which total an estimated 700,000 bbl/d.[103] PDVSA has also had to pay for massive increases to its payroll. At the time of the strike, PDVSA had about 39,300 employees (including contractors, non-oil subsidiaries, etc.), while PDVSA's 2012 annual report showed a total headcount of 145,439.[104] In 2012, PDVSA transferred $60 billion—roughly 48% of its total reported revenues—to the central government and other groups.[105] In 2011, such transfers left PDVSA with -$6.1 billion in free operating cash flow.[106]

In July 2014, an examination of financial statements showed a network of funding between PDVSA and government agencies. Since 2007, by issuing promissory notes and certificates and through borrowing, resources have been pumped into PDVSA from Venezuela's treasury; several banks, including the Central Bank; and at least one government fund.[107] With those resources being quickly drained out of PDVSA to support non-energy sectors the company's future is at risk, but to date it does not appear that Maduro's administration is willing to take the steps needed to restore foreign direct investment in Venezuela.

THE ROAD AHEAD

With the price of oil dropping 50% during the last six months of 2014—the price for Venezuela's oil basket was $46.97/barrel on December 30—what was already a serious fiscal situation for Venezuela has worsened. For every drop of $10.00 in the price of oil, the country loses an estimated $7.5 billion in annual revenues[108] and on November 26, 2014, Fitch Ratings named Venezuela among the countries most vulnerable to a downgrade if oil prices remain low. "*Venezuela appears vulnerable on all counts—it has a relatively high fiscal breakeven point (of around $110 per barrel of crude), runs a fiscal deficit equal to close to 5% of GDP, relies on oil revenues for around 40% of its fiscal revenues and depends on oil exports for close to 95% of its current account receipts*."[109] Both PDVSA and Venezuela (the bonds are linked) have experienced several downgrades and have speculative-grade ratings, with levels comparable to those during the strike of 2002–2003 and near those of the Ukraine in its present state of war. Venezuela met its obligations of $4.5 billion for October 2014, but bond prices still soared because by the end of 2016 the government and PDVSA have $21 billion in bonds maturing—an amount equal to the country's foreign reserves.[110]

The recent rapprochement between the United States and Cuba may be an acknowledgement of Venezuela's precarious situation. Cuba experienced significant economic hardship after the Soviet Union withdrew its support and an increased flow of tourists and of remittances from the US may offer quick sources of funds to help Cuba ameliorate the reduction in its supply of subsidized Venezuelan oil. The US has also indicated its willingness to reopen discussions about the territorial boundaries in the Gulf of Mexico, which are shared by Cuba, the US and Mexico but have not been delineated, potentially offering Cuba more oil of its own in the future.[111]

Oil companies from countries that follow an ideology similar to Chávez's have been favored as partners for PDVSA in the past fifteen years, however their ability to affect PDVSA's production is bound by current laws. Many of PDVSA's older oil fields require specialized expertise and resources to exploit them and the dearth of revenues for PDVSA to reinvest in its facilities has caused operational decline and damage. PDVSA has been further burdened by the massive numbers of employees that it has working outside of its core business, leading it to be critiqued as "*inefficient, inflexible and a poorly performing employment agency*."[112] The expertise (advanced technological skills and management) that once made Venezuela's oil company among the best in the world has been lost to other countries must be redeveloped over time and in partnership with technically competent firms or wooed back, which may be difficult to do given the current salaries at PDVSA and the socio-political environment in Venezuela.

Revenues from PDVSA remain, overwhelmingly, the source of government revenues. The Chávez and Maduro administrations have used the moneys for domestic subsidies and for social and international programs. The government has made great efforts to pay its foreign debt, but inside the country the shortages of basic items, power blackouts, hyperinflation, and problems with violent crime that precipitated the demonstrations in 2014 have only worsened. The bolívar has lost 90% of its value on the black market because Venezuela's central bank has quadrupled the money supply over the last two years, which has also led to record inflation.

The most recent developments in Venezuela reveal a growing dissatisfaction with the governance of President Maduro, which has produced a more distorted foreign exchange regime; much lower oil revenues; goods shortages; payments to suppliers seriously in arrears; and bond pricing at lower valuations. Maduro has quashed protests with a heavy hand and his approval rating has continued to drop since he was elected. By year-end 2014 it had fallen nearly 29 points to 22%, raising the chances of opposition victories in the 2015 National Assembly elections. Government expenditures have typically increased ahead of elections and, with its resources so limited, there are concerns that Maduro might repeat the pre-election scenario of December 2013, when he attributed the inflated prices of goods to gouging by retailers and imposed such significant price controls on them that soldiers were dispatched to the stores to manage the crowds. Still, if the 2015 elections do not produce an immediate change in government they could still significantly change the rules of Venezuela's politics and force the executive branch to accept more restrictions on its power.

The current situation gives greater urgency to the need for the government to announce measures to address some of the country's distortions and imbalances because a delayed response is more likely to have a limited effect. Maduro had announced that he would unveil a new exchange rate scheme in January 2015, but the month passed and all that he revealed was that the exchange rate of 6.3 bolívares to the dollar for food and medicine would be held constant and that a system with multiple exchange rates would be retained. Maduro has announced that there would be deep tax reforms, but he has also indicated that he intends to continue social spending and he has decreed a freeze in firing for 2015 and a 15% increase in the minimum wage. Although the fuel subsidy is an entitlement that politicians have feared to touch since the Caracazo of 1989, Maduro has proposed to begin a dialogue to "*establish a system of fair prices for fuels in the domestic market,*" but a price increase is unlikely before the next elections.[113]

Other Latin American countries have drawn on IMF resources to avoid an economic meltdown, however the conditions for such loans are usually based on the implementation of austerity measures that are counter to the stated principles of Socialism of the Twenty-first Century. Thus, Maduro has turned once again to China and on January 7, 2015 he announced that he had secured a promise of $20 billion for "*projects of the economic, energy and social kind,*" but any effect that the moneys might have on Venezuela's debt remains unclear.[114] Venezuela retains the draw of its oil resources, although to attract further foreign investments Maduro must take steps to repair—or at least stabilize—the economy and to provide a less hostile and more predictable environment. Without abandoning a left-of-center agenda, he could propose a plan to reduce the fiscal deficit by such measures as providing a single, transparent, and credible exchange rate system; returning some independence to the central bank; backing away from the present expansive monetary policy; and tackling inflation. A commitment to stop nationalizations and to respect private property would give potential investors some confidence that there are clear terms of engagement and support of private enterprise. Controlling the high crime rate and opening a dialogue with the opposition would also help.

Ironically, the way for Venezuela to solve its present crisis and to best take advantage of its great natural resources is, in my opinion, to return to the implementation of the neo-liberal policies of the last decade of the Fourth Republic while incorporating the social safety nets and social programs of the past fifteen years that have proven to be effective. Whatever Maduro chooses to implement, to restore confidence in Venezuela he must take effective steps toward repairing the economy and reestablishing PDVSA's productive capabilities. If he refuses to make such adjustments, or continues to make them primarily at a cost to the private sector, the future for Venezuela's "golden goose" and for the country's economy is bleak.

NOTES

1. Greg Palast, "Hugo Chávez Interview," *Progressive*, August 20, 2006, http://www.progressive.org/mag_intv0706. The interview with Chávez aired on the initial program of BBC Two *Newsnight* program's "Latin American Season," April 3, 2006.

2. Juan Forero, "Chávez Restyles Venezuela With '21st-Century Socialism,'" *New York Times*, October 30, 2005, http://www.nytimes.com/2005/10/30/international/americas/30venezuela.html?_r=0.

3. United States Central Intelligence Agency (CIA), "South America: Venezuela," in *The World Factbook*, last modified June 30, 2014, https://www.cia.gov/library/publications/the-world-factbook/geos/ve.html.

4. Alejandro Grisanti, "Venezuela's Oil Tale," *Americas Quarterly* (Spring 2011), http://www.americasquarterly.org/node/2436.

5. International Monetary Fund (IMF), "World Economic and Financial Surveys: World Economic Outlook Database," Data and Statistics, April 2014, http://www.imf.org/external/pubs/ft/weo/2014/01/weodata/index.aspx. Author calculations from IMF GDP constant price: 2011 data for Venezuela is estimated.

6. Alejandro Grisanti, "Venezuela's Oil Tale," *Americas Quarterly* (Spring 2011), http://www.americasquarterly.org/node/2436.

7. "Venezuelan Public Debt Estimated at USD 219 Billion in 2013," *El Universal*, August 13, 2013, http://www.eluniversal.com/economia/130819/venezuelan-public-debt-estimated-at-usd-219-billion-in-2013.

8. OPEC, *Annual Statistical Bulletin 2013* (Vienna: OPEC, 2013), http://www.opec.org/opec_web/en/publications/202.htm.

9. For a more detailed study of this topic, see Gonzalo E. Vázquez, "An Evaluation of Brain Drain in the Case of Venezuela's Petroleum Company, Petróleos De Venezuela, S.A. (PDVSA)" (Master's thesis, University of Miami, 2012), http://scholarlyrepository.miami.edu/cgi/viewcontent.cgi?article=1353&context=oa_theses.

10. Rafael Villa Duarte, "Venezuela: Political Changes in the Chávez Era," *Estudos Avançados* 19, no. 55 (2005): 154, doi: 10.1590/S0103-40142005000300011.

11. Ibid, 153–4.

12. Historians mark the start of the Fourth Republic as the time that Gran Colombia separated into the Provinces of Venezuela and the Republic of Colombia in 1831, but today the term is commonly used to refer to the forty-year period of Puntofijismo.

13. Federal Deposit and Insurance Corporation (FDIC), Division of Research and Statistics, "The LDC Crisis," in *History of the Eighties—Lessons for the Future*, vol. 1, *An Examination of the Banking Crises of the 1980s and Early 1990s* (Washington, D.C.: FDIC, 1997), 193, http://www.fdic.gov/bank/historical/history/191_210.pdf.

14. Roger W. Ferguson, "Latin America: Lessons Learned from the Last Twenty Years" (speech, Florida International Bankers Association, Inc., Miami, FL, February 11, 1999), http://www.federalreserve.gov/boarddocs/speeches/1999/19990211.htm.

15. Moisés Naím, *Paper Tigers and Minotaurs: The Politics of Venezuela's Economic Reforms* (Washington, D.C.: Carnegie Endowment for International Peace, 1993), distributed by the Brookings Institute, Washington, D.C. Moisés Naím served as Venezuela's Minister for Development and as a director of the Central Bank

16. Ibid.

17. *Economist*, "Venezuela: Fever," March 20, 1997, International [print] edition, http://www.economist.com/node/145829.

18. Julia Buxton, "The Bolivarian Revolution as Venezuela's Post-Crisis Alternative" (paper presented at the 2nd Global International Studies Conference, Ljubljana, Slovenia, 2008), draft version downloaded July 9, 2014 from www.wiscnetwork.org/ljubljana2008/papers/WISC_2008-401.doc4.

19. John V. Lombardi, "Venezuela's Permanent Dilemma," in *Venezuelan Politics in the Chávez Era: Class, Polarization, and Conflict*, eds. Steve Ellner and Daniel Hellinger (Boulder: Lynne Rienner Publishers, 2004), 5.

20. Roger Santodomingo, *La Conspiración del 98: Un Pacto Secreto para Llevar a Hugo Chávez al Poder* (Caracas: Alfadil, 1999), 16.

21. Villa Duarte, "Venezuela: Political Changes," 160.

22. Timothy P. Wickham-Crowley, *Guerrillas and Revolution in Latin America: A Comparative Study of Insurgents and Regimes Since 1956*, (Princeton University Press, 1992), 197.

23. Ibid, 199.

24. Margarita López Maya, "Hugo Chávez Frías: His Movement and His Presidency," in Ellner and Hellinger, *Venezuelan Politics*, 81.

25. Daniel Hellinger, "Political Overview: The Breakdown of Puntofijismo and the Rise of Chavismo," in Ellner and Hellinger, *Venezuelan Politics*, 33.

26. Ricardo Hausmann and Francisco Rodríguez, eds., *Venezuela Before Chávez: Anatomy of an Economic Collapse* (University Park, PA: Penn State Press, 2013), 1.

27. Rafael Di Tella, Javier Donna and Robert MacCulloch, "Oil, Macro Volatility and Crime in the Determination of Beliefs in Venezuela," in Hausmann and Rodríguez, *Venezuela Before Chávez*, cxii.

28. Allan R. Brewer-Carías, *Dismantling Democracy in Venezuela: The Chávez Authoritarian Experiment*, Cambridge University Press, 2010, 37.

29. Angela Zago, private discussion with the author, Miami, FL, June 2012. Angela Zago was an elected member of the 1999 Constituent Assembly.

30. Allan R. Brewer-Carías, "The 1999 Venezuelan Constitution-Making Process as an Instrument for Framing the Development of an Authoritarian Political Regime," in *Framing The State In Times Of Transition: Case Studies In Constitution Making*, ed. Laurel E. Miller, with Louis Aucoin, (Washington, D.C.: U.S. Institute of Peace Press, 2010), 511, http://www.usip.org/sites/default/files/Framing%20the%20State/Chapter19_Framing.pdf. Allan Brewer-Carías was an elected member of the 1999 Constituent Assembly.

31. Brewer-Carías, *Dismantling Democracy*, 19.

32. Brewer-Carías, "The 1999 Venezuelan Constitution-Making Process," 513.

33. Ibid, 516.

34. Gustavo Coronel, "The Venezuelan Elections: How Can PDVSA Recover?" *Journal of Energy Security* (July 2012), Institute for the Analysis of Energy Security, http://www.ensec.org/index.php?option=com_content&id=371:venezuela-chavez-and-PDVSAs-oil-woes&catid=128:issue-content&Itemid=402.

35. David G. Victor, David R. Hults and Mark C. Thurber, eds., *Oil and Governance: State-Owned Enterprises and the World Energy Supply* (Cambridge University Press, 2011), 491.

36. Juan Forero, "Venezuela's New Oil Law Is Seen as a Risk to Growth," *New York Times*, December 4, 2001, http://www.nytimes.com/2001/12/04/world/venezuela-s-new-oil-law-is-seen-as-a-risk-to-growth.html.

37. Coronel, "The Venezuelan Elections."

38. Gonzalo E. Vázquez, "Brain Drain from Venezuela's Petroleum Giant, PDVSA," *IMTP-Magazine on Migration Issues* (Spring 2013), http://www.trendsandperspectives.info/pdf/09.03/Gonzalo%20Vázquez%20Spring%202013.pdf.

39. Larry Spancake, "Venezuela Offers Full Market Value to Encourage Foreign Investment in Oil," U.S. Department of Energy, Energy Information Administration (EIA), March 1999, last updated April 16, 2001, http://archive.today/ftPc#selection-23.0-23.73.

40. Luis E. Giusti, "La Verdadera Situación de PDVSA," *El Espectador*, September 20, 2009, Economía, http://www.elespectador.com/impreso/negocios/articuloimpreso162486-verdadera-situacion-de-PDVSA. Luis Giusti is a former president of PDVSA.

41. Peter Wilson, "The Ugly Truth About Venezuela," *Latin Trade*, October 11, 2013, http://latintrade.com/the-ugly-truth-about-pdvsa/.

42. Brewer-Carías, "1999 Venezuelan Constitution-Making Process," 523.

43. Simon Romero, "Chávez Seeks Decree Powers," *New York Times*, December 14, 2010, Americas, http://www.nytimes.com/2010/12/15/world/americas/15venez.html?_r=0.

44. Michael Penfold-Becerra, *Clientelism and Social Funds: Empirical Evidence from Chávez's "Misiones" Programs In Venezuela* (Caracas: Inst. de Estudios Superiores de Administración [IESA], 2006), 14, http://siteresources.worldbank.org/INTDECINEQ/Resources/1149208-1147789289867/IIIWB_Conference_Clientelism_and_Social_FundsREVISED.pdf.

45. CIA World Factbook, "Venezuela," 2014. Population estimate as of July 2014.

46. OPEC, *Annual Statistical Bulletin 2013*.

47. British Petroleum (BP), *Statistical Review of World Energy*, 6, http://www.bp.com/content/dam/bp/pdf/Energy-economics/statistical-review-2014/BP-statistical-review-of-world-energy-2014-full-report.pdf.

48. Julia Buxton, "The Bolivarian Revolution."

49. Ewan Robertson, "Venezuela: An Ethical Foreign Policy?" Venezuelanalysis.com, July 14, 2013, http://venezuelanalysis.com/analysis/9852.

50. Ibid.

51. Julie M. Feinsilver, "Oil-for-Doctors: Cuban Medical Diplomacy Gets a Little Help From a Venezuelan Friend," *Nueva Sociedad* 216 (2008), http://www.nuso.org/upload/articulos/3537_2.pdf.

52. Bruce M. Bagley and Magdalena Defort, eds., "*¿La hegemonía norteamericana en declive? El desafío del ALBA y la nueva integración latinoamericana del Siglo XXI*," Facultad de Derecho y Ciencias Sociales (Cali, Colombia: Universidad Icesi, 2014).

53. Francine Jácome, *Petrocaribe: The Current Phase of Venezuela's Oil Diplomacy in the Caribbean*, trans. Juan Diego Prieto, Friedrich Ebert Stiftung: Programa de Cooperacion en Seguridad Regional, Policy Paper 40 (November 2011), 3, http://library.fes.de/pdf-files/bueros/la-seguridad/08723.pdf.

54. Ezra Fieser, "Venezuela's Regional Energy Program Petrocaribe Wobbles," *Christian Science Monitor*, November 15, 2013, http://www.csmonitor.com/World/Americas/2013/1115/Venezuela-s-regional-energy-program-Petrocaribe-wobbles.

55. Feinsilver, "Oil-for-Doctors." The referenced agreement with Cuba was made under ALBA.

56. Joshua Goodman, "Venezuelan Oil Diplomacy Curbed by Economic Crisis," Associated Press, December 2, 2013, Latin America, http://fuelfix.com/blog/2013/12/02/venezuelan-oil-diplomacy-curbed-by-economic-crisis/.

57. *Trinidad and Tobago Guardian Online*, "Venezuela Cuts Back on PetroCaribe," March 11, 2014, http://www.guardian.co.tt/business/2014-03-11/venezuela-cuts-back-petrocaribe.

58. Fox News Latino, "Chavez Reelection in Jeopardy Because of Spending and Crime in Venezuela," posted September 20, 2012, http://latino.foxnews.com/latino/news/2012/09/20/chavez-reelection-in-jeopardy-because-spending-and-crime-in-venezuela/.

59. Ibid.

60. Tina Rosenberg, "Oil Meets Socialism in Venezuela," *New York Times Magazine*, November 3, 2007, http://www.nytimes.com/2007/11/02/business/worldbusiness/02iht-wbpetro.1.8159725.html?pagewanted=all&_r=0.

61. Fox News Latino, "Chavez Reelection in Jeopardy."

62. Ricardo Hausmann and Miguel Angel Santos, "Should Venezuela Default?" Project Syndicate, September 5, 2014, http://www.project-syndicate.org/commentary/ricardo-hausmann-and-miguel-angel-santos-pillory-the-maduro-government-for-defaulting-on-30-million-citizens--but-not-on-wall-street.

63. Andreína Trujillo, trans., "Venezuela's International Reserves at USD 19.64 Billion," *El Universal*, November 12, 2014, http://www.eluniversal.com/economia/141112/venezuelas-international-reserves-at-usd-1964-billion.

64. Ibid.

65. Jhean Cabrera, trans., "Venezuela's Fiscal Deficit in 2013 Estimated at 11.5% of GDP," *El Universal*, February 22, 2014, http://www.eluniversal.com/economia/140222/venezuelas-fiscal-deficit-in-2013-estimated-at-115-of-gdp. Estimate by Ecoanalítica for year-end 2013.

66. World Bank, "Gross Fixed Capital Formation (% of GDP)," http://data.worldbank.org/indicator/NE.GDI.FTOT.ZS.

67. Andrew Cawthorne and Diego Ore, "Update 1—Venezuela Confirms Recession, Inflation Hits 63.6 Percent in Nov.," Reuters, December 30, 2014, U.S. Edition. http://www.reuters.com/article/2014/12/30/venezuela-economy-idUSL1N0UE1FY20141230.

68. Amos Irwin and Kevin Gallagher, "Chinese Finance to Latin America Tops $100 Billion Since 2005," *Inter-American Dialog*, April 2, 2014, http://www.thedialogue.org/page.cfm?pageID=32&pubID=3563

69. R. Evan Ellis, "China, Russia, India, and the Venezuelan Petroleum Industry," *Latin Business Chronicle*, December 11, 2013, http://www.jsg.utexas.edu/lacp/files/China-Russia-India-and-the-Venezuelan-Petroleum-Industry.pdf.

70. Kejal Vyas, "China Loosens Debt Terms for Venezuela," *Wall Street Journal*, November 24, 2014, World, http://www.wsj.com/articles/china-loosens-debt-terms-for-venezuela-1416858616.

71. Ellis, "China, Russia, India, and the Venezuelan Petroleum Industry."

72. Ibid.

73. Ibid.

74. United States Energy Information Administration (EIA), Independent Statistics and Analysis, "Countries: Venezuela," last modified June 30, 2014, http://www.eia.gov/countries/cab.cfm?fips=ve. The 2005 number for exports to China has been reported at up to 59,000 bbl/d. The amount of oil that China imports from Venezuela has consistently been less than the amount that PDVSA exports to China—in some years, as much as 50–70% less—and it appears that China's national oil companies

have resold the balance in local and international markets. Oil production figures also vary because some sources count the heavy oil from the Orinoco directly and others apply the 10% reduction needed to upgrade that oil to syncrude.

75. Boris Korby and Corina Pons, "Venezuela Siphoning U.S. Oil Exports to China Sinks Bonds," Bloomberg, February 4, 2014, http://www.bloomberg.com/news/2014-02-04/venezuela-selloff-worsens-as-u-s-oil-exports-sink-andes-credit.html.

76. Ellis, "China, Russia, India, and the Venezuelan Petroleum Industry."

77. Charlie Devereux, "China Bankrolling Chavez's Re-election Bid with Oil Loans," Bloomberg (Sustainability), September 26, 2012, http://www.bloomberg.com/news/2012-09-25/china-bankrolling-chavez-s-re-election-bid-with-oil-loans.html.

78. Wu Jiao and Zhang Yunbi, "Beijing, Caracas Sign Multiple Agreements," China Watch, *Washington Post*, July 22, 2014, http://chinawatch.washingtonpost.com/2014/07/post-1.php.

79. Mayela Armas H., "Venezuela Signs 22 Private Agreements with China," trans. Andreína Trujillo, *El Universal*, August 4, 2014, http://www.eluniversal.com/economia/140804/venezuela-signs-22-private-agreements-with-china.

80. Jiao and Yunbi, "Beijing, Caracas Sign Multiple Agreements."

81. *PanAm Post*, "Venezuela, China Announce 'Upgrade' to Strategic Partnership: Maduro Courts Further Chinese Investment in Crisis Economy," July 23, 2014, http://panampost.com/panam-staff/2014/07/23/venezuela-china-announce-upgrade-to-strategic-partnership/.

82. Jiao and Yunbi, "Beijing, Caracas Sign Multiple Agreements."

83. Ellis, "China, Russia, India, and the Venezuelan Petroleum Industry."

84. Jake Simpson, "Rosneft Puts Up $2.6B for Stake in Venezuelan Oil JV," Law360, May 28, 2013, http://www.law360.com/articles/445302/rosneft-puts-up-2-6b-for-stake-in-venezuelan-oil-jv.

85. Ellis, "China, Russia, India, and the Venezuelan Petroleum Industry."

86. United States Energy Information Administration (EIA), Independent Statistics and Analysis, "Data: U.S. Imports from Venezuela of Crude Oil," September 2, 2014, http://www.eia.gov/dnav/pet/hist/LeafHandler.ashx?n=PET&s=MCRIMUSVE2&f=A.

87. United States EIA, Independent Statistics and Analysis, "Countries: Venezuela."

88. Ibid.

89. "Brain Hemorrhage: Venezuela's Loss of Thousands of Oil Workers Has Been Other Countries' Gain," *Economist*, July 19, 2014, from the print edition: Americas, http://www.economist.com/news/americas/21607824-venezuelas-loss-thousands-oil-workers-has-been-other-countries-gain-brain-haemorrhage.

90. Ibid.

91. Coronel, "The Venezuelan Elections."

92. Dmitry Zhdannikov, "Venezuela Could Boost Oil Output in 5–7 Years—LUKoil," Reuters, March 8, 2013, UK Edition, http://uk.reuters.com/article/2013/03/08/venezuela-chavez-oil-lukoil-idUKL6N0C09A720130308.

93. Coronel, "The Venezuelan Elections."

94. Wilson, "The Ugly Truth."

95. EIA, "Independent Statistics and Analysis: Countries; Venezuela."

96. "Venezuela Oil Price Rises," *Latin American Herald Tribune*, April 28, 2014, http://www.laht.com/article.asp?ArticleId=2056049&CategoryId=10717.

97. *Economist*, "Brain Hemorrhage."

98. Wilson, "The Ugly Truth."

99. "PDVSA's Revenues Slump 8% in 2013," *El Universal*, June 18, 2014, News; Hydrocarbons, http://www.eluniversal.com/economia/140618/pdvsas-revenues-slump-8-in-2013.

100. William Neuman, "Venezuela May Meet New Reality, and New Price, at the Pump," *New York Times*, January 20, 2014, http://www.nytimes.com/2014/01/21/world/americas/venezuela-gasoline-prices.html?_r=0.

101. Aymara Lorenzo Ferrigni, "Inminente Aumento de Gasolina en Venezuela," *Voice of America*, August 1, 2014, http://www.voanoticias.com/content/inminente-aumento-de-gasolina-en-venezuela/1969459.html.

102. Christopher Helman, "Cheap Gasoline: Why Venezuela Is Doomed To Collapse," *Forbes,* February 20, 2014, http://www.forbes.com/sites/christopherhelman/2014/02/20/cheap-gasoline-why-venezuela-is-doomed-to-collapse/.

103. Wilson, "The Ugly Truth."

104. Ernesto J. Tovar, "PDVSA Labor Force Up 9.7% to 145,439 Employees," *El Universal*, May 8, 2013, http://www.eluniversal.com/economia/130508/pdvsa-labor-force-up-97-to-145439-employees.

105. "Fitch Rates PDVSA's Proposed US$4.5 Billion Issuance 'B+/RR4,'" MarketWatch, *Wall Street Journal*, December 6, 2013, http://www.marketwatch.com/story/fitch-rates-PDVSAs-proposed-usd45b-issuance-brr4-2013-12-06.

106. Caryn Trokie, "S&P Affirms Petroleos de Venezuela," Reuters, July 30, 2012, http://www.reuters.com/article/2012/07/30/idUSWNA225220120730.

107. Mayela Armas H., "PDVSA Gets Funding Through Six Government Agencies," *El Universal*, July 07, 2014, http://www.eluniversal.com/economia/140707/pdvsa-gets-funding-through-six-government-agencies.

108. Debbie Carlson, "Collapse of Oil Prices Leads World Economy into Trouble," *Guardian* (*London*), December 3, 2014, http://www.theguardian.com/business/2014/dec/03/oil-collapse-leads-world-economy-trouble.

109. James Kynge, "EM Oil Producers Risk Ratings Downgrade on Low Crude Price," *FT.com*, November 25, 2014, http://blogs.ft.com/beyond-brics/2014/11/25/em-oil-producers-risk-ratings-downgrade-on-low-crude-price/. Estimates for break-even are as high as $121/barrel.

110. Katia Porzecanski, "Venezuela's Got $21 Billion. And Owes $21 Billion," Bloomberg, December 12, 2014, http://www.bloomberg.com/news/2014-12-12/venezuela-s-got-21-billion-owes-21-billion-as-crisis-builds.html.

111. Mimi Whitefield, "Venezuelan Oil Might Be Behind Cuba's Pivot to the U.S.," *Miami Herald*, December 18, 2014, http://www.miamiherald.com/news/nation-world/world/americas/cuba/article4651707.html.

112. Fernando Menéndez, "After Chávez: De-socialize and Transform PDVSA," May 8, 2012, http://www.thecordobagroup.com/2012/05/08/after-chavez-de-socialize-and-transform-PDVSA/.

113. Ferrigni, "Inminente Aumento de Gasolina."

114. Juan Pablo Spinetto and Anatoly Kurmanaev, "Crude Collapse Seen Boosting China's South American Clout," Bloomberg, January 8, 2015, http://www.bloomberg.com/news/2015-01-07/venezuela-s-maduro-claims-20-billion-new-chinese-investment.html.

BIBLIOGRAPHY

Azzellini, Dario. "Constituent Power in Motion: Ten Years of Transformation in Venezuela." *Journal of Socialism and Democracy* 24 (2010): 8–31.

Bagley, Bruce M. and Magdalena Defort, eds. "*¿La hegemonía norteamericana en declive? El desafío del ALBA y la nueva integración latinoamericana del Siglo XXI*." Facultad de Derecho y Ciencias Sociales. Cali, Colombia: Universidad Icesi, 2014.

Brewer-Carías, Allan R. "The 1999 Venezuelan Constitution-Making Process as an Instrument for Framing the Development of an Authoritarian Political Regime." In *Framing The State In Times Of Transition: Case Studies In Constitution Making*, edited by Laurel E. Miller, with Louis Aucoin, 505–534. Washington, D.C.: U.S. Institute of Peace Press, 2010. http://www.usip.org/sites/default/files/Framing%20the%20State/Chapter19_Framing.pdf.

British Petroleum (BP). *Statistical Review of World Energy*. http://www.bp.com/content/dam/bp/pdf/Energy-economics/statistical-review-2014/.

BP-statistical-review-of-world-energy-2014-full-report.pdf.

Buxton, Julia. "The Bolivarian Revolution as Venezuela's Post-Crisis Alternative." Paper presented at the 2nd Global International Studies Conference, Ljubljana, Slovenia, 2008. Draft version, downloaded July 9, 2014 from www.wiscnetwork.org/ljubljana2008/papers/WISC_2008-401.doc.

Coronel, Gustavo. "The Venezuelan Elections: How Can PDVSA Recover?" *Journal of Energy Security* (July 2012). Institute for the Analysis of Energy Security. http://www.ensec.org/index.php?option=com_content&id=371:venezuela-chavez-and-PDVSAs-oil-woes&catid=128:issue-content&Itemid=402.

Corrales, Javier. "Conflicting Goals in Venezuela's Foreign Policy." In *Venezuela's Petro-diplomacy: Hugo Chávez's Foreign Policy*, edited by Ralph S. Clem and Anthony P. Maingot, 32–48. Gainesville: University Press of Florida, 2011.

Di Tella, Rafael, Javier Donna, and Robert MacCulloch. "Oil, Macro Volatility and Crime in the Determination of Beliefs in Venezuela." In Hausmann and Rodríguez, *Venezuela Before Chávez*, 407–424.

Ellis, R. Evan. "China, Russia, India, and the Venezuelan Petroleum Industry." *Latin Business Chronicle*, December 11, 2013. http://www.jsg.utexas.edu/lacp/files/China-Russia-India-and-the-Venezuelan-Petroleum-Industry.pdf.

Steve Ellner and Daniel Hellinger, eds. *Venezuelan Politics in the Chávez Era: Class, Polarization, and Conflict*. Boulder: Lynne Rienner Publishers, 2004.

Federal Deposit and Insurance Corporation (FDIC), Division of Research and Statistics. "The LDC Crisis." In *History of the Eighties—Lessons for the Future*, vol. 1, *An Examination of the Banking Crises of the 1980s and Early 1990s*,

191–210. Washington, D.C.: FDIC, 1997. http://www.fdic.gov/bank/historical/history/191_210.pdf.

Feinsilver, Julie M. "Oil-for-Doctors: Cuban Medical Diplomacy Gets a Little Help From a Venezuelan Friend." *Nueva Sociedad* 216 (2008). http://www.nuso.org/upload/articulos/3537_2.pdf.

Ferguson, Roger W. "Latin America: Lessons Learned from the Last Twenty Years." Speech given to the Florida International Bankers Association, Inc., Miami, FL, February 11, 1999. http://www.federalreserve.gov/boarddocs/speeches/1999/19990211.htm.

Grisanti, Alejandro. "Venezuela's Oil Tale." *Americas Quarterly* (Spring 2011). http://www.americasquarterly.org/node/2436.

Ricardo Hausmann and Francisco Rodríguez, eds. *Venezuela Before Chávez: Anatomy of an Economic Collapse*. University Park, PA: Penn State Press, 2013.

Hellinger, Daniel. "Political Overview: The Breakdown of Puntofijismo and the Rise of Chavismo." In Ellner and Hellinger, *Venezuelan Politics*, 27–54.

International Monetary Fund (IMF). "World Economic and Financial Surveys: World Economic Outlook Database." Data and Statistics. April 2014. http://www.imf.org/external/pubs/ft/weo/2014/01/weodata/index.aspx.

Irwin, Amos and Kevin Gallagher. "Chinese Finance to Latin America Tops $100 Billion Since 2005." *Inter-American Dialog*, April 2, 2014. http://www.thedialogue.org/page.cfm?pageID=32&pubID=3563.

Jácome, Francine. *Petrocaribe: The Current Phase of Venezuela's Oil Diplomacy in the Caribbean*. Juan Diego Prieto, trans. Friedrich Ebert Stiftung: Programa de Cooperacion en Seguridad Regional; Policy Paper 40, November 2011. http://library.fes.de/pdf-files/bueros/la-seguridad/08723.pdf.

Lander, Edgardo and Luis A. Fierro. "The Impact of Neoliberal Adjustment in Venezuela, 1989–1993." *Latin American Perspectives* 23, no. 3, (1996): 50–73. http://www.jstor.org/stable/2634107.

Lombardi, John V. "Venezuela's Permanent Dilemma." In Ellner and Hellinger, *Venezuelan Politics*, 1–6.

López Maya, Margarita. "Hugo Chávez Frías: His Movement and His Presidency." In Ellner and Hellinger, *Venezuelan Politics*, 73–92.

Naím, Moisés. *Paper Tigers and Minotaurs: The Politics of Venezuela's Economic Reforms*. Washington, D.C.: Carnegie Endowment for International Peace, 1993; distributed by the Brookings Institute, Washington, D.C.

OPEC. *Annual Statistical Bulletin 2013*. Vienna: OPEC, 2013. http://www.opec.org/opec_web/en/publications/202.htm.

Palast, Greg. "Hugo Chávez Interview." *Progressive*, August 20, 2006. http://www.progressive.org/mag_intv0706. Palast interviewed Chávez on the initial program of BBC Two *Newsnight* program's "Latin American Season," April 3, 2006.

Penfold-Becerra, Michael. *Clientelism and Social Funds: Empirical Evidence from Chávez's "Misiones" Programs In Venezuela*. Caracas: Inst. de Estudios Superiores de Administración (IESA), 2006. http://siteresources.worldbank.org/INTDECINEQ/Resources/1149208-1147789289867/IIIWB_Conference_Clientelism_and_Social_FundsREVISED.pdf.

Santodomingo, Roger. *La Conspiración del 98: Un Pacto Secreto para Llevar a Hugo Chávez al Poder*. Caracas: Alfadil, 1999.

United States Central Intelligence Agency (CIA). "South America: Venezuela." In *The World Factbook*. Last modified June 30, 2014. https://www.cia.gov/library/publications/the-world-factbook/geos/ve.html.

United States Energy Information Administration (EIA), Independent Statistics and Analysis. "Countries: Venezuela." Last modified June 30, 2014. http://www.eia.gov/countries/cab.cfm?fips=ve.

———. "Data: U.S. Imports from Venezuela of Crude Oil." September 2, 2014. http://www.eia.gov/dnav/pet/hist/LeafHandler.ashx?n=PET&s=MCRIMUSVE2&f=A.

Vázquez, Gonzalo E. "Brain Drain from Venezuela's Petroleum Giant, PDVSA."

IMTP-Magazine on Migration Issues (Spring 2013). http://www.trendsandperspectives.info/pdf/09.03/Gonzalo%20Vázquez%20Spring%202013.pdf.

———. "An Evaluation of Brain Drain in the Case of Venezuela's Petroleum Company, Petróleos De Venezuela, S.A. (PDVSA)." Master's thesis, University of Miami, 2012. http://scholarlyrepository.miami.edu/cgi/viewcontent.cgi?article=1353&context=oa_theses.

Victor, David G., David R. Hults and Mark C. Thurber, eds. *Oil and Governance: State-Owned Enterprises and the World Energy Supply*. Cambridge University Press, 2011.

Villa Duarte, Rafael. "Venezuela: Political Changes in the Chávez Era." *Estudos Avançados* 19, no. 55 (2005): 153–172, doi: 10.1590/S0103-40142005000300011.

Wickham-Crowley, Timothy P. *Guerrillas and Revolution in Latin America: A Comparative Study of Insurgents and Regimes Since 1956*. Princeton University Press, 1992.

World Bank. "Gross Fixed Capital Formation (% of GDP)." http://data.worldbank.org/indicator/NE.GDI.FTOT.ZS.

Chapter 12

Colombia and Oil

Chris Kraul

Colombia has risen over the last decade to become one of the world's 20 largest crude oil producers, its status crowned by passing the 1-million barrels per day production milestone in 2013. It's a remarkable success story whose stars are the many small wildcatters who have helped the Andean country nearly double its crude output in less than eight years. But an unforeseen decline in 2014 output, combined with the 15% slide in global crude prices has helped create a $6 billion "fiscal hole" in the Colombian government's budget. The mini-crisis has forced the government to raise taxes, provoking howls of protest from wealthy Colombians and businesses who have to shoulder most of the burden and who claim the tax system is inequitable. The experience has taught Colombia's leaders the perils of a dependence on commodity sales generally, and that oil booms can abruptly and painfully deflate.

It's a turn of events that no one would have predicted 10 or 12 years ago when the country was an oil industry pariah, its production in decline and drilling by wildcatters at a standstill because ultraviolent leftist rebels, the Revolutionary Armed Forces of Colombia (FARC), controlled the rural and jungle areas where huge untapped oil reserves were known to reside.

What changed the panorama for the better was the improvement in overall Colombian security thanks largely to $8 billion in US military and anti-terrorism aid provided under the Plan Colombia program starting in 2000. The get-tough policies of tenacious former president Alvaro Uribe who was in office from 2002 to 2010 was another important factor. The upshot was that the armed forces retook control of many remote areas, and oil companies were encouraged to return to look for oil. At the same time, the gradual rise in global oil prices made profitable the pumping of so-called heavy oil in Colombia's eastern Llanos region. Exxon and Shell previously had discovered oil there as thick as tar but could not afford to produce it due to a lack of pipeline capacity and the high cost of diluting it to make it transportable.

With expertise from Venezuelan engineers banished by Hugo Chavez who were schooled in the ways of heavy oil production in their native country's eastern Orinoco Belt, Colombian companies quickly solved the heavy oil conundrum. IN 2009, a pipeline was built to connect the Llanos to the country's main OCENSA trunk pipeline. A group of former executives at PDVSA, the Venezuelan state-owned oil producer, formed Colombia-focused Pacific Rubiales Energy, which became their adopted country's second largest producer after Ecopetrol, a former state monopoly. PRE now operates the country's largest oil field, the 200,000 b/d Rubiales field in the Llanos and its South America's largest independent producer.

From 2007 onward, Colombian oil production showed double digit percentage gains with the high point in Colombia's energy resurgence coming in January 2014 when the mining and energy ministry announced that the country averaged 1,007,000 barrels per day of crude output in 2013, nearly double the 531,000 b/d pumped in 2005.

The National Hydrocarbons Agency which was created in 2003 to independently award exploration blocks and collect government taxes and royalties, credited scores of small and daring oil companies who came to Colombia looking for a strike, a la Jett Rink, the James Dean character in the movie *Giant.*

Nowhere was the boom more welcome than at Colombia's finance ministry which by 2013 was using oil royalties and taxes to finance one fifth of its annual budget. Export sales that year of crude, natural gas and productions generated $32.5 billion in foreign currency, nearly 10 times the $3.38 billion generated in 2003. Colombia, despite its image as a coffee and

Table 12.1 PlattsEnergyEconFarcFacts. Colombia's Average Daily Oil Production

2004	5,41,956
2005	5,40,686
2006	5,48,352
2007	5,46,108
2008	6,03,674
2009	6,90,276
2010	8,05,889
2011	9,38,543
2012	9,69,054
2013	10,28,473
2014Est	9,80,000

Sources: US Energy Informaiton Adminsitration, Colombian Mining and Energy Ministry.

banana exporter, was now depending on oil for half of its total export sales: "I can't think of any downsides," said finance minister Mauricio Cardenas in a February 2014 interview with Platts Oilgram News. "The challenge is sustainability."

With the windfall expected to grow, President Juan Manuel Santos' government counted on it to finance infrastructure projects such as highways, railroads and ports that had been deferred as the government dedicated the lion's share of its budget in recent years to national defense and fighting the FARC. Santos also planned to use the oil bonanza to help fund a major initiative to improve pre-kindergarten and primary education, a category in which Colombia is among Latin America's weakest nations.

The tsunami of oil cash was also eyed as a source to finance the cost of demobilizing, educating and subsidizing thousands of FARC rebels if peace negotiations now underway in Havana are successful. In fact , for two years now the government has set aside oil cash to help finance reparations to victims of the 50-year civil conflict as well as agrarian initiatives to enable farmers displaced by violence and land-grabbers to return to their land.

But in 2014 the government experienced the flip side of an oil boom. A sudden decline in production means the country will probably produce an average of 980,000 barrels a day in 2014, according to the finance ministry, or 5% less than the 1,030,000 b/d the country was counting on this year. The revised goal is also a 2.5% decline from the previous year's record setting 1,007,000 b/d pace. In an interview in October, finance minister Cardenas said the reduced pumping combined with global softening of crude prices mean oil revenue to the government in 2014 will come in at $2 billion less than forecast, contributing to a $6 billion "fiscal hole" that the government is scrambling to fill.

The oil sector's setback has also impacted Colombia's current account or trade balance. Instead of the double digit percentage increases in oil export dollars seen over much of the previous decade, export dollar revenue from hydrocarbons sales through August were running about 3% less on a year on year basis compared with the same period in 2013, according to the DANE government statistics agency. As a result, the country anticipates a 2014 current account, or trade deficit of 4.4% of its economic output, a worrisomely high figure, Cardenas admits.

Adding to the pain is the drop in profits expected this year at Ecopetrol, the partially privatized oil company in which the government owns an 85% equity stake. Not only are revenues flat but costs are higher, which means that the oil company almost certainly will pay out smaller cash dividends next year to the Colombian government and other shareholders, another wrench in Colombia's 2015 spending plans. Expectations of a reduced dividend caused Ecopetrol's share price to drop by more than one third in 2014 to a level so low that the government was forced to kill once and for all any plan of selling part of its Ecopetrol stake to raise cash, figuring that the market is undervaluing those shares.

Meanwhile, in what is perhaps the most ominous portent of medium- and longer term production declines, foreign direct investment in Colombia's oil patch was expected to fall in 2014 on top of a 10% year on year decline in 2013 from the previous year. The investment drop comes at a time when the government has made expansion of proven and probable reserves a national priority, a goal that is hard to achieve if oil companies are not out spending money on seismic studies and drilling oil wells. Reduced exploration, if sustained, translates into reduced output. It's as simple as that, energy executives say.

The suddenly bleak energy outlook has forced the finance ministry to act quickly to try to make up for the revenue shortfall. Sticking to a target average price of $97 per barrel for all of 2014 for budget purposes, the government unveiled a tax plan that has reaped bitter recriminations from the business sector. The plan involves the raising of corporate taxes with a 3% tax surcharge on companies that report $500,000 or more in profits. Colombia is also extending two taxes that were scheduled to die or be reduced over the next year: a financial transactions charge, which was supposed to be cut in half in 2015 and a tax on wealthy individuals' assets that was supposed to die after this year. "We have always considered that we needed to keep these taxes, it was mentioned in the presidential campaigns," finance minister Mauricio Cardenas said in an interview, referring the financial transactions and wealth taxes and the President Santos' victorious campaign last Spring. "These taxes are necessary for the sustainability of public finances and without them we would have to cut investment programs the country needs in national defense and in social spending."

After near economic collapse in 1999, Colombia has clawed its way back to fiscal health and recently regained its investment grade bond rating and now enjoys high standing for the way it handles its finances. The IMF and other multinational agencies hold Colombia in high regard for its management of the economy and Cardenas, who has a UC-Berkeley Phd., wants to keep it that way.

What has happened to turn Colombia's energy panorama so bleak from the bright optimism of a year ago? The year 2014 saw a resurgence of leftist guerrilla attacks on oil installations, which observers blame on FARC rebels' desire to increase their bargaining position in Havana. Bombings on the Cano Limon pipeline over which roughly one fourth of the nation's oil is transported, coupled with an indigenous community's refusal to allow repair crews on its reservation, meant the 220,000 b/d pipeline was offline for three months this year. In addition to the FARC, the National Liberation

Army, known by its Spanish initials ELN, is also blamed for the upsurge in attacks.

In September, suspected ELN guerrillas killed two contractors for Ecopetrol who had been sent to remote North Santander province to perform maintenance on the 500-mile Cano Limon line. That pipeline, which connects Occidental Petroleum's Cano Limon oil field in eastern Arauca province to the Covenas oil port southwest of Cartagena on the Caribbean coast, has been attacked scores of times in 2014.

Also hampering production is the rising wave of community protests by farmers and rural residents protesting several issues, including alleged environmental damages, heavy traffic of tanker trucks used to haul crude over rural roads, discriminatory hiring practices and the reduction in royalties received by some communities compared with what they got in prior years. The protesters typically form blockades to obstruct vehicles entering and leaving a drilling or oil pumping operation, forcing the owners to shut the fields down.

The civil disturbances and uptick in violence have been blamed for the downturn in foreign direct investment in Colombia's oil patch so far this year when compared with 2013. FDI related to oil totaled $4.9 billion in 2013, down from $5.4 billion in 2012. Other indicators, such as exploratory wells drilled and seismic studies undertaken, also indicate a decline in foreign oil companies' commitment to Colombia.

A complicating factor for Colombia in efforts to recapture its mojo is the opening up of Mexico's oil patch after 75 years of being off limits to foreign oil companies. Late in 2013, Mexican president Enrique Pena Nieto ramrodded a constitutional change through his congress that ended the monopoly enjoyed since 1938 by Petroleos Mexicanos, or Pemex, on all oil production in that country. The first auction of exploration blocks is scheduled to be held early 2015 and companies including Pacific Rubiales have said they will be bidding. The upshot is that Colombia now competes with Mexico for oil investment dollars.

Alejandro Martinez, former president of the Colombian Petroleum Association, the nation's largest trade group of oil companies, said in an August interview that some companies are "transferring their exploration budgets" to Mexico from Colombia because the former offers better terms and "geology." A further disincentive for oil companies weighing their options in Colombia is the year-long delays that some companies have reported in receiving environmental permits for their Colombian drilling projects, a lag time that costs companies money.

The darkening panorama was blamed for the relatively poor showing at the Colombia Bid Round in July in which only 26 of 95 oil and natural gas exploration blocks being auctioned off by the nation received bids from oil companies. The National Hydrocarbons Agency had said it was hoping to sell a minimum of 38 of the blocks. A subsequent auction in August of a dozen blocks thought to contain deposits of so-called "nonconventional" reserves of crude, or those associated with coal and shale, received only one offer.

The results were especially disheartening for Colombian officials because an upsurge in exploration is what the country needs if it is to find new resources and halt the recent decline in proven crude reserves when figured as years of inventory at current production rates. The year-end 2013 total of 2.44 billion barrels equates to just 6.6 years of inventory, down from the previous year's equivalent of 6.8 years of supply.

In the October interview, finance minister Cardenas said the country is now targeting production in 2015 to reach 1,030,000 barrels per day, a sharp reduction from the 1.3 million-barrels per day goal that some industry and government officials were setting as recently as two years ago. But Cardenas was frank about the longer term outlook amid stagnant reserve levels and declining investment which he said will result in a decline in the government's oil "rent" over time, a decline one that must be replaced by a more efficient and equitable system of taxation. "As we get less revenue from oil—and we expect that to continue to happen in the future—we need more domestic sources of taxation," Cardenas said.

In what some interpreted as a symptom of the fiscal straits in which Colombia finds itself, officials in early October shocked Panama by adding it to on its list of 40 countries it considers "tax paradises." Thousands of Colombian individuals and companies own property and do business in Panama without reporting those assets and income to their home government, which deprives it of tax revenue.

Panama described the action as "unfriendly," while Colombia insisted that it was asking Panama for information that it already provides to other countries including the US, Mexico and Canada. Colombia was only acting in accordance with standards of the Organization for Economic Cooperation and Development (OECD), the club of developed nations it aspires to join. Had the designation stuck, Colombia would have begun withholding up to 33% of all payments or cash transfers sent from Colombia to Panama, a move that would have discouraged investment to put it mildly.

The dust-up ended with Colombia victorious when Panama agreed in late October to negotiate an information-sharing agreement by September 2015. Still, some observers can be excused for their impression that the timing of Colombia's action and its fiscal problems were not coincidental.

November marked the two-year anniversary of the start of peace talks between the Colombian government and the Revolutionary Armed Forces of Colombia (FARC). It is hoped the talks will bring an end to a half century of brutal and divisive conflict and the hemisphere's longest running war.

Just how much progress has been made, and how much negotiating time lies ahead is difficult to judge. Both sides play cards close to their chests. But there is cautious optimism that these negotiations, unlike several previous attempts, will eventually produce the peace deal Colombians yearn for, possibly later this year (2015).

Last November (2014), the two sides overcame a serious crisis when the FARC kidnapped Colombian army Gen. Ruben Dario Alzate as he visited an energy project in the northwest jungle region without his normal security detail. That the rebels promptly released the general was a sign that many took to mean the insurgents are committed to forging a peace deal.

The energy industry would cheer such a turn of events. Among the benefits touted by President Juan Manuel Santos as a byproduct of a deal is a peace dividend: investment in energy, manufacturing and other sectors should rise once Colombia becomes a more secure place to do business. So will Colombians' standard of living and consumption. The rebels made that point indirectly last year by stepping up oil field mayhem with pipeline bombings, kidnappings and murders, actions that have cut into crude output, oil field investment and the government "take" or portion of oil output claimed by the government in taxes and royalties.

Santos had hoped to have a deal done by the end of 2014, a notion that the FARC quickly dismissed as an effort to hurry the talks. They have a point: Santos at the start said the talks would last "months, not years," a serious miscalculation, especially given that only half of the negotiating issues had been dealt with two years into the process. Broad agreement has been announced on three of six points: agrarian reform, political participation, and drug cultivation and trafficking. The three remaining issues to be hammered out are victims, transitional justice and the mechanics of implementing an accord.

The fourth point—how both sides will treat and compensate victims—is one of the most complicated. Several victims as well as surviving family members went to Havana late last year (2014) to testify before negotiators about the horrors of the kidnapping, murder and extortion they experienced at the hands of the insurgents. The rebels aren't the only culpable ones: the Colombian armed forces have to answer for thousands of extrajudicial murders known as false positives, a practice by which military units raised their body counts by killing innocent civilians.

Recently, several FARC subcommanders have made trips to Havana, another sign that the government insists is an indication that the talks are approaching a climax. Why would rebel negotiators be talking deal points with mid-level commanders if not to clarify the plan to the rank and file, who are said to total about 10,000 fighters. Or are they simply stalling for time, as opponents to the peace talks, including former president and now senator Alvaro Uribe, insist?

President Santos was re-elected to a second term in Spring 2014 after running on a peace platform, a sign that a majority of Colombians supported his efforts. But their patience and tolerance may have limits if the talks drag on. Eventually, any peace accord probably will go before voters for approval and while Santos has said peace involves "swallowing toads," or accepting an agreement's unpleasant details in the interests of attaining an overall peace, he shouldn't underestimate the odium with which Colombians by and large regard the FARC and their brutal methods over recent decades. Unpleasant facets that voters may have to accept in a Santos peace deal is guaranteed FARC representation in Congress and minimal (or no) jail terms for rebel leaders. Those toads indeed may be tough for many to swallow.

Chapter 13

Energy Production and Armed Conflict in Colombia

Present and Future Economic and Environmental Implications

Lilian Yaffe

During the past five decades, Colombia has been immersed in an armed conflict involving multiple actors: leftist guerrillas, right-wing paramilitaries, drug traffickers, the government and its armed forces, and civil society. This armed conflict has had important implications for the energy sector: insurgent attacks on oil and gas pipelines, transmission lines and other crucial infrastructures have long plagued the country. The Caño Limón-Coveñas oil pipeline, in particular, has been extremely vulnerable to attacks by Colombia's insurgent forces because it runs mostly above ground in sparsely populated areas that have a high rebel presence. Electricity transmission towers have also proven to be easy targets; a bombing campaign has destroyed hundreds of them, causing repeated power outages in many parts of the country. Rail lines used to transport coal have been attacked as well. Besides causing millionaire economic losses, insurgent terrorist attacks on oil and gas pipelines and on electric transmission towers are also a primary cause of environmental damage to habitats in Colombia. Oil spills, primarily from the heavily bombarded Cano Limon-Covenas pipeline, have caused severe damage to some of the country's rivers and wetlands, wildlife, and farmland. The bombing campaigns over the last twenty years have resulted in the spilling of oil into watercourses, causing extreme levels of environmental destruction.[1]

Since the launching in 2012 of the Santos' government negotiation with the FARC in Norway and Havana, many analysts believe that the day will come when armed actors in Colombia abandon their frontlines to become democratic actors. The conditions for ending nearly five decades of the oldest armed conflict in the Hemisphere seem to be now more favorable than ever. What would be the consequences for Colombia's energy sector of signing a peace agreement? Will attacks to energy infrastructure end in a post-conflict scenario? Under what conditions will the country receive support from businesses and social movements for policies that promote further investment in the energy sector? And what will be the economic and social consequences of a continuous expansion of the energy sector, at the expense of other investment initiatives?

This chapter explores the relationship between armed conflict and the energy sector in Colombia. The first part presents an overall view of the Colombian energy sector, focusing on the resources most affected by the armed conflict: oil and hydrocarbons; and coal. Part two explains the different historical stages of the Colombian conflict, analyzing its consequences over the energy sector not only in terms of economic losses but also in terms of environmental costs. Part three addresses the future perspectives for the energy sector, contemplating the effects of a peace accord. A post-conflict scenario would require sustained economic growth in order to afford and stabilize peace. Will the energy sector be the main contributor to economic growth in the near future? What are the risks and implications of such strategy and the effects of an excessive reliance on exports of primary goods? What can be expected given the dramatic decrease in the international price of oil that started during the last months of 2014? The last part presents conclusions and policy recommendations. Failure to attain sustained economic growth in the post-conflict would result in the exacerbation of inequality, which could conduct to new sources of insurgency and violent contestation. As the chapter shows, the energy sector will have a key role in this scenario.

ENERGY OVERVIEW OF COLOMBIA: OIL, HYDROCARBONS AND COAL

Colombia possesses numerous fossil fuel and natural resources. The country has productive petroleum reserves, extensive coal reserves—the largest in South America—significant natural gas reserves and extensive hydroelectric resources. A large amount of potentially productive oil and natural gas areas remain unexplored. Demand for energy (petroleum, natural gas, and electricity) is expected to grow 3.5% per year through 2020.[2] In recent years, Colombia has seen a significant increase in oil, natural gas and coal production, after the implementation of a set of regulatory reforms and norms designed to increase transparency and efficiency. The enactment of a series of regulatory reforms to make the oil and natural gas sectors more competitive and attractive to foreign investors, has led to an increase in Colombian production. A partial privatization of state oil company Ecopetrol (formerly known as Empresa Colombiana de Petróleos SA) was enacted in an attempt to revive its upstream oil industry.

In 2011, the US Energy Information Administration (EIA) estimated that Colombia consumed 1.48 quadrillion British thermal units (Btus) of total energy. Oil constitutes the largest part of this amount, followed by electricity, natural gas and coal. The country relies on hydropower for the bulk of its electricity needs and uses very little coal domestically. Of the 94.6 million short tons (MMst) of coal produced, Colombia exported 94% in 2011. Natural gas consumption in Colombia has grown, rising over 54% in the past decade.[3] The most important sources of energy production are oil and hydrocarbons and coil. Their distribution and effects are discussed in the next sections.

Oil and Hydrocarbons

The infrastructure for oil production is spread throughout the country, with large oil fields and connecting pipelines. The largest producing oil field is the Rubiales heavy oil field, located in Meta department, and operated in partnership by Ecopetrol and Pacific Rubiales. Gross production at Rubiales went from 37,000 bbl/d in 2008, to 177,000 bbl/d in 2012. Other large fields include Cano-Limon, Castilla and Cupiagua. Ecopetrol, in addition to its responsibility for exploring, extracting, processing, and marketing of oil, is also responsible for its transportation through Colombia's oil pipelines.

Colombia possesses an extensive network of oil pipelines linking production areas to the main refineries and to shipping ports. There are six major pipelines, four of which connect production fields to Coveñas, where the Caribbean export terminal is located. These pipelines include the 500-mile Ocensa pipeline, which has the capacity to transport 650,000 bbl/d from the Cusiana/Cupiagua area; the 460-mile, 220,000 bbl/d capacity Cano-Limon pipeline; and the smaller Alto Magdalena and Colombia Oil pipelines. The Llanos Orientales pipeline joined the system in late 2009, connecting the Rubiales field to the Ocensa pipeline, with capacity of 340,000 bbl/d. The sixth pipeline, the TransAndino, has a capacity of 190,000 bbl/d and transports crude from Colombia's Orito field in the Putumayo basins to Tumaco, Colombia's Pacific port in the frontier with Ecuador.[4]

Important structural changes have stimulated the evolution of the oil sector in the last two decades. Before 2008, oil production had been almost constant for many years, following a period of production declines that started in 1999. The main causes of the reduction in oil production were the natural declines at existing oil fields, and a lack of discoveries of new reserves. Nonetheless, in recent years Colombia's government has implemented measures to increase the attractiveness of the investment climate to foreign oil companies. For instance, upstream sector initiatives give foreign oil companies the right to own 100% stakes in oil ventures and compete with Ecopetrol. Additionally, the government has sold shares of Ecopetrol to private investors, reducing its share to approximately 90%. These initiatives brought interested companies to participate in Colombia's oil sector, with record levels of exploration and drilling. According to the Banco de la República (Colombia's central bank), the oil sector received $5.39 billion in foreign direct investment (FDI) in 2012, which represents 34% of total FDI in Colombia.[5]

As a result of these measures, in 2008 the country reversed the decline in oil production that had begun in 1999. The oil sector is now experiencing rapid growth, with forecasts predicting that Colombian oil production will rise to more than 1 million bbl/d by the end of 2015.[6] The Ministry of Mines and Energy estimates that Colombian production is expected to reach 1.3 million bbl/d by 2020.

Coal

Colombia has the largest coal reserves in Latin America, with 5,557 MMst of probable coal reserves. It is the 11th largest global producer and 5th largest exporter of coal (after Indonesia, Australia, Russia and the United States). Coal reserves are mostly located in La Guajira, bordering the Caribbean and the Andean foothills. Consequently, the majority of Colombia's coal production and export infrastructure is located on the Caribbean coast.[7] The majority of Colombia's production is thermal coal, used for power generation.

In terms of levels of production volumes, the country produced 85.5 million tons of coal in 2013, and exported 94% of its total coal output in that year; coal is Colombia's second- largest export after oil and represented 19% of export earnings in 2012. In terms of exports destination, forecasts predict that coal exports to the US and Europe will decrease in the near future. On one hand, coal exports to the US are declining due to lower demand resulting from the increase in shale gas production. On the other hand, policies promoting renewable energy and natural gas for power generation will reduce future demand for coal in Europe as well. As consequence of these structural changes of international demand, Colombian coal will be increasingly exported to Asia.

According to the US Energy Information Administration, China and India alone will account for 80% of demand growth for coal between 2005 and 2030. Colombia has recently begun exporting significant quantities of coal to China, as Asia's share of Colombian coal exports grew from less than 1% in 2008 to 8% in 2012.[8] The causes of this expansion are multiple: a combination of higher prices in Asia, lower freight costs and falling demand from the United States have spurred this expansion into Asian markets. Talks conducted in May 2012 between Colombia and China's presidents (who met in Beijing), revealed China's interest in investing in Colombia's coal sector and in the construction of a railroad to the Pacific coast to facilitate exports to Asia. Both the expansion of the Panama Canal (currently scheduled for completion in 2015) and Chinese-backed transportation infrastructure projects could generate greater exports of Colombian coal to Asia in the future.[9] Overall, by 2019, coal production is expected to increase to 143.9 Mt with exports rising to 120 Mt from 50 Mt in 2003.[10]

THE ENERGY SECTOR AND ARMED CONFLICT IN COLOMBIA

From the sixties Colombia has been immersed in an armed conflict involving multiple actors: leftist guerrillas; right-wing paramilitaries; drug traffickers; the Colombian government with its armed forces; and the civil society. There is no clear consensus on which is the exact origin of the Colombian armed conflict, but the most widely accepted explanation[11] points out that the origins of the current insurgency go back to the period known as "La Violencia," a civil war that took place between 1946 and 1966.

During this bloody period the country was radically divided between support for the Liberal and Conservative parties. Since then, Colombia has witnessed successive emergence of guerrillas, drug traffickers and paramilitary groups. The guerrillas emerged in the 60s, being the most important the Revolutionary Armed Forces of Colombia (FARC), the Army of National Liberation (ELN); the Movement April 19, (M-19) and the People's Liberation Army (EPL). Drug trafficking, meanwhile, started with the cultivation and trafficking of marijuana in the early 70s, followed by cocaine in the 80s. Paramilitary groups, in turn, emerged in the early 80s as a form of retaliation by drug dealers, ranchers and some soldiers, who formed groups of "self-defense" to protect themselves from the violence perpetrated by the guerrillas. The proliferation and expansion of these groups—different in nature but all generators of violent conflict—has plunged the country into a wave of violence that has lasted more than five decades.[12]

Colombia's armed conflict has had important implications for the Energy sector: insurgents' attacks on oil and gas pipelines, transmission lines and other crucial infrastructures have long plagued the country. Electricity transmission towers have also proven to be easy targets; a bombing campaign has destroyed hundreds of them, causing repeated power outages in many parts of the country. Rail lines used to transport coal have been attacked as well. Besides causing millionaire economic losses, these insurgent terrorist attacks on oil and gas pipelines and on electric transmission towers are also a primary cause of environmental damage to habitats in Colombia. Oil spills, primarily from the heavily bombarded Cano Limon-Coveñas pipeline, have caused severe damage to some of the country's rivers and wetlands, wildlife, and farmland.

The bombing campaigns over the last twenty years have resulted in the spilling of oil into watercourses, causing extreme levels of environmental destruction.

Criminal actions perpetrated by the guerrillas against the energy sectors also included the practice of kidnapping and extorting major oil companies (such as the foreign companies in charge of building the Caño Limon-Coveñas oil pipeline). This allowed the guerrillas to occupy areas of high economic potential in the energy and mining sectors, and to extract of these highly significant resources. Within these areas are the oil and coal producing zones of the country, such as Arauca, Casanare, Barrancabermeja, La Guajira and Cesar.[13]

These terrorist actions against energy targets were substantially decreased through the implementation of Democratic Security Policy (2002–2010), a strategy based on a concerted military offensive against guerrillas and paramilitaries. Through its implementation, FARC was ousted from central Andean departments and lost its urban presence as well as control over strategic corridors and was forced to retreat to hideouts in ever more remote rural areas.[14] Over 12,000 FARC members were killed by security forces between 2002 and May 2009; another 12,000-plus were captured between 2003 and 2009; and over 17,000 were demobilized between 2002 and August 2012. The number of fighters decreased to some 9,000 in 2012. As a result, the number of municipalities exposed to offensive military operations by FARC shrank from 377 (of some 1,100) in 2002 to 142 in 2010.[15] At the same time, security forces extended their presence across the territory. By 2006, the police had at least some presence in all municipalities, up from 932 in 2002, even though territorial control in peripheral areas remains often restricted to urban centers.[16]

Finally, after the government's investment in technology and significant military resources to protect infrastructure, criminal actions against oil pipelines and electricity towers were reduced significantly in the last decade. For instance, pipeline attacks had significantly declined from 155 in 2005, to 31 in 2010. However, the integrity of energy assets is still threatened by illegal armed groups, with harsh increases in the number of attacks to energy infrastructure in the last year. Beginning in 2011 there were 84 pipeline attacks and there has been continuous increase in the number: in 2013 there were more than 150 attacks, which have led to significant increases in unplanned production disruptions in Colombia.[17] According to the Ministry of Defense,[18] there were 259 attacks on oil pipeline infrastructure in Colombia last year, the highest figure in a decade.

There are many possible explanations for this new increase of guerrilla actions against energy infrastructure. Some experts claim that the success of Colombia's oil industry in recent years has made it a target of attacks by the country's two largest rebel groups, the Revolutionary Armed Forces of Colombia (FARC) and the National Liberation Army (ELN). Territorial control, extortion, kidnappings, contraband and "political messages" drive these groups. While the stated motive of both leftist groups is to broaden citizens' access to natural resource royalties, the attacks point toward the rebel groups' desperate attempts to maintain leverage during ongoing peace negotiations. For the ELN, the attacks and kidnappings are a way to demonstrate that it poses a serious enough threat to be included in the negotiations. In the case of the FARC, the attacks also signify a change in the groups' tactics following a series of high-level losses in recent years and the appointment of a new commander, Timochenko. The attacks in early 2014 against state-owned Ecopetrol's Caño Limón-Coveñas and Transandino pipelines accrued significant losses.[19]

Attacks on the Colombian pipeline system by FARC rebels over the last year also speak about the ongoing struggle of the FARC insurgency in the country. Though the rebel group is said to be at its weakest point in decades and has been in peace talks with the government since 2012, it continues to use tactics such as the oil pipeline attacks as a way to apply pressure on the state and remain a threat. The rebels specifically go after state or commercial entities of strategic importance to the Colombian government. Usually, the pipelines are back online within a matter of days, but the losses are adding up, particularly given the growing frequency and severity of such attacks. For example, the recent months-long closure of the Caño Limon- Coveñas pipeline run by the state entity Ecopetrol reportedly cut the daily Colombian production by 7%. The government claims that such attacks do not affect output, but it is a setback for Colombia, which is seeking to increase its oil exports.[20] Furthermore, and particularly affecting the mining sector, there is an additional challenge in illegal mines operated by paramilitary groups and artisans. A lack of regulation has increased environmental hazards and generated illicit flows of mining profits in Colombia.

Despite the plurality of causes, the truth is that security issues and social unrest have led to short-term production interruptions, increased transportation costs and waning investor confidence. Shares of several local energy companies are down, and Colombia's overall oil output fell to an average 935,000 barrels per day in April, the lowest level since August 2012 and well below the record of 1.12 million barrels per day reached in November 2013. Experts estimate Colombia averaged 35,000 bbl/d of unplanned production disruption in 2013, more than a 115% increase from 2012.[21]

FUTURE PERSPECTIVES FOR THE ENERGY SECTOR: ENVIRONMENTAL AND ECONOMIC IMPLICATIONS IN A POST-CONFLICT SCENARIO

This section addresses the future perspectives for Colombia's energy sector, contemplating the potential effects of signing a peace accord. Although the results of the peace talks currently conducted in Havana are still uncertain, the peace process seems to be the way forward agreed upon by both sides in the conflict—and an important alternative for finally bringing an end to more than fifty years of internal conflict. At the time of publication, negotiations had resumed in Havana and three of the six points in the negotiating agenda have been agreed (integral agricultural development policy; political participation and solution to the problem of illicit drugs). If a complete agreement is reached and peace accords are signed, a post-conflict scenario would require sustained economic growth in order to afford and stabilize peace. Will the energy sector be the main contributor to economic growth in the near future? What are the risks and implications of such strategy and the effects of an excessive reliance on exports of primary goods? What can be expected given the dramatic decrease in the international price of oil that started during the last months of 2014?

Economic Challenges

Several elements have positively changed in Colombia's economic scenario during the last decade, particularly regarding foreign direct investment and portfolio capitals. On one hand, the country's investment climate has substantially improved, and is currently ranked 42nd of 183 countries, according to the World Bank's indicators on ease of doing business. Additionally, the improved security situation has propelled Colombia's status among Latin American economies, though concerns persist in the short term over FARC attacks on railways transporting coal, oil pipelines and other infrastructure. Furthermore, in an effort to fight corruption, the government has increased institutional commitment, establishing a special division under the Executive Office to monitor procurements for large infrastructure projects.

However, despite Colombia's great economic performance of the past four years (average GDP Growth of 4.5%) the recent sharper decline in commodity prices caused by weaker demand from some major commodity-importing economies, especially China, poses a downside risk for the country.[22] Also, oil and gas production in the United States and Canada has increased considerably since 2008, as technological advancements have driven development of unconventional resources such as shale and tight oil. This expansion of world's supply will dramatically affect the energy outlook for Latin America, although the extent and nature of its impact varies widely between countries. Colombia will be among those greatly affected by the price decline, given that in the last five years the oil sector has contributed to approximately 5.6% of GDP and about 40% of export earnings, therefore being a key economic driver.[23]

Regarding peace prospects and economic forecasts, evidence is accumulating of a contradiction between security goals and wider economic policies. In particular, the expansion of extractive industries, partly caused by improved security conditions under DSP, is proving counterproductive for conflict resolution. Extractive industries have become income sources for all illegal armed groups, including FARC, mostly via extortion.[24] Given their illegal nature, there is no direct evidence for payments, but attacks against oil and mining sectors reportedly mainly motivated by failure to pay extortion are rising. FARC has repeatedly struck the railway tracks of Cerrejón, the largest coalmine, and pipelines such as the Transandino and Caño Limón Coveñas have again become frequent targets. Despite military protection, FARC has repeatedly attacked trucks carrying oil from San Vicente del Caguán and even produced a production halt. Kidnapping risks for oil workers have increased. This is costly not just for oil companies and the state. Pipeline attacks also cause substantial environmental and collateral social and economic damage, for instance water contamination.[25]

The conjunction of these events (the uncertainty of the peace talks, the decreased price of oil, and the bureaucratic delays in Colombia for granting drilling rights) has created concern among foreign investors. It has not helped that FARC attacks to oil infrastructure have continued, and the attacks were a main factor in the falling production of oil. The predictable consequence has been that energy companies are now more reluctant to invest in Colombia. Oil investment, at $2.8 billion through June 2014, was the lowest for that half-year period since 2011. An oil auction in July lured commitments of $1.4 billion, just more than half of what the government had intended. [26]

Oil is a major revenue generator that the Colombian state counted on to fund the post-conflict projects, to build infrastructure, roads, tunnels and waterways, as well as schools and hospitals. It is undoubtedly the main source of financing the governments' promises if the peace accord is reached. At the same time that the FARC bombed oil infrastructure, experts calculated post-conflict rebuilding—solely for the countryside—would require between 80 and 100 billion pesos in the next ten years.[27] Policy makers hope increased output will blunt any lingering price declines. But producers face disruptions related to the conflict with the FARC, as well as an uncertain investment climate. These two situations constitute a clamp pressing the government, and particularly the energy sector.

Finally, another element that has to be considered is the volatility of the exchange rate, which undoubtedly will affect Colombia's economic scenario. The fluctuations of the past year have brought the country from the extremes of fearing Dutch disease (from extreme revaluation of the Colombian peso), to the current devaluation caused by the fall in oil prices. This erratic and unpredictable behavior of the exchange rate will certainly affect the countries' finances and disrupt estimations of peace construction.

Environmental and Infrastructure Challenges

In order to increase production and simultaneously extend oil reserve life the country has to expand exploration of oil and other resources. However, this strategy comes at a cost, not only financial—at current dropping prices—but mostly environmental too. Public opposition and conflicts over social and environmental impacts are on the rise. Colombia has the world's second highest number of environmental conflicts after India and leads the Latin America ranking with about half of conflicts related to coal and gold mining, says the Environmental Justice Organizations, Liabilities and Trade project.[28]

In response to strong public opposition to mining due to social and environmental concerns, the government has enforced stricter permitting rules, delaying projects. In an initiative to show support to environmental protection in 2011 the National Authority for Environmental Licensing (ANLA) was created. The main purpose of this entity is to manage environmental licensing. However, ANLA has been slow to award permits and has a huge backlog. As part of additional institutional initiatives, the government also sought to improve regulations by establishing the National Mining Agency (ANM) to manage the regulatory process. Currently most mining concessions are increasingly encroaching on indigenous reserves, which cover approximately 30 million hectares of the country. Because of this, prior consultations must be carried out with indigenous and other ethnic minority communities that will be impacted by mining projects. Consequently, the process is often slow or communities oppose projects altogether.[29]

Road and rail infrastructure are major challenges for Colombia's energy sector (and for the economy overall) to improve competitiveness. The shortage of roads and rail to bring coal cargos from the country's interior to its ports leads to high transportation costs that make exporting products from the interior very expensive. The government estimates resolving the infrastructure deficit could raise annual GDP growth by a full percentage point. Accordingly, the government has designed plans to improve roads and railways to reduce travel time and improve cargo capacity at ports. President Santos created the National Infrastructure Agency (ANI) in 2011, initiating a plan to award at least 25 concessions by 2014 and dedicating $23.2 billion to infrastructure projects.[30]

In addition, the private sector has already worked with officials by kicking in $7 billion in capital spending so far in 10 approved public-private partnerships, with the government organizing the deals and accepting much of the risk. The hope is that the infrastructure spending to improve the Andean nation's underdeveloped roads, tunnels and waterway will also add to future growth.[31]

CONCLUSION

After two years of peace talks and negotiations between the Colombian government and the FARC in Havana, the country is still divided awaiting the results of the talks. Optimists believe there has never been a more propitious time than now to achieve peace, given the particular characteristics of the negotiations and the current conditions of the country, the guerrillas and the Colombian government. Pessimists, in turn, oppose the process passionate and viscerally, accusing the negotiators of handing the country to the guerrillas.

Nonetheless, given the pace and progress of the peace conversations, it seems highly likely that the meetings in Havana will conclude with the signature of a peace treaty. A crucial step in this direction was recently announced by FARC negotiators. As part of the progress made so far in the peace negotiations underway in Havana, the FARC guerrillas declared an indefinite and unilateral ceasefire, starting December 20, 2014. They warned that the ceasefire would be terminated if attacks occur by the security forces to any of the different guerrilla columns.[32]

It is clear that if peace accords are signed, the successful implementation of the agreement will depend strongly on an injection of large amounts of funds to ensure and stabilize peace. The source of these funds was always identified as future income from exploitation and investments in natural resources—being oil the main one. With the current deep and fast slide in oil prices, the expected royalties and revenues generated by oil exports seem uncertain for years to come, putting the government in a difficult and unstable situation vis-a-vis the signees of the peace treaty and the rest of the Colombians, who have followed the process fearing that it will result in additional and heavier tax burdens. Santo's recent tax reform, budgeted counting on oil prices that reflected the boom of the past 3 years (around US 100 per barrel), is now clearly insufficient to finance Colombian post-conflict given the sharp fall in oil prices (U$ 60 per barrel).

Another irony of the situation is that it was precisely the guerrillas who constantly targeted some of the most important sources of financing for post-conflict reparation and reconstruction—oil exploration and production—for many years. To the losses caused by those attacks, the country will have to add the ones resulting from lower oil and commodity prices. Overall, there will be less financial resources to satisfy the construction of a post-conflict scenario. This could, in the long run, reduce the possibilities of long-lasting peace for Colombia.

Regarding security, a likely successful outcome of the peace talks would certainly not lead to an immediate improvement in the security situation. Especially extortion related crimes, which potentially include small- scale terrorist attacks, will likely increase in the period following a peace agreement. It seems now evident that the situation will not improve until economic development, application of the rule of law and improvements in social indicators are fostered more effectively in rural Colombia. The failure to do so encourages the existence of illegal armed groups and lies at the heart of Colombia's conflict. The longer the state takes to actively combat and eradicate this situation, the more complex security issues will become. With criminal gangs emerging, the security panorama in Colombia will remain a dynamic, moving target that will continue to affect the energy sector.[33]

Finally, many countries (Colombia among them) struggle with competing interests of developing natural resources while protecting the environment and local populations and distributing oil wealth efficiently. The government must set a clear energy strategy and build support from industry as well as social and environmental activists. It's crucial to keep in mind that success in Havana depends on a strong government with public support as well as a thriving economy.

NOTES

1. Richard Lynch, "An Energy Overview of Colombia," Washington: U.S. Department of Energy, 2003, accessed September 3, 2014, http://www.geni.org/globalenergy/library/national_energy_grid/colombia/EnergyOverviewofColombia.shtml.

2. Richard Lynch, "An Energy Overview of Colombia," Washington: U.S. Department of Energy, 2003, accessed September 3, 2014, http://www.geni.org/globalenergy/library/national_energy_grid/colombia/EnergyOverviewofColombia.shtml.

3. U.S. Energy Information Administration, "Colombia has seen a dramatic increase in oil, natural gas, and coal production in recent years after the implementation of a series of regulatory reforms," Washington: U.S. Energy Information Administration, 2014, accessed August 14, 2014, http://www.eia.gov/countries/cab.cfm?fips=CO.

4. U.S. Energy Information Administration, "Colombia has seen a dramatic increase in oil, natural gas, and coal production in recent years after the implementation of a series of regulatory reforms," Washington: U.S. Energy Information Administration, 2014, accessed August 14, 2014, http://www.eia.gov/countries/cab.cfm?fips=CO.

5. U.S. Energy Information Administration, "Colombia has seen a dramatic increase in oil, natural gas, and coal production in recent years after the implementation

of a series of regulatory reforms," Washington: U.S. Energy Information Administration, 2014, accessed August 14, 2014, http://www.eia.gov/countries/cab.cfm?fips=CO.

6. U.S. Energy Information Administration, "Short-Term Energy Outlook (STEO)," Washington: U.S. Energy Information Administration, 2013, accessed August 16, 2014, ww.eia.gov/forecasts/steo/archives/Nov13.pdf.

7. U.S. Energy Information Administration, "Colombia has seen a dramatic increase in oil, natural gas, and coal production in recent years after the implementation of a series of regulatory reforms," Washington: U.S. Energy Information Administration, 2014, accessed August 14, 2014, http://www.eia.gov/countries/cab.cfm?fips=CO.

8. U.S. Energy Information Administration, "Colombia has seen a dramatic increase in oil, natural gas, and coal production in recent years after the implementation of a series of regulatory reforms," Washington: U.S. Energy Information Administration, 2014, accessed August 14, 2014, http://www.eia.gov/countries/cab.cfm?fips=CO.

9. U.S. Energy Information Administration, "Colombia has seen a dramatic increase in oil, natural gas, and coal production in recent years after the implementation of a series of regulatory reforms," Washington: U.S. Energy Information Administration, 2014, accessed August 14, 2014, http://www.eia.gov/countries/cab.cfm?fips=CO.

10. Lisa Viscidi, "Colombia Extractive Industries and the Economy," Washington: Inter-American Dialogue April 29, 2014, accessed September 20, 2014, http://www.thedialogue.org/Energy.

11. Gonzalo Sánchez and Peter Bakewell, "La Violencia in Colombia: New Research, New Questions," *The Hispanic American Historical Review* 65, no. 4 (1985): 789–807.

12. Lilian Yaffe, "Colombia, Prospects for Peace in the 21st Century: Past Failures, Present Challenges and Future Opportunities," In *Re-conceptualizing Security in the Western Hemisphere in the 21st Century*, edited by Bagley, Kassab & Rosen. (Forthcoming 2014)

13. Fabio Sánchez y Mario Chacón. "Conflicto, Estado y Descentralización: del Progreso Social a la Disputa Armada por el Control Local, 1974–2002," En *Nuestra Guerra Sin Nombre: Transformaciones del Conflicto en Colombia*, editado por Gutiérrez Sanín, Francisco, María Emma Wills O., Gonzalo Sánchez G., Universidad Nacional de Colombia, y el Instituto de Estudios Políticos y Relaciones Internacionales, 361. Bogotá: Grupo Editorial Norma, 2006.

14. International Crisis Group, "Colombia: Peace at Last?" International Crisis Group, *Latin America Report* N°45, September 25, 2012, accessed October 3, 2014, http://www.crisisgroup.org/en/regions/latin-america-caribbean/andes/colombia/045-colombia-peace-at-last.aspx.

15. International Crisis Group, "Colombia: Peace at Last?" International Crisis Group, Latin America Report N°45, September 25, 2012, accessed October 3, 2014, http://www.crisisgroup.org/en/regions/latin-america-caribbean/andes/colombia/045-colombia-peace-at-last.aspx.

16. Lilian Yaffe, "Conflicto Armado en Colombia: Análisis de las Causas Económicas, Sociales e Institucionales de la Oposición Violenta," *CS* No. 8 (2011): 187–208.

17. U.S. Energy Information Administration, "Colombia has seen a dramatic increase in oil, natural gas, and coal production in recent years after the implementation of a series of regulatory reforms," Washington: U.S. Energy Information Administration, 2014, accessed August 14, 2014, http://www.eia.gov/countries/cab.cfm?fips=CO.

18. U.S. Energy Information Administration, "Colombia has seen a dramatic increase in oil, natural gas, and coal production in recent years after the implementation of a series of regulatory reforms," Washington: U.S. Energy Information Administration, 2014, accessed August 14, 2014, http://www.eia.gov/countries/cab.cfm?fips=CO.

19. Lisa Viscidi, Juan Mauricio Ramirez, Andrés Mejía Vergnaud, "How Should Colombia Handle Oil and Mining Royalties?," Washington: Inter-American Dialogue July 3, 2014, accessed December 1, 2014, http://www.thedialogue.org/page.cfm?pageID=32&pubID=3635.

20. Jack Devine, "What is the security outlook for Colombia's oil sector?," Latin America Advisor Blog, Inter-American Dialogue, May 23, 2014, accessed September 4, 2014, http://latinamericaadvisor.org/2014/05/23/what-is-the-security-outlook-for-colombias-oil-sector/.

21. U.S. Energy Information Administration, "Colombia has seen a dramatic increase in oil, natural gas, and coal production in recent years after the implementation of a series of regulatory reforms," Washington: U.S. Energy Information Administration, 2014, accessed August 14, 2014, http://www.eia.gov/countries/cab.cfm?fips=CO.

22. Lisa Viscidi, Juan Mauricio Ramirez, Andrés Mejía Vergnaud, "How Should Colombia Handle Oil and Mining Royalties?," Washington: Inter-American Dialogue July 3, 2014, accessed December 1, 2014, http://www.thedialogue.org/page.cfm?pageID=32&pubID=3635.

23. Lisa Viscidi, "A Difficult Balancing Act," Washington: Inter-American Dialogue July 11, 2014, accessed December 1, 2014, http://www.thedialogue.org/page.cfm?pageID=32&pubID=3639.

24. International Crisis Group, "Colombia: Peace at Last?" International Crisis Group, *Latin America Report* N°45, September 25, 2012, accessed October 3, 2014, http://www.crisisgroup.org/en/regions/latin-america-caribbean/andes/colombia/045-colombia-peace-at-last.aspx.

25. International Crisis Group, "Colombia: Peace at Last?" International Crisis Group, *Latin America Report* N°45, September 25, 2012, accessed October 3, 2014, http://www.crisisgroup.org/en/regions/latin-america-caribbean/andes/colombia/045-colombia-peace-at-last.aspx.

26. William Mauldin, "Fast-Growing Colombia Faces Challenges in Oil Price, Conflict Related Spending," *The Wall Street Journal*, October 8, 2014, accessed November 14, 2014, http://www.wsj.com/articles/fast-growing-colombia-faces-challenges-in-oil-price-conflict-related-spending-1412809300.

27. Paula Delgado-Kling, "Santos more likely suspended talks due to the economy, not the army general's kidnapping," Talking about Colombia, November 17,

2014, accessed December 9, 2014, http://talkingaboutcolombia.com/2014/11/17/santos-more-likely-suspended-talks-due-to-the-economy-not-the-army-generals-kidnapping/.

28. Lisa Viscidi, "A Difficult Balancing Act," Washington: Inter-American Dialogue, July 11, 2014, accessed December 1, 2014, http://www.thedialogue.org/page.cfm?pageID=32&pubID=3639.

29. Lisa Viscidi, Juan Mauricio Ramirez, Andrés Mejía Vergnaud, "How Should Colombia Handle Oil and Mining Royalties?," Washington: Inter-American Dialogue, July 3, 2014, accessed December 1, 2014, http://www.thedialogue.org/page.cfm?pageID=32&pubID=3635

30. Lisa Viscidi, Juan Mauricio Ramirez, Andrés Mejía Vergnaud, "How Should Colombia Handle Oil and Mining Royalties?," Washington: Inter-American Dialogue, July 3, 2014, accessed December 1, 2014, http://www.thedialogue.org/page.cfm?pageID=32&pubID=3635

31. William Mauldin, "Fast-Growing Colombia Faces Challenges in Oil Price, Conflict Related Spending," *The Wall Street Journal*, October 8, 2014, accessed November 14, 2014, http://www.wsj.com/articles/fast-growing-colombia-faces-challenges-in-oil-price-conflict-related-spending-1412809300.

32. Redacción Paz, "Farc decretan cese al fuego por tiempo indefinido," *El Espectador*, December 2014.

33. Sergio Torres, "What is the security outlook for Colombia's oil sector?," Latin America Advisor Blog, Inter-American Dialogue, May 23, 2014, accessed September 4, 2014, http://latinamericaadvisor.org/2014/05/23/what-is-the-security-outlook-for-colombias-oil-sector/.

BIBLIOGRAPHY

Delgado-Kling, Paula. "Santos more likely suspended talks due to the economy, not the army general's kidnapping." Talking about Colombia, November 17, 2014, accessed December 9, 2014, http://talkingaboutcolombia.com/2014/11/17/santos-more-likely-suspended-talks-due-to-the-economy-not-the-army-generals-kidnapping/.

Devine, Jack. "What is the security outlook for Colombia's oil sector?" Latin America Advisor Blog, Inter-American Dialogue, May 23, 2014, accessed September4,2014,http://latinamericaadvisor.org/2014/05/23/what-is-the-security-outlook-for-colombias-oil-sector/.

International Crisis Group. "Colombia: Peace at Last?" International Crisis Group, *Latin America Report* N°45, September 25, 2012, accessed October 3, 2014, http://www.crisisgroup.org/en/regions/latin-america-caribbean/andes/colombia/045-colombia-peace-at-last.aspx.

Lynch, Richard. "An Energy Overview of Colombia." Washington: U.S. Department of Energy, 2003, accessed September 3, 2014, http://www.geni.org/globalenergy/library/national_energy_grid/colombia/EnergyOverviewofColombia.shtml.

Mauldin, William. "Fast-Growing Colombia Faces Challenges in Oil Price, Conflict Related Spending." *The Wall Street Journal*, October 8, 2014, accessed November

14, 2014, http://www.wsj.com/articles/fast-growing-colombia-faces-challenges-in-oil-price-conflict-related-spending-1412809300.

Redacción Paz, "Farc decretan cese al fuego por tiempo indefinido." *El Espectador*, December 2014.

Sánchez, Gonzalo and Peter Bakewell. "La Violencia in Colombia: New Research, New Questions." *The Hispanic American Historical Review* 65, no. 4 (1985): 789–807.

Sánchez, Fabio, y Mario Chacón. "Conflicto, Estado y Descentralización: del Progreso

Social a la Disputa Armada por el Control Local, 1974–2002." En *Nuestra Guerra Sin Nombre: Transformaciones del Conflicto en Colombia*, editado por Gutiérrez Sanín, Francisco, María Emma Wills O., Gonzalo Sánchez G., Universidad Nacional de Colombia, y el Instituto de Estudios Políticos y Relaciones Internacionales, 361. Bogotá: Grupo Editorial Norma, 2006.

Torres, Sergio. "What is the security outlook for Colombia's oil sector?" Latin America Advisor Blog, Inter-American Dialogue, May 23, 2014, accessed September 4, 2014, http://latinamericaadvisor.org/2014/05/23/what-is-the-security-outlook-for-colombias-oil-sector/.

U.S. Energy Information Administration."Short-Term Energy Outlook (STEO)." Washington: U.S. Energy Information Administration, 2013, accessed August 16, 2014, ww.eia.gov/forecasts/steo/archives/Nov13.pdf.

U.S. Energy Information Administration. "Colombia has seen a dramatic increase in oil, natural gas, and coal production in recent years after the implementation of a series of regulatory reforms." Washington: U.S. Energy Information Administration, 2014, accessed August 14, 2014, http://www.eia.gov/countries/cab.cfm?fips=CO.

Viscidi, Lisa. "Colombia Extractive Industries and the Economy." Washington: Inter-American Dialogue, April 29, 2014, accessed September 20, 2014, http://www.thedialogue.org/Energy.

Viscidi, Lisa. "A Difficult Balancing Act." Washington: Inter-American, Dialogue July 11, 2014, accessed December 1, 2014, http://www.thedialogue.org/page.cfm?pageID=32&pubID=3639.

Viscidi, Lisa, Juan Mauricio Ramirez, and Andrés Mejía Vergnaud, "How Should Colombia Handle Oil and Mining Royalties?," Washington: Inter-American Dialogue July 3, 2014, accessed December 1, 2014, http://www.thedialogue.org/page.cfm?pageID=32&pubID=3635.

Yaffe, Lilian. "Conflicto Armado en Colombia: Análisis de las Causas Económicas, Sociales e Institucionales de la Oposición Violenta." *CS* no. 8 (2011): 187–208.

Yaffe, Lilian. "Colombia, Prospects for Peace in the 21st Century: Past Failures, Present Challenges and Future Opportunities."In *Re-conceptualizing Security in the Western Hemisphere in the 21st Century*, edited by Bagley, Kassab & Rosen. (Forthcoming 2014)

Part IV

HYDROCARBONS VS. RENEWABLES IN THE FUTURE OF GLOBAL ENERGY PRODUCTION

Chapter 14

Energy and Systemic Change

The Emerging Power of Coal

Hanna Samir Kassab[1]

In July 2014, the world collectively held its breath for three reasons. The first was in suspense over the escalating conflict in the Ukraine. The second was in shock over the discovery of a Chinese spy submarine off the coast of Hawaii.[2] The third reason was literally for want of clean air. As reported at the time in *Foreign Policy,* a BRICS (Brazil, Russia, India, China and South Africa) development bank is now underway, aimed at financing coal development. The author of the piece, Keith Johnson, posited that the new bank ". . . boost financing for coal-fired power plants around the world . . ."[3] The Global South logic is that since BRICS countries depend on coal for cheap energy, weak states would seek to borrow funds to develop their own coal industries (Ibid). It stands to reason that, along with the new development bank may come further political demands from creditor countries. Nevertheless, emergence of this new multilateral lender suggests it will assist in overturning status quo norms, rules and institutions, upheld by Global North institutions like the World Bank, which advocates the use of clean energy (as well as provide their own developmental loans to weak states). To what extent is a process like this one a foregone conclusion?

This study aims to answer this question by discussing patterns of state behavior using the theory of Structural Realism espoused by Robert Gilpin in his book *War and Change in World Politics* (1981) (hereafter *War and Change*). Gilpin posits that states go about creating structures based on their particular social, political and economic interests.[4] In *War and Change*, Gilpin focuses on patterns of state behavior during times of systemic transition or change. He studies 2,400 years of human history and concludes that "the distribution of power among states constitutes the principal form of control in every international system."[5] If there are certain rules that serve the interests of a set of states, then changes in the distribution of capabilities may change the rules of the system. As does Gilpin, we can predict whether or not energy as a capability, and therefore coal development, is part of such systemic considerations. Therefore, the purpose of this paper is to analyze coal as a material capability, its use as well as its ability to influence systemic change in relation to rising powers.

This chapter discusses the future of coal within the context of international systemic change and its ramifications. By applying Structural Realism to current international conditions, one could potentially gain some insight into the future of energy, the environment and the struggle for domination between status quo powers and emerging powers in the twenty-first century. The paper will begin with a short discussion of Structural Realism as the theoretical perspective adopted to help us analyze and explain coal use during periods of transition. Secondly, a short overview on this history of coal and coal use must be undertaken to establish social, political and economic context. Thirdly, I will apply the theory to the period of systemic transition while underscoring its process. I will focus on Chinese investment and external assistance practices as well as Russia's use of oil and gas as a weapon of foreign policy.

THEORETICAL PERSPECTIVE

This paper applies the theory of Structural Realism to explain the continuities and processes which occur alongside systemic change, or perceptions of such change taking place. This section hopes to briefly summarize Structural Realism as posited by both Kenneth Waltz in *Theory of International Politics* (2010) and Robert Gilpin in *War and Change* (1981). To make sense of the claim that the rise of emerging powers will result in the increases in coal use, a theoretical foundation must first be set. This particular theory was built taking into consideration patterns of great power behavior (Waltz) as well as those behavioral patterns during times of transition (Gilpin). If emerging powers do indeed want to change the international system to suit their particular interests, then their practices, economic growth and development models in particular, will become the agenda for other states within their spheres of influence. The fact remains that emerging powers are the main consumers and producers of coal; this is a fundamental part of their economic growth and development model. It is to be expected, therefore, that coal and coal use will increase, and cause greater discord between status quo powers given their supposed dedication to the environment and clean energy.

To begin, the theory of Structural Realism sees the world and world politics as constructed by a structure of the world system. This structure shapes the behavior of states. Simply put, states rationally pursue their interests of security and survival through a system of self-help in the absence of an overarching order. This system consists of a structure, and ". . . the structure is the systems-level component that makes it possible to think of the units as forming a set as distinct from a mere collection."[6] The system creates this state behavior. The structure is defined by three factors. First, by anarchy, that is the absence of an overarching authority. Second, by the functions and then the capabilities of interacting units, more specifically, states.[7] In this environment, states seek to survive by any means, either through war or isolation. This world system of anarchy is permanent unless the structure changes. No state, or unit of that structure, can alter this framework. This, according to the theory, is the cause of war.[8] Nothing then can alter a state's behavior unless the system itself transforms. Thus, Waltz sees the world as afflicted by a determinant structure of anarchy that cannot be mitigated. Structural Realists like Waltz see power as ". . . defined in terms of the distribution of capabilities."[9] This makes up the third factor that forms the structure of the international system. The structure deviates with fluctuations in the distribution of capabilities among nations.[10] These capabilities can be economic, military and other factors like: size of population and territory, political stability and competence. Capabilities provides a tangible and material formulation of a state's ability to defend its security to survive.

Gilpin furthers Waltz's argument but frames systemic change as a function of asymmetrical economic growth. Such asymmetrical growth facilitates state economic and military power. Gilpin is particularly interested in identifying trends in behavior within the structure of the international system.[11] He sees the international system as similar to any other system: ". . . actors enter social relations and create social structures in order to advance particular sets of political, economic or other types of interests . . . the particular interests that are most favored by these social arrangements tend to reflect the relative power of the actors involved."[12] In other words, the nature of the system and all its rules, norms, economic relations, and energy use, are determined by ". . . whose interests are being served by the functioning of the system."[13] Hence, systems are created to serve the interests of great powers who construct such a system for their own benefit.

Further, political change arises with ". . . the efforts of individuals or groups to transform institutions and systems in order to advance their interests . . . the political system will be changed in ways that will reflect these underlying shifts in interests and power."[14] Hence, we see states engaged in a struggle for dominance given conflicting interests and systems of governance. Gilpin describes the structure as existing in five particular phases, determined by conflicting types of structures:

1. An international system is stable (i.e., in a state of equilibrium) if no state believes it profitable to attempt to change the system.
2. A state will attempt to change the international system if the expected benefits exceed the expected costs (i.e., if there is an expected net gain).
3. A state will seek to change the international system through territorial, political, and economic expansion until the marginal costs of further change are equal to or greater than the marginal benefits.
4. Once the equilibrium between the costs and benefits of further change and expansion is reached, the tendency is for the economic costs of maintaining the status quo to rise faster than the economic capacity to support the status quo.
5. If the disequilibrium in the international system is not resolved, then the system will be changed, and a new equilibrium reflecting the redistribution of power will be established.[15]

This process of behavioral transition is exacerbated during periods of unequal economic growth between competing powers. Differences in growth can take many forms but may result in differences in political, economic, technological innovations that drive economic growth and development. If a state is in status quo, its place in the structure is determined. If it is not growing in power terms vis-a-vis its potential challengers, the result could be disequilibrium.[16] This disequilibrium as posited in point 5 above, may encourage competing powers to change the system; replacing it with a system that suits their interests, and advances further economic growth. This may lead to conflict if status quo powers find such a transition untenable.

Gilpin's explanation is slightly different from that of Waltz, but the main emphasis on the capability of structure to shape state behavior remains constant. The international system is more fluid than Waltz's conception allows, and control over methods of transaction defines the Gilpinian system. Gilpin borrows from Robert Mundell and Alexander Swoboda who define the international system as ". . . an aggregation of diverse entities united by regular interaction according to a form of control."[17] Hence, according to Gilpin, the international system is one of control and competition over relationships; of relative struggle and of a constant "seeking to control" of one state system over another.[18]

In such an environment of competition, weak states choose between Gilpinian systems. Holding aside the notion of opportunism, weak state form parasitic relations with competing powers and their systems.[19] Schweller (1994) posits that weak states bandwagon with great power speculated as the "wave of the future." This bandwagoning is for the purposes of economic gain and not for security. Gilpin's research relied on the foreign policy of weak states during the Cold War. Many states reasoned that the Soviet Union was the "wave of the future" and, for that reason, created closer ties to the Soviets than to the Western world. "Bandwagoning for Profit," as Schweller calls it, reflects the Gilpinian international system.[20] Two competing rules and frameworks of operation obtain between two power groupings, as in the case of the socialist world versus the capitalist world during the Cold War. Weak states, motivated by economic growth and development, adopt the international system as a set of social, political and economic arrangements determined by one great power. Inevitably, such adoption fulfills the great power's own efforts at perpetuating its power. Such patterns involve certain practices and development models, which of course includes structures of production and, as a consequence, types of energy developed and consumed.

Today's international environment is, arguably, witness to competition between one set of powers, that is the status quo as led by the United States, and a complex amalgam of competing powers, that is the BRICS states. The United States would prefer that all states serves its strategic interests by working within its rules, regulations and institutions. However, the BRICS states are going forward with their own endeavors; their development bank plan presents an interesting example. The BRICS development model (to be explained in section # to follow), is one that requires strong ties among the BRICS countries. This presents a new competing system for the world, which can be seen as an opposing system to the Western world. There is an existing fear, and that ". . . fear is that a new Beijing or Chavez consensus will replace the long-hallowed Washington consensus."[21] Given the theoretical propositions here, and given repeating patterns of behavior as posited by Gilpin, could the following theoretical expectation regarding coal be predicted? I argue yes. China sees it necessary to use economic power to facilitate coal production which serves to only further enhance its economic and political power position.

COAL AND COAL USE IN THE WORLD

People have been using coal since the days of Aristotle; and it was more widely used during the heyday of the Roman Empire, when Emperor Hadrian used the mineral to keep warm during cold Scottish nights.[22] Demand for coal increased with time, especially during times of timber scarcity. Coal could be extracted from the earth and easily transported throughout the city, greatly increasing its popularity compared to wood.[23] With increases in demand came increases in production. Mining techniques moved from small endeavors during the 1500s to large scale operations of the 1800s forward.[24] Because of its combustion capabilities, it was naturally the fuel of the Industrial Revolution.

Industrial centers the world over, in Great Britain, the United States and France not only adopted coal as their backbone of economic growth, they further built their industries around coal.[25] Coal helped power the international trade which defined the 1800s until the outbreak of the First World War. After the Second World War, Europeans understood the importance of coal, not only to their economy, but their national security and defense. This realization about coal led to the creation of the European Coal and Steel Community which attempted to bind the success and survival of European states together. Creating such a community under coal and steel cooperation would create spillover and dependence, tying the states closer together in the spirit of European brotherhood.[26]

Today, coal is still widely used by many countries; moreover, since 2000, its use grew faster than other energy sources. Seventy-six percent of world coal production is consumed by five countries: China, the USA, India, Russia and Japan; three of these are BRICS countries. Estimations conclude there is 861 billion tons of coal worldwide.[27] Given estimates and current use, there is enough coal to last 112 years as compared to oil and gas reserves: between 46 and 54 years.[28]

There are two main types of coal, each with their different uses. Steam or thermal coal is mainly used to generate power and electricity. Coking or metallurgical coal is used to produce steel. Contemporary coal use is primarily for the purpose of electricity, steel production, in the manufacturing of cement and as a liquid fuel.[29] Coal is also used for a multiplicity of other reasons such as ". . . alumina refineries, paper manufacturers, and the chemical and pharmaceutical industries."[30] The by-products of coal are seemingly endless: soaps, aspirins, solvents, dyes, plastics, filters, fertilizers, fibers and a plethora of chemicals, just to name a few.[31]

Coal's inexpensive nature, its wide versatility, linkage potential (industries resulting from byproducts), and abundance present many advantages to today's economies. The countries making up the BRICS make full use of coal and all its benefits. In the next section, I discuss coal use in the BRICS countries as a way to introduce the main section of the paper: the proliferation of coal as a global energy source as a result of the rising power and importance of the BRICS countries.

COAL AND THE BRICS

The BRICS is a term to contextualize the loose alliance of emerging countries growing in power and influence in both absolute and relative terms after the 2008 financial crisis. These states, Brazil, Russia, India, China and South Africa, all need access to energy supplies to pursue economic growth in a sustainable and dynamic way.[32] Coal is a fundamental to the success of the countries. As an important part of their economy, these states must continue to invest in coal production and ensure that sources of coal are open for business, even if it means supporting other countries through aid and investment.

China in particular has the most vociferous appetite for coal of the BRICS bloc and indeed the entire world. China depends on coal for 69% of its energy production; other sources like oil (18%) and hydroelectric power (6%) are dwarfed by coal.[33] China's demand for this resource far outstrips all the others: 50% of both the forces of global aggregate supply and global aggregate demand originate from within its borders.[34] Hence, China consumes nearly as much coal as all the other countries in the world combined.[35]

In terms of global aggregates in 2012, China produced 46% of the world's coal, almost four times as much coal as the United States, its second largest producer. Consumption data is more astounding: China consumes 49% of world coal. To fill this gap, China imports coal. As a consequence, there is significant pressure on China to secure energy and continue to grow; not to mention pressure to avoid an energy crisis which would serve only to arrest growth.

More importantly for future prospects, China's consumption of coal is expected to grow. There are efforts by China's leadership to slow-down its coal usage in the next decade.[36] China must, however, continue to mine and consume coal for the purpose of growing the economy in the short-term. The economic environment at this very precarious time in their development and progress as an emerging power is palpable. In human terms, they need ready supplies of coal to provide energy for 20% of the world's population. A shift to gas is already taking place. On September 1, 2014, Russia and China began construction on a gas pipeline that would link the two countries.[37] However, until the pipeline is fully functional, China will continue to make good use of coal; Chinese analyst Xiao Yunhan estimate that China's consumption of coal will likely double over the next ten years.[38] Hence, constant access to coal is fundamental to the success of China.

The other BRICS members are also some of the top producers of coal. Russia and India join China among the top five coal producers.[39] While the United States' coal production decreased (indeed, coal production in the United States is declining especially domestically[40]), China, India and Russia are all expanding their production of coal. The need for cheap and available energy is a vital to fuel continuing economic growth. One must conclude then, that the BRICS model of development is based on the use of cheap energy and the construction of its infrastructure. This may bring about a coal revolution, as poorer countries seek to build their own coal projects with BRICS investment. The same countries could also seek aid and loans for the purpose of development and may have to expand their own coal infrastructure to be considered attractive by the mentioned bloc.

SYSTEMIC CHANGE, THE BRICS CHALLENGE AND COAL[41]

In the previous section, I describe the importance of coal to BRICS countries. In particular, I highlight the importance of coal to China. While all of the BRICS countries produce and consume coal to advance their economic development, China best illustrates the paper's proposition of systemic change because it is China that consumes more coal than it produces. To sustain its recent economic growth of, for example, 7.7% in 2012,[42] and avoid an energy crisis, then China must continue to import coal from reliable sources. China must therefore continue to give other countries incentives to produce coal and it will likely do so by applying its economic and political influence. We currently see China exerting such influence in the underdeveloped countries of world such as in sub-Saharan Africa.

With regard to coal, we are looking at changing the system to favor the production and consumption of coal in the twenty-first century; we are talking about an energy revolution (really a return to coal) and China is the primary author of this systemic change Systemic change, as previously mentioned, is defined as ". . . the efforts of individuals or groups to transform institutions and systems in order to advance their interests . . . the political system will be changed in ways that will reflect these underlying shifts in interests and power."[43] In the context of coal and its role in the international system, there are several factors that may contribute to systemic change: Chinese external aid and investment are key to the overall process.

Today, the practice of Chinese aid is a major driving force of systemic change. Chinese aid takes the form of concessional grants, zero-interest loans, and low interest loans.[44] By concessional is meant that the state and state institutions are in full control of the external assistance process. To elaborate, the foreign aid loan program is operated by China's state run Eximbank which combines economic matters with diplomacy, developmental practice and business objectives.[45] For example, to gain aid, states must first recognize the Peoples' Republic of China over the Republic of China (or Taiwan). Therefore, the political motivations behind China's economic aid practices are made apparent even before the external assistance is paid out.

Borne out of this concessional approach, Chinese assistance is made contingent upon its Eight Principles:

1. Aid from China to foreign states is for mutual gain rather than based on one-sided grants.

2. A "no-strings attached" relationship that respects the sovereignty of the aid recipients.
3. China will provide interest free or low interest loans as economic aid with the option of delaying payment if necessary.
4. Aid is a means of assistance to assist countries develop independently rather than become reliant on China.
5. Projects are to provide quick results through small investments in order to give nations the ability to grow revenue.
6. China will provide Chinese manufactured goods at a competitive market price.
7. The Chinese will provide technical assistance and training.
8. Chinese citizens sent to work in various nations receiving aid will be paid at the level of their host country nationals working in the same field.[46]

While #2 states that there would be a "no-strings attached" approach, there are certain expectations that stem from #'s 5, 6, 7 and 8. Chinese aid projects are conducted through bi-lateral negotiations with little to no transparency. In exchange for assistance, China gains access to resources through preferential agreements. Further, they gain through special trade agreements and commercial investments. The end results are large infrastructure projects which use Chinese manpower and cater to the interests of China: for instance, the production of coal. This is obvious in many countries, especially those of the African continent.

To further look into the political motivations of aid, Ngaire Woods, in her article *Whose aid? Whose influence? China, emerging donors and the silent revolution in developmental assistance* (2008) illustrates standard Chinese aid practice. She argues that China, along with other emerging donors, are "giving aid on terms of their own choosing."[47] Much of this aid goes to countries in dire need of it because of economic stagnation. Other countries, like Zimbabwe, need aid because of political isolation due to lack of democratic institutions.[48] While OECD countries deny Zimbabwe aid because of its anti-democratic practices, China is not afraid to embrace the African country to gain from its vast coal mines. Many therefore posit that China is undermining global democracy to achieve its own goals. Moises Naim labels this "rogue aid" which is defined as ". . . development assistance that is nondemocratic in origin and nontransparent in practice; its effect is typically to stifle real progress while hurting average citizens."[49] Naim's conclusions here can be borrowed to further understand the motivations of China with its global coal campaign.

While not taking a Neo-Marxist perspective, it is important to note Raul Prebisch's Dependency Theory (1959). This theory seeks to explain underdevelopment by looking at economic production of rich and poor countries.

This structural argument posits that elites of core or developed countries dictate the production and exportation of poor and underdeveloped countries through supply and demand. To apply this to the current international political economic context, China will use its capacity for demand to encourage states to produce coal for its own sake. Using this perspective, we can predict the growing importance of coal in the next few years. Currently, China is involved in a number of coal mining projects globally: Bangladesh, Bosnia and Herzegovina, Burma, Cambodia, Canada, India, Indonesia, Iran, Malawi, Malaysia, Mongolia, Pakistan, Romania, Russia, Serbia, South Africa, Tanzania, Turkey, the Ukraine, the United States, Vietnam, Zambia, and Zimbabwe.[50] In these cases, China has played a vital role in improving the coal industries of these many countries; even developed countries like the United States and Canada. The growing demand for coal is changing the world and the way it produces energy. China's action plays an important role in facilitating coal's rise, making coal part of the systemic force of change.

AIDING COAL: THE CASE OF PAKISTAN, ZIMBABWE AND ZAMBIA

To empirically demonstrate the theoretical connection I assert between coal and systemic change, in this section I briefly mention three cases of Chinese external assistance which have served to facilitate coal production in these countries. Pakistan, Zimbabwe and Zambia (and much of Africa as Johnson points out) all have difficulty producing energy for domestic use. Coal as a cheap and readily available fuel source becomes the answer; and China is there to assist these countries with coal development as it seeks to further its own national interests. Through these three examples, we see the dynamic of coal as a growing influence on world energy use.

To begin, Pakistan is a country experiencing an energy crisis. There are severe power outages, sometimes for up to 10 to 15 hours in a day. This is attributed to the spike in the demand for energy given increases in economic growth and development.[51] To help meet energy demands, the Pakistani leadership has chosen to develop its coal mines, especially the coal mining fields of Thar. According to estimates "Thar coal fields have estimated reserves of 175 billion tons. These reserves could be utilized to produce 100,000 MW of power for many decades."[52] In August 2013, Pakistani scientist Samar Mubarakmand argued that such an endeavor would be costly and that China, with its China Coal and Mining Technology University, would be an excellent collaboration opportunity. Mubarakmand also suggested partnership with the Chinese government given the tremendous expense).[53] In October of the same year, China Power International Holding announced that it would begin construction of ten coal-fired power plants in at the Thar coal fields, a project that would cost $7.2 billion.[54] Currently, there are plans for even more coal based power plants, up to fifteen.[55]

There are, of course, environmental concerns over these matters, but thus far, such concerns have fallen on deaf ears. Coal has been deemed the solution to the energy crisis and Pakistan. The economic benefits that come from coal production, especially the replacement of oil imports (USD$14 billion annually), has proven vital to the economic success of the country.[56] In Pakistan, therefore, we see China is filling a need through coal production. While China does not currently import a major percentage of its coal from Pakistan (Pakistani coal production currently fuels 6% of the country's energy needs), but this percentage is expected to increase as Pakistan's coal production increases with time.

Zimbabwe is also seeing a major developments in its coal industry. Speculators and experts alike conclude that the country holds the second largest coal reserves in the southern African region (Ibid). It, however, lacks the capability to produce coal for the global market. Its rampant underdevelopment, underscored by terrible hyperinflation is worsened by a dependence on oil imports. To correct this, Zimbabwean minister of Mining and Mine Development announced the beginning of an energy revolution, selling 20 mining licenses for coal production. However, given economic difficulties and a lack financial resources, rapid growth will take some time. For this, Zimbabwe reached out to the world. Given animosity toward the Western world as well, due largely to Western-backed sanctions designed to punish President Mugabe's anti-democratic tendencies, other states had to be considered by Zimbabwe.[57]

Currently, there is competition between Japan (a major coal consumer) and China for access to the Zimbabwe's massive coal deposits.[58] Both China and Japan, through their state enterprises, began courting Zimbabwe in hopes of securing coal for future imports. In particular, it is said that Japan is seeking 15 million tons a year; but Zimbabwe can only produce 2 million tons a year. Hence, Japan has promised Zimbabwe to help them produce the resource by financing the construction of infrastructure, like railroads.[59]

Like Japan, China is also looking to gain a firm foothold in the country. Chinese coal firm *China Africa Sunlight* injected USD $600 million to help aid the industry. In return, it was granted a 100,000 hectare concession.[60] From this, a coal project partnership between China's China Africa Sunlight and Zimbabwe's Oldstone Investments named "China-Africa Sunlight Exploration" was born.[61] This large-scale, ambitious joint venture combines the Zimbabwean concession and Chinese financing of more than USD$2 billion as of 2013. The project should earn USD$1 billion a year upon completion. In all, the project hopes to assist Zimbabwe with its energy needs but also serve Chinese interests, securing a ready supply of the substance for many years to come.

Considering Zambia, the country has produced coal for many years now, since 1967.[62] It has a proven reserve of 20 million metric tons. Prior to 2000, the country produced on average 214,000 tons of coal. However, by 2010, that figure withered to a mere 1000 tons. The mines needed investment to keep the resource moving.[63] To add to this, Zambia at the time was suffering from an energy crisis. Mostly run by hydroelectricity, the state is looking to diversify its energy resources. To add to the urgency, other industries depend on coal to light their furnaces. The country reached out to global investors, like Singapore and China.

Currently, China enjoys a robust presence in Zambia especially in the mining sectors of cobalt, copper and coal.[64] Naturally, this linkage injected some much needed financial capital, approximately USD$1.2 billion in 2010, from China to Zambia.[65] This infusion helped the country produce 238,000 tons in 2011. By the end of this year, estimated coal production is set for 600,000 tons. Moreover, this production helped provide 25,000 jobs for Zambians.[66]

Zambian production of coal through Chinese investment has not been free of challenges. Zambian workers accuse the Chinese of excessive cruelty and hazardous working conditions: "'we do not have support timbers everywhere they need to be, and we have no masks to protect us from the coal dust,' said Boston Sikalamba, 21, who was buried for several minutes by a cave-in this month. 'After the dynamite is set, there's nothing to do about the dust but breathe it, and if you are slow at your work, the Chinese beat you.'"[67] Matters such as this led to a riot which ended in tragedy when hundreds of workers rioted against two Chinese managers. The managers, who did not speak English or Tonga, the indigenous language of Zambia, fired into the crowd wounding 13.[68] While such events are quite tragic, the official statement from the Zambian government is: "'we're an economy in transition, and we can't afford to lose the cow that gives us milk today,' said Labor Minister Austin Liato."[69] Today, Zambia's coal industry continues to thrive from Chinese investment. Like much of Africa, Zambia needs Chinese investment to fund its wide array of projects for the sake of development. China knows this and will continue its activities to gain access to coal and other resources (Zambian copper and cobalt benefit from Chinese investment) for its own economic growth and development.

The power of Chinese aid to bring about the systemic change is indeed something to study in depth. It is only natural for China to pursue its interests in such a manner. Pakistan, Zimbabwe and Zambia all gain from these relationships as well: their coal has a customer who holds a stake in their success. Systemically though, these three cases only serve to illustrate examples of what is occurring in the international system. In the next section, I will discuss the importance of Chinese aid in ushering in a world order funded by China and fueled by its demand for coal and other resources.

BANDWAGONING FOR PROFIT AND SYSTEMIC CHANGE

As discussed, Chinese aid is predicated on preferential trade and treatment coupled with, most importantly for this paper, special access to resources, specifically coal. Given aid, the demands associated with this aid will be welcomed by recipients given underdevelopment. This then, as Schweller argues, is a fundamental part of bandwagoning for profit. If China is perceived as the "wave of the future," then it would be wise for weak and underdeveloped states to begin creating good, working relationships with China. Many of the world's countries now make full use of Chinese aid in return for making the needed concessions. The table below document's all Chinese aid recipients (87 countries in total) for 2011.

Chinese aid is now global in reach. Of the 194 independent countries in existence today, 87 of them accept aid from China. In turn, each of these countries supply concessions in one way or another to the Chinese, as per established practice. In many instances, this is coal for the purpose of securing energy sources for China to ensure continual growth. Such cases of bargaining are transforming the international system as certain practices, such as the use of a specific energy source, is taking hold.

Aid is not as benign as one might think. Rather, Chinese aid is perceived by the United States and the rest of the OECD as a threat to established norms and practices. Naim poignantly describes the perception of this activity:

What we have here—in states like China, Iran, Saudi Arabia, and Venezuela—are regimes that have the cash and the will to reshape the world into a place very different from where the rest of us want to live. Although they are not acting in concert, they collectively represent a threat to healthy, sustainable development. Worse, they are effectively pricing responsible and well-meaning aid organizations out of the market in the very places where they are needed most. If they continue to succeed in pushing their alternative development model, they will succeed in underwriting a world that is more corrupt, chaotic, and authoritarian. That is in no one's interests, except the rogues.[70]

The activities of China and the other powers mentioned here do indeed undercut the development of global democracy; the Zimbabwean example is key. Zimbabwe's undemocratic ruling regime and the power of Robert Mugabe is given strength to survive another day because of these relations.

The inherent purpose of this discussion was to illustrate the systemic power of China. Seeing this power, and understanding its activities in terms of energy acquisition, I make the argument that coal will itself become a

Table 14.1 Chinese Aid Recipients in 2011

*Country**	*Country**	*Country**
Afghanistan	Equatorial Guinea	Mozambique
Algeria	Eritrea	Myanmar
Angola	Ethiopia	Namibia
Argentina	Gabon	Nepal
Bangladesh	Ghana	Niger
Benin	Grenada	Nigeria
Bolivia	Guinea	Pakistan
Botswana	Guinea-Bissau	Peru
Brazil	Haiti	Philippines
Burundi	Honduras	Rwanda
Cambodia	India	Senegal
Cameroon	Indonesia	Seychelles
Cape Verde	Iran (Islamic Republic of)	Sierra Leone
Central African Republic	Iraq	Singapore
Chad	Jordan	Somalia
Chile	Kenya	South Africa
Colombia	Lao People's Democratic Republic	Sri Lanka
Comoros	Lesotho	Sudan
Congo	Liberia	Thailand
Costa Rica	Libyan Arab Jamahiriya	Togo
Cuba	Madagascar	Tunisia
Côte d'Ivoire	Malawi	Turkey
Democratic People's Republic of Korea	Malaysia	Uganda
Democratic Republic of the Congo	Mali	United Republic of Tanzania
Djibouti	Mauritania	Uruguay
Dominica	Mauritius	Venezuela
Ecuador	Mexico	Viet Nam
Egypt	Mongolia	Zambia
El Salvador	Morocco	

*Many thanks to Wenyuan Wu for finding this information taken from: Wolf, C, Wang, X and Eric Warner 2013 "China's Foreign Aid and Government Sponsored Investment Activities: Scale, Content, Destinations and Implications" The RAND Corporat*ion*, Washington DC.

systemic force. As China facilitates its production it for its own purposes, others will follow.

Considering loans, and bringing in the other members of the BRICS alignment, the creation of a BRICS development bank will only serve to assist countries construct their own coal mines.—of *Foreign Policy* argues that "the new bank would likely finance projects similar to those backed by the institutions it is modeled on."[71] He cites Professor Kevin Gallagher of Boston University who suspects that the new bank would ". . . put a premium on infrastructure and energy, which was the World Bank's model in the 1970s . . ."[72] However, given the failure of the International Monetary Fund (IMF) to secure pragmatic economic growth due to the demands of the dreaded structural adjustment programs, the predicament of underdeveloped countries is understandable. For example, to deal with rogue states on a bilateral basis while making concessions is less difficult than to cut taxes and government expenditures.

In summation, weak countries are in need of energy and China is willing to help them develop their own coal industries; and in return, gain resource. Systemic change in energy use is on the horizon. In this section, we looked at the use of coal as a tool of an emerging power's policy. However, such reliance on coal of this nature is short-sighted. China has a serious and mounting pollution problem which affects the health and wellness of the world's population. If we adopt a broadened approach to the study of security that is threats, objects or complications that impede the autonomy of states and the degradation of human life,[73] then China's coal use is indeed a threat to its own national security. There will be a time when the stifled populations of China will rise up against these policies. The demand for clean air will become very real once it becomes absolutely intolerable, when the costs (pollution) attributed to economic growth begin to outweigh its benefits. Until this time, coal use will continue because of the economic gains which, as explained, will lead to domination of the international structure. However, human suffering will eventually come to a tipping point resulting in protest and even revolution.

In the following section, coal use as a reaction to an emerging power's policy is further demonstrated. I shall discuss the growing power of Russia and its use of its oil and gas supplies as a weapon that punishes countries for disobedience. In 2014, we find Ukraine and much of the European continent worried about the coming winter due the lack of energy supplies. As a result, analysts are now discussing the possibility of coal as a short-term solution to the conflict.

THE EU AND RUSSIA AND COAL: EUROPE NEED ENERGY DIVERSIFICATION NOW

The continuing conflict in Ukraine has proven to be a source of instability for the region and the world. At the time of writing, August 27, 2014, Ukraine voiced its concern over Russia shutting off its gas taps and pipelines to Ukraine and the countries of Europe. If this threat proves true, it would come as a response to growing European Union opposition against Russia due to its annexation of Ukrainian Crimea and continued Russian interferences in breakaway parts of eastern Ukraine. Many members of the European Union are dependent on Russia for much of their gas.[74]

Table 14.2 Percentage of Countries' Dependence for Russian Energy*

Countries	Percentage of Russian Dependence (approximations %)
Bulgaria	90 %
Czech Republic	66 %
Hungary	80 %
Lithuania	100 %
Poland	90 %
Slovakia	100 %
Slovenia	100 %

*"Factbox: How EU's eastern members depend economically on Russia" reuters.com, March 24, 2014 http://www.reuters.com/article/2014/03/24/us-ukraine-crisis-east-sanctions-factbox-idUSBREA2N0Q020140324

There are many more indices to study these countries' dependence on Russia such as in terms of exports and finance. Reliance of this magnitude creates a tenuous situation within this political context. However, energy dependence in this magnitude also presents enormous difficulty for these countries as the biting cold of the fall and winter months are approaching. If Russia does act in this manner, there is the threat of humanitarian disaster.

The threat is real as it would not be the first time Russia demonstrated its power in such a manner. In June 2014, Russia cut gas supplies to Ukraine. The reason was that Ukraine failed to pay its gas bill of $4.5 billion. However, this occurred within the context of the current conflict. Such acts would not just hurt Ukraine; it would hurt the European Union as pipelines from Russia feed into Ukraine and divert into other countries.[75]

Further, it must be noted that 2014 was not the first time Russia halted gas deliveries to other states as a show of force. In 2010, Lithuanian gas supplies fell drastically after a conflict over payment between Russia and Belarus led to Russia cutting gas supplies by 60%.[76] There have been other occasions as well; below is a list of some instances, as illustrated in Table 14.3.

Such action given dependence on Russia calls for drastic diversification of energy sources. The challenge for policy-makers is the avoidance of civilian suffering and death from freezing (most of the above events take place in winter months).

Interestingly, analysts are already drawing such conclusions: that coal and its use could be a viable alternative to Russian confrontational energy policy. Today, more than half of all coal exports from the United States go to Europe.[77] Given close ties to the United States, analysts expect coal exports to rise to fill energy gaps. Ernie Cecilia, chief investment officer at Bryn Mawr Trust stated in an interview with Reuters that "export demand will certainly increase, with the situation in Russia and Ukraine having a big impact on Europe with respect to natural gas."[78]

Table 14.3 Cases Russian Pipeline Disputes*

Country(ies) Affected	*Date*	*Summary*
Ukraine	January 2006	Moscow halts supply of gas due to pricing dispute with Ukraine
Ukraine	March 2008	Pricing dispute with Moscow results in a halving of gas supply
Ukraine/EU	January 2009	Pricing dispute with Moscow results in complete stoppage, including Europe
Ukraine	April 2010	New pricing deal: lower price in exchange for extended naval base lease in Ukrainian Black Sea
Belarus	January 2010	Month-long quarrel results in negotiation
Belarus/EU	June 2010	Moscow cuts supply by 15% due to debt; EU supplies threatened
Belarus	November 2011	100% ownership of Belarussian pipeline in exchange for 40% price cut and $10 billion for nuclear plant

*"Factbox: Russia's energy disputes with Ukraine and Belarus" reuters.com, December 21, 2012 http://www.reuters.com/article/2012/12/21/us-russia-gas-disputes-idUSBRE8BK11T20121221

Threatened countries, and citizens of these countries with their energy supplies under threat, must begin investing in alternatives to oil and gas. In the short-term, given the availability of coal as a cheap and versatile fuel source, states and peoples may begin a program of coal accumulation to prepare for the uncertain months ahead. Industries that run on oil and gas will ultimately be harmed as electricity is produced by these resources. Coal cannot hope to replace oil and gas unless the right infrastructure is installed. In the long-term, these states must begin to invest in energy diversification. It is necessary that these be sustainable fuel sources: wind, solar and other forms of natural, readily available or kinetic energies. Being 90–100% dependent on any power, especially a former oppressor, is both irresponsible and foolish.

To end this section, we must understand the context of this energy crisis. Relating back to the subject matter of this essay, this conflict is simply a part of systemic transition as Russia is growing in power and seeking recognition as a great power. The United States, the European Union and other major states must acknowledge Russia and respect its spheres of influence. This means appeasing Russia, allowing the aggressor to gain territory and influence in its desired areas, even at the expense of other states. This may staunch President Putin's appetite for power and it may not. Russia may not cease its desire to expand, but for the sake of peace in Ukraine, a compromise must be found. As Mearsheimer argues "appeasement accomplishes this end . . . by allowing the appeaser to demonstrate its good intentions and by shifting the military balance in the appeased state's favor, thus making it less vulnerable and more secure, and ultimately less aggressive."[79]

While the sanctions by the United States and the European Union against Russia targets specific people, Russia's weaponization of energy actions target the people of Europe. Russian action of this sort does nothing but degrade the soft power of Russia, ruining its reputation in the world and reducing Russian economic growth during global recovery. In this light, European states must adopt drastic and unorthodox measures in the short-term (stockpiling coal) and engage in pragmatic energy development and diversification in the long-term, especially of the clean and renewable variety. Countries of Europe who depend on Russia for fuel must begin to diversify; and do so quickly.

CONCLUSION

In the first few months of 2013, China imported 110 million tons of coal which represents a 25.6% increase compared to the entire length of 2012.[80] China must continue to develop the coal industries of other countries to secure the resources needed to continue growth and to avoid an energy crisis. The end result supports this paper's main argument: coal energy is a systemic force in the world. As coal use increases, then its importance will become an essential part of the new world order. China facilitates the use of coal mines in other countries for its own purposes. This is occurring especially in Zambia, Zimbabwe and Pakistan countries and many others experiencing energy crises. As a cheap and reliable fuel source compared to expensive alternatives, coal is a vital part of the economic geography. In this sense, underdeveloped countries can begin to grow their economies by way of coal development. China has proven itself to be a ready partner in helping these countries develop their coal resources. Unlike Western powers who advocate use of clean but unreliable and expensive technologies, China is willing to act quicker to produce energy more efficiently for these countries. However, as argued, this is not because of China's benevolence; rather, it is because China is acting in its own interests given its own energy needs. China needs to import the coal it needs, and aid will help encourage coal production in coal rich countries.

The world is entering multi-polarity and so status-quo powers must begin to expect and adapt to such change. The conflict in the Ukraine is only the beginning. Hence, I recommend that Western powers and their policymakers must now begin to recognize certain spheres of influence of BRICS countries. The West must begin to compete for the affections of weak states and other middle powers; they must make it easier to acquire loans and produce clean energy more affordably, more reliably and therefore more attractively than coal. This means using economic power and soft power as a means to influence. The only way to fight the seduction of coal as a cheap and ready resource is to continue developing sustainable energy sources. Once alternatives are provided to developing states as a feasible alternative to coal through a "carrot and stick," soft power approach, then the West can expect to curb, not only coal use, but the political influence of BRICS countries. The benefits will be priceless and may include energy independence, clean air, the stabilization of states' economies through the balance of payments and a reduction of Russian ability to use oil and gas as a geopolitical weapon. The world will indeed be a more stable place to live.

NOTES

1. The author would like to thank Elias Samire Kassab, M.A. for his kind contributions to the argument and research of this paper.
2. "China defends dispatch of spy ship to monitor US-led naval drills off Hawaii" *South China Morning Post*, July 21st, 2014 http://www.scmp.com/news/world/article/1556956/chinese-spy-ship-monitors-us-led-naval-drills-hawaii.
3. K Johnson, "Coke Brothers" Foreign Policy, 2014 http://www.foreignpolicy.com/articles/2014/07/22/coke_brothers_brics_bank_coal_china_ccs.
4. R Gilpin, *War and Change in World Politics*, (Cambridge: Cambridge University Press, 1981), 25.
5. Gilpin, *War and Change*, 29.
6. Kenneth Waltz, *Theory of International Relations*, (New York: Columbia University Press, 2010), 40.
7. Waltz, *Theory of International Relations*, 88.
8. Waltz, *Theory of International Relations*, 118.
9. Waltz, *Theory of International Relations*, 192.
10. Waltz, *Theory of International Relations*, 97.
11. Gilpin, *War and Change*, x.
12. Gilpin, *War and Change*, 90.
13. Gilpin, *War and Change*, 10.
14. Gilpin, *War and Change*, 10.
15. Gilpin, *War and Change*, 10–11.
16. Gilpin, *War and Change*, 13.
17. R. Mundell and A. Swoboda, *Monetary Problems of the International Economy*, (Chicago: University of Chicago, 1968), 343.
18. Gilpin, *War and Change*, 28.
19. For an in-depth look at weak state behavior, see: Kassab, Hanna S. *Weak States and Major Powers in the 21st Century*. Miami: University of Miami Dissertation, 2014.
20. Schweller, R. "Bandwagoning for Profit: Bringing the Revisionist State Back In." *International Security*, 35(1), (1994), 74.
21. N. Woods, "Whose Aid? Whose Influence? China, Emerging Donors and the Silent Revolution in Development Assistance," *International Affairs*, 84(4), (2008), 4.
22. C Kernot, *The Coal Industry*, (Cambridge: Woodhead Publishing Limited, 2000), 1.

23. Kernot, *The Coal Industry,* 2.
24. Kernot, *The Coal Industry*, 2.
25. Kernot, *The Coal Industry,* chapter 2, 1.
26. A. Warleigh-Lack, *European Union: The Basics*, (New York: Routledge, 2009), 4.
27. "Where is cold found?" *World Coal Association*, accessed September 2, 2014, http://www.worldcoal.org/coal/where-is-coal-found/.
28. "Global Power and Global CO2 emissions" *Thomson Reuters,* accessed September 6, 2014 http://blog.thomsonreuters.com/index.php/tag/united-states/.
29. "Uses of coal" *World Coal Association*, accessed August 28, 2014 http://www.worldcoal.org/coal/uses-of-coal/.
30. "Uses of Coal" *World Coal Association.*
31. "Uses of Coal" *World Coal Association.*
32. "Fueling Growth" *Geospatial World, accessed September 6, 2014* http://geospatialworld.net/images/BRICS-Mining-Production-Facts.jpg.
33. "Total energy consumption in China by Type, 2011" *US Energy Administration,* 2011 http://www.eia.gov/countries/cab.cfm?fips=ch.
34. "Fueling Growth" *Geospatial World,* accessed September 6, 2014 http://geospatialworld.net/images/BRICS-Mining-Production-Facts.jpg.
35. "China produces and consume almost as much coal as the rest of the world combined" *US Energy and Information Administration,* accessed September 4, 2014 http://www.eia.gov/todayinenergy/detail.cfm?id=16271.
36. J. Watts 'China's coal addiction" *Foreign Policy,* 2010 http://www.foreignpolicy.com/articles/2010/12/02/china_s_addiction_to_coal.
37. "Russia and China launch gas pipeline" *BBC News,* accessed September 2, 2014 www.bbc.com/news/business-29011092.
38. Watt "China's coal addiction."
39. "World's Largest Coal Producers" *Oil Price*, accessed September 6, 2014 http://oilprice.com/Energy/Coal/Peak-Coal-And-Chinese-Economic-Growth.html.
40. A. Otani, "U.S. coal stocks could gain on Russia tension" *Reuters,* 2014 www.reuters.com/article/2014/08/30/us-usa-stocks-weekahead-idUSKBN0GT26V20140830.
41. The BRICS can hardly be considered a solid, unbreakable alliance. On September 1, 2014, India's new Prime Minister, Narendra Modi, announced his concern over China. Although not specifically mentioning China by name, it is obvious: "everywhere around us, we see an 18th century expansionist mindset: encroaching on another country intruding in others' waters, invading other countries and capturing territory" (Modi quoted in Obe and Mandhana 2014, wsj.com). In classical balance of power fashion, India is reaching on to another concerned power: Japan. Along with the United States leading a conglomeration of other Asian states, India and Japan hope to counter any aggressive Chinese behavior. However, this still does not discount the power of coal as a systemic force. Individually, both China and India make good use of coal and they will continue to do so for the new decade.
42. "Total energy consumption in China by Type, 2011" US Energy Administration, 2011 http://www.eia.gov/countries/cab.cfm?fips=ch.
43. Gilpin, *War and Change*, 10.

44. D. Brautigam, "Aid 'with Chinese characteristics': Chinese foreign aid and development finance meet the OECD-DAC aid regime," *Journal of International Development,* 23(5), (2011), 752; 755.

45. Brautigam, "Aid with Chinese characteristics," 755.

46. He Wenping, "Moving Forward with the Time: the Evolution of China's African Policy" 2006, 5.

47. Woods, "Whose Aid? Whose Influence," 1.

48. Woods, "Whose Aid? Whose Influence," 3.

49. M. Naim, "Rogue Aid" *Foreign Policy,* 2007 http://www.foreignpolicy.com/articles/2007/02/14/rogue_aid.

50. "International Chinese Coal Projects," *sourcewatch.org,* accessed September 4, 2014 http://www.sourcewatch.org/index.php/International_Chinese_coal_projects.

51. S. Shaikh and S. Tunio, "Pakistan surges into coal-fired power plants to meet energy demand," *trust.org,* 2014 http://www.trust.org/item/20140611093028-fa051/?source=jtDontmiss.

52. "China firm to set up 10 coal fired power plants costing USD$7.2 billion" October 29, 2013 http://www.brecorder.com/pakistan/industries-a-sectors/142083-china-firm-to-set-up-10-coal-fired-power-plants-costing-us$-72-bln.html.

53. Z. Bhutta, "Chinese assistance sought for coal gasification," *The Express Tribune,* 2013 http://tribune.com.pk/story/592848/chinese-assistance-sought-for-coal-gasification/.

54. "China firm to set up 10 coal fired power plants costing USD$7.2 billion" October 29, 2013 http://www.brecorder.com/pakistan/industries-a-sectors/142083-china-firm-to-set-up-10-coal-fired-power-plants-costing-us$-72-bln.html.

55. Shaikh and Tunio "Pakistan surges."

56. Shaikh and Tunio "Pakistan surges."

57. "Zimbabwe and Coal" *sourcewatch.org,* accessed September 6, 2014 http://www.sourcewatch.org/index.php/Zimbabwe_and_coal.

58. "Japan and China to compete for Zimbabwean coal," *The Africa Report,* January 3, 2013 http://www.theafricareport.com/Southern-Africa/japan-and-china-to-compete-for-zimbabwean-coal.html.

59. "Japan and China to Compete for Zimbabwean coal," *The Africa Report.*

60. "Japan and China to Compete for Zimbabwean coal," *The Africa Report.*

61. "Coal exploration project by China-Africa Sunlight Exploration" *China Aid Data,* accessed September 6, 2014 http://china.aiddata.org/projects/30306.

62. KMPG International 2014 *Zambia Mining Guide* http://www.kpmg.com/Global/en/IssuesAndInsights/ArticlesPublications/Documents/zambian-country-guide.pdf.

63. "Zambian coal mine looking to increase productivity" *miningweekly.com,* September 13, 2013 http://www.miningweekly.com/article/zambian-coal-mine-looking-to-increase-productivity-beyond-2013-09-13.

64. KMPG International, *Zambia Mining Guide.*

65. Bearak, B. "Zambia Uneasily Balances Chinese Investment and Workers' Resentment," *The New York Times,* 2010 http://www.nytimes.com/2010/11/21/world/africa/21zambia.html?pagewanted=all&_r=2&.

66. "Zambian Coal Mine Looking to Increase Productivity," *miningweekly.com,* September 13, 2013 http://www.miningweekly.com/article/zambian-coal-mine-looking-to-increase-productivity-beyond-2013-09-13.

67. B. Bearak, "Zambia, Uneasily Balances."

68. Wonacott P. "China Investment Brings Jobs Conflict to Zambia Mines," *The Wall Street Journal,* 2012 http://online.wsj.com/news/articles/SB10000872396390444082904577609161447720808.

69. B. Bearak, "Zambia, Uneasily Balances."

70. Naim, "Rogue Aid."

71. Johnson, "Coke Brothers."

72. Johnson, "Coke Brothers."

73. Richard Ullman, "Redefining Security," *International Security,* 8(1), (1983), 133.

74. K. Friar, "Russia plans to halt gas flow to the EU, says Ukraine PM" Marketwatch.com, 2014 http://www.marketwatch.com/story/russia-plans-to-halt-gas-flow-to-eu-says-ukraine-pm-reuters-2014-08-27?dist=countdown.

75. K. Friar, "Russia Plans to Halt Gas Flow" .

76. "Lithuania Gas Supply Falls After Russia-Belarus Dispute," *BBC News,* June 23, 2010 www.bbc.co.uk/news/10389250.

77. Otani, "U.S. Coal Stocks Could Gain on Russia Tension."

78. Otani, "U.S. Coal Stocks Could Gain on Russia Tension."

79. John J. Mearsheimer, *The Tragedy of Great Power Politics*. (New York: W.W. Norton and Company, 2001), 163.

80. D. Juan "China coal imports continue to affecting global prices: Platts" *China Daily,* 2013 http://www.chinadaily.com.cn/business/2013-05/22/content_16518558.htm.

BIBLIOGRAPHY

Bearak, B. "Zambia Uneasily Balances Chinese Investment and Workers' Resentment," *The New York Times,* 2010 http://www.nytimes.com/2010/11/21/world/africa/21zambia.html?pagewanted=all&_r=2&.

Bhutta, Z. "Chinese Assistance Sought for Coal Gasification," *The Express Tribune,* 2013 http://tribune.com.pk/story/592848/chinese-assistance-sought-for-coal-gasification/.

Brautigam, D. "Aid 'With Chinese Characteristics': Chinese Foreign Aid and Development Finance Meet the OECD-DAC Aid Regime," *Journal of International Development,* 23(5), (2011): 752–764.

Carr, E. H. *The Twenty Years' Crisis, 1919–1939: An Introduction to the study of International Relations*. London: Macmillan, 1978.

"China Defends Dispatch of Spy Ship to Monitor US-led Naval Drills off Hawaii," *South China Morning Post,* July 21st, 2014 http://www.scmp.com/news/world/article/1556956/chinese-spy-ship-monitors-us-led-naval-drills-hawaii.

"China Firm to Set Up 10 Coal Fired Power Plants Costing USD$7.2 Billion," October 29, 2013 http://www.brecorder.com/pakistan/industries-a-sectors/142083-china-firm-to-set-up-10-coal-fired-power-plants-costing-us$-72-bln.html.

"China's Foreign Aid and Government Sponsored Investment Activities: Scale, Content, Destinations and Implications," *The RAND Corporation*, Washington D.C.

"Coal exploration project by China-Africa Sunlight Exploration," *China Aid Data,* accessed September 6, 2014 http://china.aiddata.org/projects/30306.

"Factbox: Russia's Energy Disputes With Ukraine And Belarus," reuters.com, December 21, 2012 http://www.reuters.com/article/2012/12/21/us-russia-gas-disputes-idUSBRE8BK11T20121221.

"Factbox: How EU's Eastern Members Depend Economically on Russia," reuters.com, March 24, 2014 http://www.reuters.com/article/2014/03/24/us-ukraine-crisis-east-sanctions-factbox-idUSBREA2N0Q020140324.

Friar, K. "Russia Plans to Halt Gas Flow to the EU, says Ukraine PM," Marketwatch.com, 2014 http://www.marketwatch.com/story/russia-plans-to-halt-gas-flow-to-eu-says-ukraine-pm-reuters-2014-08-27?dist=countdown.

"Fueling Growth," *Geospatial World,* accessed September 6, 2014 http://geospatialworld.net/images/BRICS-Mining-Production-Facts.jpg.

Gilpin, R. *War and Change in World Politics.* Cambridge: Cambridge University Press, 1981.

"Global Power and Global CO_2 Emissions," *Thomson Reuters,* accessed September 6, 2014 http://blog.thomsonreuters.com/index.php/tag/united-states/.

"International Chinese Coal Projects," *sourcewatch.org,* accessed September 4, 2014 http://www.sourcewatch.org/index.php/International_Chinese_coal_projects.

"Japan and China to Compete for Zimbabwean Coal," *The Africa Report,* January 3, 2013 http://www.theafricareport.com/Southern-Africa/japan-and-china-to-compete-for-zimbabwean-coal.html.

Johnson, K. "Coke Brothers," *Foreign Policy,* 2014 http://www.foreignpolicy.com/articles/2014/07/22/coke_brothers_brics_bank_coal_china_ccs.

Juan, D. "China Coal Imports Continue to Affecting Global Prices: Platts," *China Daily*, 2013 http://www.chinadaily.com.cn/business/2013-05/22/content_16518558.htm

Kassab, Hanna S. *Weak States and Major Powers in the 21st Century.* Miami: University of Miami Dissertation, 2014.

Kernot, C. *The Coal Industry*. Cambridge: Woodhead Publishing Limited, 2000.

Khan, A. "The Tyranny of Small Decisions: Market Failures, Imperfections, and the Limits of Econometrics," ed. Bruce M. Russett, *Economic Theories of International Relations*, Chicago: Markham, 1968.

KMPG International 2014 *Zambia Mining Guide* http://www.kpmg.com/Global/en/IssuesAndInsights/ArticlesPublications/Documents/zambian-country-guide.pdf.

"Lithuania Gas Supply Falls After Russia-Belarus Dispute," *BBC News,* June 23, 2010 www.bbc.co.uk/news/10389250.

Mearsheimer, John, J. *The Tragedy of Great Power Politics*. New York: W.W. Norton and Company, 2001.

Mundell, R and Swoboda, A. *Monetary Problems of the International Economy.* Chicago: University of Chicago, 1968.

Naim, M "Rogue Aid," *Foreign Policy,* 2007 http://www.foreignpolicy.com/articles/2007/02/14/rogue_aid.

Otani, A. "U.S. Coal Stocks Could Gain On Russia Tension," *Reuters,* 2014 www.reuters.com/article/2014/08/30/us-usa-stocks-weekahead-idUSKBN0GT26V20140830.

Prebisch, R. "Commercial Policy in Under-developed Countries," *American Economic Review, Papers and Proceedings,* 492, (1959): 251–273.
"Russia and China Launch Gas Pipeline," *BBC News,* accessed September 2, 2014 www.bbc.com/news/business-29011092.
Shaikh, S and Tunio, S. "Pakistan Surges into Coal-Fired Power Plants to Meet Energy Demand," *trust.org,* 2014 http://www.trust.org/item/20140611093028-fa051/?source=jtDontmiss.
Schweller, R. "Bandwagoning for Profit: Bringing the Revisionist State Back In.," *International Security,* 35(1), (1994): 72–107.
"Total Energy Consumption in China by Type, 2011," *US Energy Administration,* 2011 http://www.eia.gov/countries/cab.cfm?fips=ch.
Ullman, Richard. "Redefining Security," *International Security,* 8(1), (1983): 140.
"Uses of Coal," *World Coal Association*, accessed August 28, 2014 http://www.worldcoal.org/coal/uses-of-coal/.
Waltz, Kenneth. *Theory of International Relations.* New York: Columbia University Press, 2010.
Warleigh-Lack, A. *European Union: The Basics.* New York: Routledge, 2009.
Watts, J. 'China's Coal Addiction," *Foreign Policy,* 2010 http://www.foreignpolicy.com/articles/2010/12/02/china_s_addiction_to_coal.
Wenping, He. "Moving Forward with the Time: the Evolution of China's African Policy," 12 November 2006. http://www.cctr.ust.hk/materials/conference/china-africa/papers/He,Wengping.pdf.
"Where is Cold Found?" *World Coal Association*, accessed September 2, 2014, http://www.worldcoal.org/coal/where-is-coal-found/.
Wonacott P. "China Investment Brings Jobs Conflict to Zambia Mines," *The Wall Street Journal,* 2012 http://online.wsj.com/news/articles/SB10000872396390444082904577609161447720808.
Woods, N. "Whose Aid? Whose Influence? China, Emerging Donors and the Silent Revolution in Development Assistance," *International Affairs*, 84(4), (2008): 1–17
"World's Largest Coal Producers," *Oil Price*, accessed September 6, 2014 http://oilprice.com/Energy/Coal/Peak-Coal-And-Chinese-Economic-Growth.html.
"Zambian Coal Mine Looking to Increase Productivity," *miningweekly.com,* September 13, 2013 http://www.miningweekly.com/article/zambian-coal-mine-looking-to-increase-productivity-beyond-2013-09-13.
"Zimbabwe and Coal," *sourcewatch.org,* accessed September 6, 2014 http://www.sourcewatch.org/index.php/Zimbabwe_and_coal.

Chapter 15

Russian Re-Engagement with Latin America

Energy and Beyond

Vladimir Rouvinski

During the first week of September 2014, the news headlines in Russia continued to be dominated by the topics related to the conflict in Ukraine and the new round of sanctions imposed on the country by the United States and the European Union. Taken into account the importance given by the Kremlin to the use of mass media for keeping high the level of public support for the Russian stand in the conflict, it seemed very unlikely that any other news—unrelated to the troubled Eastern Ukraine—would make it to the headlines. Nothing indicated that only a few days later, on September 16th, the Russian media would explode with the news of the detention of one of the top Russian businessmen and the member of the Forbes short-list, Vladimir Yevtushenkov.

For many of the Russian populace, the placing of one of the richest men in the country under a house arrest was difficult to explain. Only a few trusted the official story that the arrest was related to some crimes committed by Mr. Yevtushenkov in the past. However, no other version was easily coming to mind since Yevtushenkov was not known as a supporter of the Russian opposition, which would be a more reliable explanation of the cause of the arrest. After all, a strong opponent of Vladimir Putin and a former oil tycoon Mikhail Khodorkovsky paid ten years in prison for giving a financial support to the opposition parties without the Kremlin's permission[1].

Almost simultaneously, as the general public in Russia was looking for these answers, the man—now believed to be behind the events that happened in Moscow in mid-September—was far from Russia: the head of Rosneft, the biggest state-owned oil company in the world, Igor Sechin, was busy planting trees in Sabaneta, Venezuela, the birthplace of Hugo Chavez. At the main square, he was accompanied by the President of Venezuela, Nicolas Maduro, and they jointly announced a donation, equivalent to 1 million

US dollars, issued by Rosneft to the government of Venezuela and destined to support the construction of a museum dedicated to Chavez. Whereas the experts in Moscow eventually came to a conclusion that the real "guilt" of Yevtushenkov—that caused the arrest—was his refusal to pass one of his company's important oil assets, Bashneft, to Rosneft,[2] the head of the latter had a prefect alibi: in Moscow, it is said that Sechin is spending more time in Venezuela then visiting the company's key oil fields in Siberia.[3] Hence it was nothing unusual that Sechin was paying a visit to his friend Maduro.

It is important to note here that being one of Russia's oldest oil companies, Bashneft has been producing oil since 1932. Today, it is amongst the country's fastest growing oil companies, which ranks in the top-10 Russian companies in terms of oil production and in the top-5 in terms of oil refining. The Bashneft owns over 170 fields of commercial production that exceeds 16 million tons of oil per year and it has a considerable research potential and extensive experience in developing and introducing state-of-the-art technologies for oil exploration and production[4]. When Yevtushenkov was arrested, he owned 85 per cent of Bashneft. The Russian prosecutors claimed that the businessman obtained the company illegally. At the same time, it is well-known that in the recent past Igor Sechin has attempted several times to annex Bashneft to Rosneft.[5]

Yevtushenkov was suddenly released on December 17. Right after his liberation was announced, he joined a reception hosted by Vladimir Putin in the Kremlin. Again, at first sight, it seemed the release of Yevtushenkov could be linked to a severe Russian currency crisis occurring at the time, and the desire of the Russian President to ensure the support by the Russian business community. However, almost at the same time, it became known that the oil asset in question was taken away from the company by the Russian government.[6] Soon, Yevtushenkov publicly announced that he would not fight to get it back.[7]

The above events in Moscow were an indication of the increasing political power of Igor Sechin, who nowadays is one of the most powerful figures in the country.[8] The case is also evidence of the special place that oil-related issues play in the Russian politics today. If it would not be for oil, it is highly unlikely that Sechin would risk going after Yevtushenkov thus violating the unwritten rules for businessmen in Putin's Russia: "safety from prosecution in exchange for no involvement in politics."[9]

Moreover, it is interesting to note that only a few people in Russia raised questions over the fact that the state-owned company was giving away a huge amount of money to a country in Latin America, which has been in the midst of a dangerous political turmoil and accumulating a significant debt for the Russian supplied arms and machinery. But rather than being a case of public inattentiveness, it is an indication of a high degree of public approval in

Russia of a comradeship built by Russian and Venezuelan leaders in recent years that has been nurtured by intensive bilateral political contacts and attention given to the topic by the government-controlled mass-media. From this perspective, it is worth mentioning here that as a member of Vladimir Putin's "inner circle," Igor Sechin is considered one of the most powerful figures in Russia. In addition, he is believed to be in charge of Moscow's policy toward Latin America.[10]

At the same time, there are evidences that the official discourse, which emphasizes the equally attractive benefits of the Russian trade and energy collaboration with Venezuela, is far from telling the whole story. It seems that, in reality, some of the Russian companies, with already established presence in Venezuela, fear any further involvement and even try to leave the country because of the political instability and variety of other business risks. However, they are under a heavy pressure from the Kremlin power men who want them to remain in Venezuela.[11] But why does Moscow want them to stay?

This question cannot be answered in a proper way if taken outside of a much broader context of the Russian re-engagement with Latin America and its global energy policies. To better understand the latter, it would be necessary to first examine a number of issues that are related both to the transformation of the Russian energy companies following the collapse of the Soviet Union *and* of the Russian foreign policy toward Latin America. The fact is that it is very difficult to properly identify the interests behind the continuing presence of the Russian businesses in Venezuela, and in some other countries in the region, without taking into consideration the entire evolution of Russian foreign policy in Latin America following the collapse of the Soviet Union almost a quarter of century ago.

This process is complex in nature, and often not properly understood by many observers. Whereas some of these observers regard the presence of the Russian energy companies in Latin America as no more than a huge cover operation to hide Moscow's intention to establish a long-term presence for spying and implementing other kind of hostile activities toward the United States,[12] other experts do not see many differences between the way in which Russia is doing business in Latin America and, say, China, although with much limited resources.[13] In my view, neither interpretation is accurate since—as it will be shown in this paper—the Russian interests, with regard to its energy sector presence in the Americas, is by no means limited to simply bothering Washington in its "near-abroad" in response to the US support of Georgia or Ukraine.[14] Moscow does not copy the strategy adopted by Beijing either: for the latter, the energy cooperation with Latin America is the priority of its policy toward the region[15] whereas for Moscow it is one of the dimensions related to the other objectives discussed below in this text.

Hence, the Russian case in Latin America is rather a unique one. Any attempts to demystify the nature of energy cooperation between Russia and Latin America, must examine the following: (1) How has the changing Russian domestic rules of the game, in politics and business, shaped the transformation of strategies adopted by the Russian energy companies for its activities outside Russia in the 1990s and during the first 15 years of the new century? (2) What are the changes in defining the short, middle, and long-term goals of the Russian foreign policy in Latin America during the same period? (3) What is the role given to the energy cooperation with Latin America by the Russian government between 1992 and 2014?

RUSSIAN ENERGY COMPANIES IN CHANGING BUSINESS AND POLITICAL ENVIRONMENTS

The emergence of the companies that dominate today's Russian energy sector began in the 1990s, a time remembered by many Russians as an era of the mafia wars and the dominance of organized crime groups in many aspects of economic and daily life. It was back then when a number of individuals had started making unbelievable fortunes by taking advantage of the opportunities arising during the unprecedented transformations of post-Soviet Russia, thanks to ambiguous regulations and corrupted officials.[16] Perhaps, we will never know the true story of how those fortunes were really made. However, what is important to note is that the group of "Russian oligarchs"—most of whom were related to the energy sector—in the Russian political process, cannot be overestimated, especially during the second term of President Yeltsin (1996–2000). This group was getting what it wanted and when it wanted it. Hence, these companies were quite satisfied with the status-quo that existed in Russia during that time, and, as yet, were not very much into opportunities outside the Russian borders. The other two reasons for the shortage of noticeable activities by Russian energy companies in other countries was the absence of national qualified and experienced personnel, capable of doing business outside the country, and, basically, no resources to hire expensive foreign staff, much less to invest outside the country. It was the time when joint ventures in Russia with foreign companies were booming: the companies from Europe and the United States had had both qualified personnel and resources to invest.[17]

New scenarios appeared after Vladimir Putin took power in 2000. On the one hand, Putin made it more difficult for the Russian energy tycoons to continue dealing with the Russian government on the same terms as during the times of Boris Yeltsin. On the other hand, the rising oil prices, and a larger number of qualified staff and new business strategies adopted by companies like Yukos and Lukoil made possible the beginning of the expansion of the Russian energy companies beyond the Russian borders.[18] It was back then when, for the first time, a Russian private energy company Lukoil made its presence noticeable in Latin America by signing a contract with Ecopetrol in Colombia.[19] However, the quest of a number of important privately owned Russian energy companies for new assets outside Russia had intensified following the destruction of the Yukos Company and the imprisonment of Michail Khodorovkosky in 2003. The rules of the game had changed, and, following the nationalization of Yukos, it was not entirely clear whether or not the government may attempt to adopt similar measures for other private companies. Furthermore, in some cases, for these companies, making business abroad turned out to be a more profitable and safer than at home;[20] at least until 2014, when the sanctions imposed on Russia in the context of the ongoing crisis in Ukraine threatened both state and private Russian companies. It is important to note here the companies that have been active participants of the Russian-Latin American interactions during the last two decades, namely, Gazprom, Lukoil, and Rosneft.

In the above list, Lukoil occupies a special place since it was the first Russian private energy company, which started its operations abroad. The company was founded in 1991, got its current name in 1993 and, today, it is the second largest oil producer in Russia, with 182 branches in 43 countries, including the United States and Venezuela. The company's key figure is Vagit Alikperov, who started his career as an oil engineer and was eventually promoted to the post of deputy minister of oil and gas production of the Soviet Union. While back in 2003, the company's revenue was approximately 3,701 billion US dollars, ten years later (2013), the number grew to almost 140 billion. The company currently employs some hundred and fifty thousand people.[21]

Another company, Rosneft, was founded in 1993, with 100% participation reserved to the government of Russia. During the 1990s, the company almost went bankrupt. Many experts believed that it was due to the management problems. When the government appointed Serguei Bogdanchikov as its CEO, Rosneft started improving its financial situation, but the real success began in 2004, when the company acquired almost all of the former assets of the Yukos company. Today the largest Russian oil producer, Rosneft employs more than 170,000 people, and its revenue reached 102 billion US dollars in 2013.[22] From May, 2012, the Company's president is Igor Sechin, and he is actively pursuing new markets and new assets for Rosneft. Currently, besides Venezuela, the company has its interest in a dozen of countries and holds a 20-year contract to export 15 million tons of oil to China.

No description of the key Russian energy companies can be complete without mentioning of Gazprom, the largest extractor of natural gas in the world and one of the world's largest companies. The Gazprom was founded back in the 1980s, when the Soviet ministry of gas industry was converted into a corporation, and today the Russian government owns 50,01% of it. By 2013, the Gazprom's revenue reached 153 billon US dollars, and it employs almost 400,000 people.[23] Before the 2014 crises, the profits earned by Gazprom surpassed the combined income of all the companies of about a hundred of sovereign nations.

Besides these three giant corporations, Russia has other relatively smaller energy companies, which fully took advantage of favorable scenarios during the last decade. At the same time, it is necessary to mention an important characteristic regarding the evolution of the Russian energy sector, especially when it comes to the changes in the domestic politics during the second term of President Putin, when the state-owned companies benefitted more from the governmental support and had a better understanding of the changing rules of the game inside the country. When the state-owned companies venture outside the country, it normally takes place either with an exceptionally strong governmental support or as a part of a Russian foreign policy strategy toward a particular country or a region. As I show herein, nowhere can this be seen more clearly than in the case with the Russian policy toward Latin America. At the same time, in order to better understand the Russian energy policy particularly in Latin America, it is very useful to take into account the following characteristics that were common to many Russian energy companies, which intended to run their business abroad following the collapse of the Soviet Union.[24]

Despite the fact that the Russian energy companies are well aware of the experiences, strategies, and best practices used by the well-known energy companies from the other countries, they often fail to apply this knowledge to design their own strategies in their conquest of the markets. As I discuss throughout the paper, in more than just a few cases, the companies from Russia have been making painful mistakes in their attempts to enter Latin American markets. These mistakes most likely would have been possible to avoid had the companies better calculated the related risks.

With regard to the international expansion of the Russian companies, the most frequently used and familiar business strategies relate to investment, above all through organic growth strategy—that is placing emphasis on the company's own resources—, along with mergers and acquisitions, and through building international strategic alliances.

The investment strategies used by the Russian oil and gas companies are often different from the ones used by other Russian companies with expanding international presence. Whereas in recent years the majority of the Russian companies outside of the energy sector have focused on the strategy of mergers and acquisitions, the oil and gas companies preferred the implementation of organic growth strategy in the form of exploration, mining projects, joint ventures and some others. These strategies became a priority for a number of reasons. On the one hand, during the last decade, many big Russian energy companies have accumulated significant financial resources as a result of successful exports of hydrocarbons and high energy prices, making significant investments in projects "from scratch" possible without extensive external borrowing. On the other hand, the adoption of the organic growth strategy when doing business abroad by Russian private energy companies was also due to the prevalence of negative domestic factors, such as: limited availability of proven domestic fields, prevention, by the government, from the participation in shelf projects in Russia, and decreasing performance of the existing projects. When it comes to the use of the strategy of mergers and acquisitions of foreign companies, Russian energy companies mainly pursue the goal of increasing their market share. Russian energy companies cultivate international strategic alliances in order to gain access to advanced technology, as well as to use the commercial and business networks of the foreign partners. Attempting to diversify and expand their business abroad, the Russian companies are increasingly using a whole variety of marketing strategies. However, branding and the promotion of a positive image of the company are the least developed elements.[25]

The Russian energy sector recognizes the importance of adopting the innovation strategies as the key to improve its international competitiveness and sustainability. Indeed, there are examples when the Russian companies are pioneering the use of innovative technologies.[26] And yet, both the Russian companies and the Russian government make no mistake that the innovative strategies remain the Achilles' heel of the Russian energy sector. With a few exceptions, the integrated innovation strategies have not been adopted by the majority of the Russian players in the field. In comparison with its foreign counterparts, there has been no major investment in research and industrial design by the key Russian companies. Russia lacks its own production of important technological equipment for gas liquefaction and production of synthetic fuels. The Russian energy companies have only a very limited experience in the production of hard hydrocarbons. There is a noticeable lag in terms of possessing modern technological equipment and know-how with regard to the refining segment.[27] In this context, whereas in the beginning of its business activities in the international markets—above all in the developing countries—, it was possible for the Russian companies to rely on the Soviet era technological expertise, today, the weakness of the Russian energy sector with regard to technologies and know-how is alarming, and it had an inevitable impact on the companies' performance everywhere even before the sanctions introduced by the United States and its allies in 2014.

Very importantly, for the most part of the post-Soviet era, it seems that in many cases the Russian private energy companies have failed to coordinate its strategies with the Russian Embassies whereas the strategies of the state-owned companies—as in the example of Latin America discussed here—rapidly became too dependent on the government's changing political stakes with regard to one or another country.[28] Summing up, it is possible to outline the following fundamental characteristics of the transformation of strategies adopted by the Russian energy companies for its activities outside Russia, including Latin America, in the 1990s and during the first 15 years of the new century.

The process of the Russian energy sector abroad can be divided into two phases. The first one starts in the mid-1990s and lasts till the beginning of the 2000s. Back then, it was the private sector that was leading the Russian efforts to establish the presence in foreign markets. The next phase—the current one—started after Vladimir Putin came to power and it is heavily marked by the direct involvement of the Russian government interested in the use of the energy strategy as a useful tool to reach its policy goals in and outside the country. Both periods, however, have some things in common, namely, the disadvantages of the Russian private energy companies in terms of business strategies, know-how, technologies and often the lack of the government support comparing with the capacities of the rival companies from North America, Europe, and China. In addition, since the changing political settings in and outside of Russia turned out to be a key driver of the Russian involvement in Latin America, before embarking upon the story of the Russian energy sector presence in this part of the world, it is necessary to describe the changes with regard to short, middle, and long-term goals of the Russian foreign policy Latin America during the same periods.

THE RUSSIAN "RETURN": WHEN AND WHY

When it comes to Russian policy in Latin America following the collapse of the Soviet Union, it is possible to talk about the early twenty-first century as the beginning of a new era in relations between Russia and Latin America. Back in 1997, the Russian prime-minister, Yevgeny Primakov, undertook his famous tour of several Latin American countries. Primakov's trip can be regarded as the first attempt by post-Soviet Russia to look into opportunities to build stronger ties with Latin America through trade and energy cooperation since the Russian foreign minister actively promoted the interests of the Russian energy and military-industrial sector. And yet, it would have been difficult for Mr. Primakov to imagine that soon after his journey to the subcontinent, top-level Russian officials would become frequent visitors to the region: between 2000 and 2014, Russian Presidents would fly eight times to Latin America, and the Minister of Foreign Affairs would visit Latin America a dozen of times during the same period. It would have been equally difficult to imagine in 1997 that Russian trade with this part of the world—quite unimpressive in the 1990s—would be counted by billions US dollars by the end of 2014.

In 2015, it is impossible to deny the existence of strong evidence of the Russian multidimensional advance into the formerly taboo territory.[29] This story is not only about rocketing growth of Russian arms sales to Latin American countries. It is also about the visits by Russian navy ships and strategic bombers to Venezuela, Cuba, and Nicaragua; the talks about possible Russian military bases in Central and South America as well as about the re-opening of the surveillance facilities on Cuba. And yet, although these developments can be considered as quite alarming by Washington, it is difficult to agree that they really mark the beginning of Russia implementing a policy similar to the one the Soviet Union had in the recent past. The Russians indeed have returned to the region, but they are back because of the reasons quite different from those that brought the Russian missiles to the island of Cuba in 1962.

It is important to note that since the 2008 version of Russian Foreign Policy Concept, the Latin American vector of its foreign policy is considered as a strategic partnership, aiming at broadening "its political and economic cooperation with . . . Latin American and Caribbean countries and their associations relying on the progress achieved in relations with the States of this region in recent years," enhancing "its interaction with these States within international organizations," promoting "export of Russia's high-technology products to Latin American countries" and implementing "joint energy, infrastructure and high-tech projects."[30]

However, the placing of emphasis on energy and infrastructure does not mean that Moscow would support any project in the energy sector of Latin America. In order to be approved, it would likely to have both a substantial political value for the Kremlin and be attractive for the Russian energy companies in terms of potential profits. This is because it is difficult to characterize contemporary Russian politics in the region as the Russian "return." Today's Russian interests in Latin America are multidimensional and thus very different from those behind the Soviet foreign strategy during the Cold War years, when the Kremlin's acting in the Western Hemisphere was doomed by the logics of the rivalry between two superpowers, USSR and USA. For the most part of the second half of the twentieth century, the Soviet Union would look for an opportunity to challenge the United States anywhere in the world. In Latin America, this can be illustrated by the case of the Soviet-Cuban relations when Moscow supported Castro's Cuba and helped Havana to resolve its energy demands even though it was not brining any tangible financial profits to the USSR.[31] But during the early post-Soviet period and up until the 2008 war with Georgian, the Russian policy toward Latin America was designed to take into account primarily the commercial interests of Russian private and state-owned companies. Therefore, to be precise, one should not talk about a Russian "return" to Latin America but about the new Russian politics in this region, and since the Primakov's Latin American tour, the commercial interests of Russia Inc. have been the major driving force of these politics, including after 2008, when Russian foreign policy in Latin America sought, for the first time, to respond to the challenges brought about by the events that were taking place in Europe. Back then, Moscow urgently needed the political support by Latin American countries of the Russia's stance and the recognition of the breakaway republics of Abkhazia and South Ossetia.[32] As a result of the ongoing crisis in Ukraine, there have been increased attempts at achieving a broader political cooperation between Russia and some of the governments in Latin America. However, Russia continues to rely on trade, commerce, and energy cooperation as the fundamentals of its cooperation with this part of the world.[33]

Along with those discussed earlier, it is possible to identify several important developments related to the Russian presence in Latin America during the years after the collapse of the Soviet Union in 1991. Firstly, by 2008 the Russian trade with the countries like Brazil, Argentina, Chile, and Ecuador was already counted by hundreds of millions of US dollar. Perhaps even more significantly, the key partners of Russia were not the former Soviet allies and sympathizers but ones of the most dynamically growing countries, including Brazil, the South American emerging power.

At the same time, 2008 was also the year the ties that Moscow was starting to build with the countries of the Bolivarian Alliance for the Americas (ALBA) were being tested, in particular, with Chavez' Venezuela and Nicaragua in the hands of Daniel Ortega. In the case of the former, since 2005, Russia has supplied Caracas with a hundred thousand Kalashnikov automatic rifles, two dozen on modern fighter jets, and approximately 50 helicopters, at a total cost estimated at four billion dollars. This constituted a dramatic breakthrough by Russia into the Latin American arms market and placed Venezuela on the top of the Kremlin's favorites list. However, that year, the Russian government had a chance to show its appreciation to Chavez and Ortega for another reason as well. In August 2008, having won, in military terms, the five-day war with Georgia over the support of the separatists republics in the Caucasus, Russia was losing the media war, and urgently needed some kind of international backing. Nicaragua became the first nation in the world, besides Russia, which officially recognized Abkhazia and South Ossetia as independent countries just a few days after the war ended. Venezuela soon followed. Although no other country in the Western Hemisphere gave its recognition to the break-away republics, for Moscow, the case became an indication that it is possible to deal with Latin America beyond trade and commerce.[34] Hence, when a similar scenario was developing in Europe during 2014, that is Moscow was facing an international isolation because of its acting in Ukraine, Russia was eagerly looking for the support in the region, which it obtained from a number of governments.[35] These two cases—the political support of Russia in the Georgian 2008 crisis and in the Ukrainian 2014 crisis—convinced Vladimir Putin and the Russian elite, on the one hand, that Latin America can be trusted and, on the other hand, have shaped the way in which Moscow has been doing business in the Latin American energy sector.

RUSSIAN ENERGY COMPANIES IN LATIN AMERICA: BETWEEN PROFITS AND POLITICS

The interest of the Russian energy companies toward the possibilities that were opening up in the Western Hemisphere, and, especially in South America became noticeable as early as the middle of the 1990s. Back then, the young Russian companies were full of high expectations based on the optimistic forecasts related to the new opportunities in Venezuela, Argentina, Brazil, Mexico, and Cuba. As the title of an article published by one of the Russian most important business community newspapers stated, "Latin Americans cannot cope with their oil on its own."[36] Yet, at the moment that the above-mentioned article was published, the only Russia's presence in the region was in the southern part of Peru, where one of the Yukos subsidiaries Yuganskneftegaz that established a joint venture with Petro Andes S. A. It looks like the Russian companies thought to take advantage of the Soviet-era networks in order to build their relations with Latin American partners of the Soviet energy companies.[37] However, soon it became clear that the conquest of the Latin American energy sector would not be an easy ride.

The mistakes made in the designing of business strategies, security issues, lack of experience and/or government support caused troubles to many companies. The example of the presence of the Russian oil companies in Colombia is an illustrative one. To begin with, it is important to mention that besides a short-lived rapprochement between Moscow and Bogota during the government of Ernesto Samper (1994–1998) and a few other episodes, the political relations between these two countries could be characterized as rather "dormant."[38] Hence, there was little political support from the Kremlin for the companies that ventured to explore the possibilities in the Colombian energy sector.

In 2001, Rosneft became the first of the Russian companies to enter the Colombian market when it acquired 45 per cent of the Colombia Energy (the other partners were Petrotesting Colombia S.A. (27.5%) and Holsan Chemicals Ltd (27.5%). On 25 December 2002, oil was first obtained in the southwestern part of Colombia, and it was expected that by mid-2003, the oil production would reach 6 thousand barrels per day.[39] Further developments would include the modernization of the pipelines and the facilities at the Colombian port of Tumaco on the Pacific coast. However, the project failed to fully meet the expectations of the Russian partners while the security risks turned out to be too high in the midst of the internal conflict Colombia, and Rosneft was forced to sell its share in 2004.[40]

Another company that eventually left Colombia was Lukoil. The company started its explorations in the Condor block, located in the Eastern Cordillera foothills in the western part of the Llanos oil and gas basin,[41] approximately 70 kilometers to the southeast of Bogota, by entering into an alliance with Ecopetrol and acquiring 70 per cent of the shares. The alliance expected to last nearly 30 years. However, in the end of almost a decade of exploration, Lukoil quit the Condor project in the end of 2012 having invested some 200 million US dollars and forced to pay a fine for the environmental damage a few years earlier.[42] The official explanation given by the company justified the decision to quit by a low level of production and the changing business strategy.[43]

At the same time, the failures to obtain the expected results were not sufficient causes to quit in those cases when Moscow was interested in the strengthening of its ties with the political leadership of particular Latin American countries. The Russian energy companies' involvement in Venezuela—the country neighboring Colombia—is one of these cases, which serves as a good illustration of the new Russia approach to Latin America. Russia's political stand and support in the international arena, it seems, was placed above the pursuit of commercial profits in the changing international context after the war with Georgia in 2008, and especially during the tensions with the United States and its allies as a result of Moscow's policy in Ukraine and the annexation of the Crimean peninsula in 2014.

In Venezuela, amid the powerful Russian privately and state-owned corporations that were gaining access to this South-American market in the 2000s, one could find all three companies mentioned earlier, namely Gazprom, Rosneft, and Lukoil. Already during Chávez's first visit to Moscow in 2001, it had been decided that a high-level intergovernmental commission should be created in order to promote economic links between two nations. Since October of 2004, the commission was initially chaired, on the Russian side, by technocrat and vice-prime minister, Alexander Zhukov to be later replaced by Igor Sechin. On the Venezuelan side, the commission was headed by the Vice-President José Vicente Rangel who, during his visit to Moscow shortly before Chávez's second visit in November of 2004, met with a number of Russian businessmen, including those involved in the energy sector.[44]

Just one month after the Venezuelan delegation visited Moscow, in December of 2004, Alexey Miller, the CEO Gazprom paid a personal visit to Caracas. Later on, in 2006, Gazprom won a tender for the exploration of some fields in the Venezuelan territorial waters. However, in the end, having invested some 300 million and having drilled the first and the only well at Urumaco I, Gazprom failed to reach any plausible results. However, the company was not allowed to leave Venezuela following the example of Rosneft in Colombia, and the head of the company publicly manifested that Gazprom would remain in the country. That was the price to pay for the continuing Venezuelan support of the Moscow's stand on international arena. Moreover, Gazprom's experience, although notable, was not as impressive in the long-term. A major contract concerning the start of a joint venture to explore the Venezuelan oil reserves in the Orinoco river area, specifically within the framework of the Junin-6 project, was agreed upon prior to Putin's visit to Caracas in 2010 between the Russian National Petroleum Consortium (NNK) and Petroleos de Venezuela SA (PdVSA). It was expected that the total investments in this project would reach between twenty and thirty billion dollars by 2035.[45]

It is appropriate to mention here that until March 2011, combining positions in the government with participation on the governing boards of the biggest state-owned enterprises like Gazprom continued to be a common practice in Russia: for instance, Igor Sechin served on the governing boards of important Russian state-owned companies for several years, while occupying key positions in government. Hence, the institutional arrangements were already in place when Moscow decided to award its allies in Latin America in the context of the 2008 crisis caused by the war in Georgia by means of energy cooperation. After Daniel Ortega government's recognition of Abkhazia and South Ossetia, in November 2008, Igor Sechin visited Managua with some specific proposals for Nicaragua. Soon after, concluding his visit to Moscow in December 2008, Ortega signed an agreement with the Russian state-owned company Inter RAO EES to build several small—and medium-sized hydroelectric and geothermal plants in Nicaragua. Conveniently, at that time Igor Sechin was not only the Russian vice prime-minister, but also the Chair of the Board of Directors at Inter RAO EES.[46]

After Sechin formally left the Government and became head of Rosneft in May 2012, he took advantage of his knowledge of the Russian energy sector while maintaining his influence in the power circles, in order to further adjust the Russian global energy strategy as one of the key tools in promoting the Russian political interests abroad. From this perspective, it is quite indicative that by the end of 2014 Rosneft was in control of the NNK group whereas other partners like Lukoil, TNK-BP, and Surgutneftegas eventually left the consortium.[47] At the same time, the year of 2014 turned out to be full of very important developments related to the Russian energy strategy not only in Venezuela but in other Latin American countries as well.

Following the announcement of the sanctions imposed on Russia and determined to show that Washington failed to isolate Russia, Vladimir Putin undertook a long trip to Latin America visiting several countries. In the context of the discussions herein, it is particularly interesting to look at the cooperation strategies discussed by the Russian leader with Cuba and Argentina.

When approaching the issue of Cuban relations with Russia, it should not be forgotten that the Caribbean island was the reason the region became open to Moscow from the time of the Cold War. Nowadays, the Soviet legacy is perhaps most visible precisely in the energy sector of Cuba: a part of the fact that Havana was the biggest importer of the Soviet crude oil in the Western Hemisphere, most of its energy capacity was built with the Soviet assistance, including the biggest power plant "Mariel." When visiting Cuba in July 2014, President Putin not only confirmed that Russia has written off the long-disputed debt of Cuba estimated in 35 billion US dollars, but also promised important investments in the energy sector of the island. In particular, a contract with a price tag of 1 billion US dollars was signed to supply "Mariel" with four new generators and to further modernize the plant. The discussions were on hold with regard to the possibility of re-opening the Soviet-era project of building a nuclear power plant on the island. Rosneft has signed an agreement with the state-owned La Unión Cuba-Petróleo (CUPET) that opens the possibility for the Russian oil company to participate in oil projects in the Cuban territorial waters beginning in 2015.[48]

With regard to Argentina, in recent years Moscow has managed to build an impressive comradeship with la Casa Rosada. In July 2014, while visiting Buenos Aires, the Russian President reached an agreement with the government of Argentina concerning the participation of Russian companies in the exploration of the Vaca Muerta fields which proved reserves are estimated in 927 million barrels.[49] In addition, Inter RAO EES—the same company which was called upon to supply Nicaragua with its energy needs back in 2008–on this occasion, is expected to contribute 2 billion US dollars in the construction of Chihuido 1 hydroelectric plant.[50]

Summing up, Russian involvement in the energy sector of Colombia, Venezuela, Cuba, and Argentina show a great degree of dependence between the incidence of the Russian companies and the quality of political links that Moscow established with the governments of particular nations.

CONCLUDING REMARKS: BEYOND ENERGY

Considered as one of today's most important energy powers on the global scale, Russia has been involved in the energy sector of Latin America since the mid-1990s, even before the beginning of the so-called Russian "return" to the area. However, until the early 2000s, it is difficult to talk about any kind of a solid energy strategy promoted by Moscow. Rather, soon after the disappearance of the Soviet Union, Russian energy companies looking for minor opportunities and chances to acquire experiences abroad began a modest international expansion in the Latin American region. However, in the middle of 2000s, the sector was already marked by the presence of the key Russian energy players bidding for the participation in the most demanding energy projects of the time.

By the middle of the second decade of the new century, Russia began fully advancing a number of important projects in several Latin American countries and many more projects on a smaller scale everywhere in the region. However, the visits of Vladimir Putin to the Caribbean, Central and South America in 2014 reveals that the priorities are set to promote energy cooperation with those countries that support the Russian political stand on the international arena. While profits still matter for Russia Inc., in Moscow, when it comes to the designing of the Russian international energy strategy of the second decade of the twenty-first century, the political sympathies and the degree of the willingness of Latin American leaders to back, even if only symbolically, Russia in its tensions with the United States and its allies, are ones of the most important factors in consideration by the Kremlin. This is possible because nowadays the government in Moscow can effectively exercise a large degree of control not only over the Russian state—but also privately owned energy companies' strategies in this region.

In this context, the sustainability of the strategy chosen by Russia in their Latin American expansion greatly depends on the availability of tangible resources that would enable it to meet the obligations assumed. In the beginning of 2015, some of the observers were already casting doubts that the crisis-driving Moscow would be able to do so because of the dramatic drop of oil prices. If this would be the case, the failure to comply with the multibillion promises made in Havana, Caracas, and Buenos Aires would be extremely damaging for the Russian reputation in the region. For now, however, the role given by the Russian government to the energy cooperation with Latin America during the last two and a half decades that passed after the collapse of the Soviet Union is that of a driving force behind of the Russian re-engagement with the countries in this part of the Western Hemisphere.[51]

NOTES

1. As reported by Interfax News Agency. See "Situation with Yevtushenkov is similar to the Yukos No. 2—Shokhin." In Russian. *Vedomosti.ru*. September 16, 2014. Accessed January 10, 2015). Mikhail Khodorkovsky—considered the richest man of Russia in the beginning of the 2000s—was arrested on October 25th, 2003 accused of tax fraud. Soon, his company was nationalized, and the main assets transferred to the Rosneft Company. See Gessen, Masha. "The Wrath of Putin," *Vanity Fair*. April 2012. Accessed January 11, 2015.

2. Vinogradova, Elena et al. "An Investigation: How the Owner of AFK Sistema ended up under the criminal investigation." In Russian. *Vedomosti.ru*, September 22, 2014. Accessed January 10, 2015.

3. Vedomosti. "Tango is Back in Fashion." In Russian. *Vedomosti.ru*, September 16, 2010 Accessed January 10, 2015.

4. According to the information provided by the company: http://www.bashneft.com/company/. Accessed 15 February 2015.

5. Vinogradova et al, Ibid.

6. Melnikov, Kyrill et al. "Bashneft": there and back. As the company has gone from the State and returned to the State." In Russian. Kommersant. 8 December 2014.

7. Gamov, Alexander. Interview with Vladimir Yevtushenkov. In Russian. *Komsomolskaya Pravda*. 10 January 2015.

8. Peleshuk, Dan. "Think it's just Putin who runs Russia? Guess again." *CNBC.com*, 24 December 2014. Accessed January 14, 2015.

9. See Stoner-Weiss, Kathryn. Russia Report. Freedom House. 2007. Available online: https://freedomhouse.org/report/countries-crossroads/2007/russia#.VOD5NuaUeSo. Accessed January 15, 2015.

Kutuzov, Roman. "The Oligarchs." In Russian. *Forbes.ru*. 20 October 2010. Accessed January 10, 2015.

10. Gabuev, Alexander at al. "Igor Sechin opened an Allied America. Russia mobilizes Venezuela and Cuba to fight the US." In Russian. *Kommersant* 168 (2008): 9;

Melnikov, Kyrill. "The Curfew." In Russian. *Kommersant Vlast* 14 (2013): 30.

11. Melnikov Ibid.

12. See, for example, Blank, Stephen. "Russia's Goals, Strategy and Tactics in Latin America." Paper Delivered as part of the LACC/ARC/U.S. Southern Command Policy Roundtable Series 28 April 2014 Miami, Florida.

13. See Rumer, Eugene. "Russia in Latin America." Paper Delivered as part of the LACC/ARC/U.S. Southern Command Policy Roundtable Series 28 April 2014, Miami, Florida.

14. As emphasized by Mattern, Benjamin. "Russia and Latin America: Geopolitical Posturing or International Partnership" Council on Hemispheric Affairs. 20 June 2014, available online at http://www.coha.org/russia-and-latin-america-geopolitical-posturing-or-international-partnership/, Accessed January 14, 2015.

15. See Ellis, Robert Evan. *China on the Ground in Latin America*. New York: Palgrave Macmillan, 2014.

16. Gustafson, Thane. *Wheel of Fortune: The Battle for Oil and Power in Russia.* Cambridge: Harvard University Press, 2012.

17. Zaslavskiy, Alexander. "Foreign Companies and the Russian Oil." In Russian, *Pro et Contra* 53 (2011): 40–50.

18. Liuhto, Kari and Peeter Vahtra. "Foreign operations of Russia's largest industrial corporations—building a typology," *Transnational Corporations* 16 (2007). Accessed January 14, 2015. http://unctad.org/en/docs/iteiit20071a5_en.pdf.

Vahtra, Peeter and Liuhto, Kari. "An Overview of Russia's Largest Corporations Abroad" In *Expansion or Exodus: Why Do Russian Corporations Invest Abroad?* Edited by Kari Liuhto, 23–40. Binghamton: Haworth Press, 2004.

19. Volcheck, Grigory. "Interview with Isaak Yanovich." *Oil News*, October 17, 2003.

20. Liuhto, Kari and Peeter Vahtra. "Foreign operations of Russia's largest industrial corporations—building a typology," *Transnational Corporations* 16 (2007). Accessed January 14, 2015. http://unctad.org/en/docs/iteiit20071a5_en.pdf.

Vahtra, Peeter and Liuhto, Kari. "An Overview of Russia's Largest Corporations Abroad" In *Expansion or Exodus: Why Do Russian Corporations Invest Abroad?* Edited by Kari Liuhto, 23–40. Binghamton: Haworth Press, 2004.

21. Lukoil. Annual Report. 2013. Accessed January.

10, 2015. http://www.lukoil.com/materials/doc/AGSM_2014/GO_Book_eng.pdf.

22. Rosneft. Annual Report. 2013. Accessed January 10, 2015 http://www.rosneft.com/attach/0/58/80/a_report_2013_eng.pdf.

23. Gazprom. Annual Report. 2013. Accessed January 11, 2015 http://www.gazprom.com/f/posts/60/660385/gazprom-annual-report-2013-en.pdf.

24. I am grateful to Rosali Archieva, of Luikoil Company, for pointing out some of these characteristics, which reflects her personal opinion and by no means express the official or in any other way endorsed view of Lukoil.

25. This is particularly true in the case of Latin America: in a conversation held by the author in 2008 with then the head of the public relations office of Lukoil Overseas in Colombia, the company's employee stated that the company preferred not to attract attention to its Russian identity rather than to promote or taking advantage of it. Similar attitude of the Russian companies can be observed in other countries as well.

26. For example, as reported at the 21st World Petroleum Congress held in Moscow in 2014, in Colombia, Lukoil was the first in the world to use an innovative technology of extracting oil at the altitude of 1.500 meters above the sea level by applying the method of a directional drilling of large diameter in order to prevent fluid loss.

27. Konończuk, Wojciech. *Russia's Best Ally. The Situation of the Russian Oil Sector and Forecasts for Its Future.* Warsaw: Centre for Eastern Studies, 2012. Ministry of Energy of the Russian Federation (Minenergo). Energy Strategy of Russia for the period until 2035 (basic principles). A proposal. In Russian. Moscow: Minenergo. 2014. Available online http://minenergo.gov.ru/upload/iblock/665/665a6512e64ffd5e3d30d9448d7b7fff.pdf Accessed January 14, 2015.

28. Reports of the Russian trade missions to Moscow available at the open sources such as polpred.ru provide plenty of evidences that Russian diplomats are often

frustrated with the lack of coordination of their efforts with the Russian private energy companies and loss of opportunities by both state-owned and private enterprises.

29. Farah, Douglas and Liana Eustacia Reyes. "Russia in Latin America: The Strategic Challenge," *Perspectives on the Americas.* January 15, 2015. Accessed January 15, 2015 https://umshare.miami.edu/web/wda/hemisphericpolicy/Perspectives_on_the_Americas/Farah%20-%20Final.pdf.

30. The Foreign Policy Concept of the Russian Federation. 12 July 2008. President of Russia. Official Web Portal. Accessed January 15, 2015 http://archive.kremlin.ru/eng/text/docs/2008/07/204750.shtml.

31. Benjamin-Alvarado, Jonathan. "Investment and International Cooperation in Cuba's Energy Sector," *Cuba in Transition*: Volume 8. Papers and Proceedings of the Eighth Annual Meeting of the Association for the Study of the Cuban Economy, Miami, Florida, 6–8 August 1998.

32. Rouvinski, Vladimir. "Rusia y ALBA: Un Matrimonio Por Conveniencia?" In ¿La Hegemonía Norteamericana en Declive? *El Desafío del ALBA y la Nueva Integración Latinoamericana del Siglo XXI*, edited by Bruce M. Bagley and Magdalena Defort, 469–496. Cali: Icesi University, 2014.

33. Trenin, Dmitri. "Putin in Latin." Carnegi.ru. Available online at http://carnegie.ru/eurasiaoutlook/?fa=561352014 Accessed January 10, 2015.

34. Rouvinski, Ibid.

35. Sanchez, Alejandro. Latin America's mixed response to the Ukraine crisis. *Voxxi.com.* 21 March 2014. Available online: http://voxxi.com/2014/03/21/latin-america-reaction-ukraine-crisis/ Accessed 15 January 2015.

36. Paniev, Yuri. "Latin Americans cannot cope with their oil on its own." In Russian. *Kommersant*, Feburary 3, 1996. Accessed January 14, 2015. http://www.kommersant.ru/doc/126116.

37. Ibid.

38. Rouvinski, Vladimir. "Colombia Frente a Las Potencias Extra-Hemisféricas: el Caso de Las Relaciones Colombo-Rusas," *Revista CS* 14 (2014): 77–110.

39. Vedomosti. Rosneft obtains first oil in Colombia." In Russian., *Vedomosti* 236 (799), 27 December 2002.

40. RosBiznesKonsalting. Rosneft Comments on Decision to Leave Colombia Energy Consortium.5 July 2004. Available online at http://www.shana.ir/en/newsagency/25605/Rosneft-Comments-On-Decision-To-Leave-Colombia-Energy-Consortium. Accessed January 10, 2015.

41. The Llanos basin is the largest oil and gas basin in Colombia, with an area of 227,780 square kilometers.

42. Minsiterio de Ambiente, Vivienda y Desarrollo Territorial. Resolution 1339. In Spanish. 27 July 2007.

43. Portafolio. "Rusos abandonan contrato petrolero en Colombia." In Spanish. 12 December 2012. Accessed January 15, 2015 http://www.portafolio.co/negocios/contrato-petrolero-rusos-y-colombianos.

44. Rouvinski, Vladimir. "Rusia y ALBA: Un Matrimonio Por Conveniencia?" In *¿La Hegemonía Norteamericana en Declive? El Desafío del ALBA y la Nueva*

Integración Latinoamericana del Siglo XXI, edited by Bruce M. Bagley and Magdalena Defort, 469–496. Cali: Icesi University, 2014.

45. According to the information provided at the Consortium website: http://consorcio.ru/ Accessed 15 January 2015.

46. Ibid.

47. "Lukoil" sold "Rosneft" 20% in the Venezuelan National Petroleum Consortium. As reported by *Vesti Economica* on 24 December 2014. Available online at http://finance.rambler.ru/news/economics/155574119.html. Accessed 15 January 2015.

48. Shvetz, Evgeniya. Russia and Cuba: energy cooperation. In Russian. Available online http://analyticsmz.ru/?p=609 Accessed 15 January 2015.

49. Dinatale, Martín. The Mission of the Russian Oil Companies. In Spanish. En *La Nación*, 10 November 2014.

50. Bidegaray, Martín. the Government relies on the Russian loans to finance the construction of the Chihuido dam. *El Clarin*. 8 December 2014.

51. The arms trade is another important factor of the Russian relations with Latin America. However, it is not clear to what extent this business can be sustainable, especially, in the case of Venezuela. See Safronov, Ivan and Pavel Tarasenko. "After Chavez. Will Russia remains an ally of Venezuela? " In Russian. *Kommersant Vlast* 16 (28 April 2014): 11.

Chapter 16

The Macroeconomic Determinants of Fuel Subsidies in Brazil, 1990s to Present

John Twichell

This case study examines the context of fuel subsidy reform in Brazil, from the 1990s to present. It describes the liberalization reforms of the 1990s, and discusses the impact of the reforms on energy development. It also touches upon the more recent national debate over the distribution of hydrocarbon rents, and on the subsequent politicization of price subsidies on hydrocarbon derivatives. These subsidies were put in place and upheld by two consecutive Workers Party (PT) executive administrations. The analysis finds that, due to a confluence of exogenous and macroeconomic constraints spawned by the 2008 global financial crisis, Brazil's leadership has had minimal policy space available to roll back fuel subsidies at a time when more of its citizens have achieved middle class status, complete with the power to own automobiles.

The study begins with the moment in the 1990s when the center-right Social Democrat (PSDB) party implemented a series of IMF-sponsored economic and structural reforms through 2001.[1] These reforms included measures to reduce fuel subsidies as part of IMF conditionality. The elimination of these subsidies produced a turnaround from a cost of approximately 1% of GDP in subsidies in the mid-1990s to a revenue generating scenario since 2002. The analysis then looks at the 2000s, when, during times of macroeconomic expansion and growth in the oil sector, the PT government restored subsidies. It reflects an observed pattern in Brazil whereby subsidies were reduced, but have reemerged, and remain a political and policy issue.

THE BRAZILIAN ENERGY SECTOR:

Under authoritarian rule, and a state-led development strategy also known as import substitution industrialization (ISI), Brazil transformed its national energy sector in the mid-twentieth century. Since the end of World War II, Brazil's energy matrix has been consciously shaped by statist policy making. Brazil is one of the few countries that responded clearly, consistently and effectively to the international oil shocks and energy crisis that occurred throughout the 1970s. The energy crisis of the 1970s gave Brazil the incentive it needed to diversify. Statist policies were implemented to summon hydropower and ethanol to compensate for a lack of domestic carbon-based resources.

The severe challenges of the 1970s crisis put Brazil on a path that now makes it one of the few countries that have developed alternatives to carbon-based energy for both the power generation and transportation sectors. Brazil's more recent approach to development of its energy sector reflects a broader even-handed and pragmatic approach to macroeconomic policy, which corresponds to the country's route of democratic institutional development since the late-1980s. Today, in the Latin American context, its pragmatism in economic policy making separates Brazil from its similarly endowed neighbors such as Argentina and Venezuela.

Through the energy crisis of the 1970s and the debt crisis of the 1980s, Brazil's leadership resisted growing international pressure to implement neoliberal reforms. Brazil took a pragmatic approach to reforms and subsidized development of its domestic sugarcane and hydrocarbon sectors. These measures were part of a broader energy policy strategy focused on diversification of supply. In the area of hydrocarbons, this meant greater investment in Petrobrás, the state oil company. Sugarcane production is traceable back to 1532, and considered to be the first organized economic activity in Brazil. A profile of the sugarcane sector valued it as a $25 billion industry in 2010. Sugarcane is now produced as a source to generate electricity and fuel automobiles. Moreover, between ethanol and hydropower, 30% of Brazil's energy matrix is now renewable while in most countries only 10% is renewable.

In the late 1990s, under democratic PSDB leadership, the state monopoly on hydrocarbon sector development was lifted, and a regulatory framework was established for both statist and private investment. In 1997, Petrobrás was partially privatized and made to compete in the newly created market for hydrocarbon development. The internationally competitive hydrocarbon industry Brazil has today, with 15 billion barrels of proven reserves, is the result of significant government and private investment to fund deep-water technology, exploration and well drilling. This transformation is widely attributed to continued pragmatism and consistency in energy policy, even in the face of changing governments and international shocks. Again, the partial and balanced approach to energy development in Brazil draws a stark contrast to the extreme deep broad and abrupt cases carried out during the same time period in Argentina and Venezuela.

Today, Brazil's energy matrix is relatively environment-friendly and it is among a unique group of countries that have delinked from traditional twentieth century economies in terms of the forms of energy consumed to power industrialization. For example, one quarter of all transport in Brazil is powered by biofuels. Accordingly, Brazil is expected to take the lead in Latin America on energy-related economic, political and climate change issues. On the other hand, all of these prospects could change completely if, as expected, Brazil becomes a major oil producer in the near future.

Most recently, Brazil has gained global recognition for its hydrocarbons revolution. In 2006, Brazil transformed from being the largest oil importer in Latin America into a self-sufficient oil producing country. Brazil is now capable of supplying enough of its own oil resources to its large and ever growing domestic market, and it has now even surpassed Mexico and Venezuela as the largest holder of petroleum reserves in Latin America. Including the ultra-deep water discoveries of the last decade, proven reserves have been estimated as high as 150 billion barrels. The development prospects inspired by these monumental discoveries have placed federal hydrocarbon policy on the legislative agenda to decide the distribution of *pre-sal* oil and gas revenues. In addition, the federal government has established a sovereign wealth fund to reinforce state commitments, assure investors, and to control the value of the Real, in an effort to stave off the secondary economic effects of Dutch disease.

BACKGROUND MACROECONOMIC AND POLITICAL HISTORY

In 1985, Brazil was a new democracy with a state-led economic development model. Brazil was known in the 1980s and 1990s for patterns of fiscal insolvency, high levels of unemployment, hyperinflation, and indebtedness. In 1994, President Fernando Henrique Cardoso (PSDB) oversaw a series of market-friendly economic reforms that, in less than a decade, transformed Brazil into an attractive country for trade and investment at the turn of the century. When, following Cardoso, Luiz Inácio Lula da Silva of the center-left Workers Party (PT) was elected president and took office in 2003, there were doubts that he would uphold Cardoso's market-friendly development model. In fact, da Silva did do so, and he even capitalized politically at home and on the world stage as a result of related macroeconomic dividends. The quintessential charismatic leader widely referred to simply as "Lula," da Silva brought a sense of optimism to Brazil. Emblematic of his charisma was this statement made by da Silva in his October 2006 runoff reelection speech delivered in São Paulo: "We have the foundations for an extraordinary quality leap in this coming four years; we solved the macroeconomics problem, but now we're tired of being an emerging power . . . If in my first mandate we advanced at 80 kilometers per hour, in these next four years we'll be advancing at 120 kilometers per hour."[2] Almost in lockstep with his prediction, in addition to new oil discoveries and elevated commodity prices in world markets, Brazil benefited from foreign direct investment (FDI) and from portfolio investment throughout most of da Silva's second term as president.

By 2008, the global financial crisis did threaten to impact Brazil's financial, trade and investment sectors with potentially problematic slowdowns. However, at the time, macroeconomic conditions allowed da Silva to respond with counter-cyclical fiscal and monetary policies. Also, da Silva was able to draw down international reserves in order to bolster trade finance, and reduce reserve requirements on bank deposits. Aided by burgeoning commodities exports to China, these policy responses by the da Silva government enabled Brazil to recover from the crisis in rapid fashion. These policies were kept in place up until the presidential election in October 2010. In 2010, Brazil achieved annual growth of 7.5% while US growth measured less than 3%. By all accounts, at the end of da Silva's second term, Brazil had transformed into a politically stable, economically competitive and geopolitically influential country that seemed to have dodged the effects of the 2008 crisis.

Brazil has in fact emerged, and the world is now witnessing, in Brazil, the maturation of a process that took root in the 1990s, namely democracy and economic stability. In 1994, Brazil's economy was twice that of Argentina. By 2010, it was five times larger than that of Argentina. The Brazilian economy has also surpassed both Canada and Mexico in terms of GDP, making it second only to the US in the Western Hemisphere. Brazil currently ranks fifth in the world in land mass and in population level, and it is predicted to become the fifth largest economy in the world within the next decade, thanks to its recent discovery of major oil reserves. From an international relations standpoint, these material capabilities make Brazil a powerful pillar in what is now a multipolar world. Moreover, Brazil is a dominant regional power in Latin America, and a driver of economic growth in South America through Mercosur.

Nevertheless, a series of short-run and structural challenges associated with rapid development were not addressed by the da Silva administration, and for that reason, the likelihood that Brazil could sustain economic growth given global economic conditions became less certain by 2011. These challenges include the combination of an expanding middle class, high levels of domestic consumption, inflation, and an overvalued currency. Fiscally, the federal government has more often been on the verge of spending more than it is collecting in tax revenue. The federal taxation system is unevenly structured and in need of reform. Government spending on social welfare and programs for the poor majority are not being adequately funded. Brazil is sorely lacking the type of airports, deep-water ports, and highways required for a country with its size and GDP to sustain growth. More than half of Brazil's fifteen-year-old youth are illiterate and only one quarter of its young adults are college educated. Decrepit infrastructure and an overwhelmed public education system threaten long-term competitiveness.

TWENTY-FIRST CENTURY STRUCTURAL FACTORS

The 2008 global financial crisis originated in the United States. Brazil, among the other larger Latin American countries, initially found itself in a better macroeconomic position to weather the current downturn than in past crises. This is, in part, the result of decades of loan conditionality and neoliberal reforms imposed upon Latin American countries by the developed world and international financial institutions (IFIs). Conditionality required countries in the region to effectively manage their fiscal accounts, and therefore, solid fiscal management has been institutionalized. As a result, Brazil was transformed into an economy that accrued budgetary and trade surpluses. These surpluses represent a welcome change and it has been suggested these policies have actually become an item of "political machismo" among leaders of Latin American countries.[3] The Brazilian Central Bank has also had welcome success controlling inflation.

Under the above set of exogenous constraints, and in the current post-crisis cycle, the Lula government, and its successor had to apply an unprecedented set of policy levers to restore short-term growth. Deficit spending has not been an option. The post-Brady Plan global financial structure makes a policy of deficit spending very difficult for developing countries to pursue, and the memory of hyperinflation and its political costs remains embedded in the minds of Brazil's leadership.[4] Under these constraints, the Lula government applied counter-cyclical measures in response to the 2008 global crisis, to the extent possible. Macroeconomic trends in Brazil, leading up to the 2008 global crisis and afterward are shown below in Figure 16.1.

THEORETICAL EXPLANATIONS

The neoliberal economic view is that policies and reforms implemented by Brazil since the 1990s have provided security against more severe and long-lasting effects. This school of thought recommends that: (1) public debt is kept at a structured minimum; (2) the exchange rates is allowed to float and

	2000	2003	2008	2010	2011
GDP per capita ($US)	3,751	3,104	8,729	10,816	12,917
GDP growth (percent)	4.31	1.15	5.16	7.49	3.77
Inflation (percent)	6.18	13.72	8.33	8.23	6.97
Overall fiscal balance (percent of GDP)	-3.37	-5.31	-2.34	-5.93	-3.57
Gross public debt (percent of GDP)	51.1	59.6	58.5	63.7	62.2
Net public debt (percent of GDP)	47.7	54.9	38.1	40.2	38.6
Current account balance (percent of GDP)	-3.76	0.76	-1.71	-2.21	-2.12
Oil imports (percent of GDP)	1.19	1.16	1.84	1.21	1.25
Oil exports (percent of GDP)	0.16	0.33	0.33	0.19	0.23
Oil consumption per capita (liters)	412	394	482	624	n.a.
Poverty headcount ratio at $1.25 a day (PPP) (percent of population)	11.82	11.21	6.01	6.14	n.a.

Figure 16.1 Brazil's Economic Indicators since 2000.

stay competitive; (3) fiscal resources are conserved; (4) banks should be highly capitalized and maintain strict lending regulations; and (5) governments should act with continuity, uphold property rights and not allow political motivations to hinder private enterprise.

The more skeptical structuralist view takes a contrary perspective. Structuralists argue that improved fiscal positions in Latin America in general and Brazil in particular, have been the result of temporary factors such as high commodity prices and lenient international credit markets. Latin American countries best positioned to weather the recession have export sector diversification, adequate reserves of foreign exchange, and limited foreign debt obligations. More industrialized and economically diversified countries like Brazil, Chile, Colombia, Mexico, and Peru have been more apt to show the lowest declines in growth as a result of this recession. Countries highly dependent on commodity exports like Argentina, Bolivia, Ecuador, and Venezuela have faced the greatest economic challenges since the 2008 global financial crisis.

Latin America's financial openness makes countries in the region such as Brazil sensitive to the vagaries of highly liquid international capital finance, which can be detrimental to investment and growth. Other regions are financed by domestic savings whereas foreign savings funds a considerable percentage of Brazil's development. Structural forces in the global economy have persisted even after the 2008 global financial crisis. For that reason, there has been no noticeable shift in this region-wide neoliberal policy of financial openness through capital account liberalization. This would be a difficult policy to reverse because it has positively impacted economic growth and also because there are no practical financing alternatives available to countries like Brazil, over the short-term. But at the same time, capital account liberalization is a dangerous policy because of the risks associated with interdependence of international and domestic constraints. Case in point is that sustained reliance on global demand for its commodity exports puts Brazil's economy at tremendous risk, in the event of a commodities bust.

BRAZIL UNDER PRESIDENT DILMA ROUSSEFF

Brazil's ability to address its challenges, achieve its development potential, and secure its unprecedented global stature, have depended, to a considerable extent on the policies of da Silva's successor, Dilma Rousseff (PT), who was elected as Brazil's first female president in October 2010. Rousseff took office in January 2011, less than two years after the world economy had just suffered a financial collapse. By then, despite the efforts of the d Silva government, the effects of the global crisis were being felt in the Brazilian economy.

Rousseff, who had never previously held an elected position, has had to prove to critics that she is qualified for the job, and that she is more than the puppet of her former boss and predecessor who was not able to run for reelection a third time due to term limits. As president, Rousseff has surprised both her critics and her supporters, in part because her stance on many topics was difficult to read during the campaign. There had been a broad perception that Rousseff was more ideologically to the left than da Silva, more committed to state-led capitalism, and less pragmatic in general. One of Rousseff's first moves after being elected was to reassure Brazilians and the world that she would maintain continuity with Lula, particularly in terms of market friendly macroeconomic policy and property rights. Rousseff indicated her intentions to uphold the same macroeconomic fiscal and monetary policies as had been overseen by da Silva. This recipe was made up of the same set of policies put in place by Cardoso, including inflation-targeting, a floating exchange rate, and fiscal targets, collectively referred to as the "policy tripod." However, the newly elected Rousseff did signal a more interventionist stance on matters of microeconomic policy, which was not a surprise coming from a career technocrat and former Energy Minister.

Rousseff has presided over a political coalition that enjoys a powerful three-fifth's majority in both houses of Congress. But unlike da Silva, who had firm control over all factions of the PT, Rousseff is more indebted to those who put her in power. She has had to manage the far left, the social democrats, as well as the PMDB party, which the PT relies upon to maintain its coalition. It is under this complex set of domestic political circumstances that Rousseff also has had to deal with numerous matters of foreign policy, including: trade, foreign investment, regional security, the proliferation of illegal narcotics, human rights, and geopolitical issues of terrorism, war and peacekeeping unfolding in other regions of the world. There is an unprecedented degree of global level attention presently focused on Brazil. The pressure has therefore been on Rousseff to prove she is capable of sound decision making and policy making on behalf of Brazil as a unitary actor in the international system, as well as in Brazil's capacity as a member of the UN, the G-20, the WTO, and now as a lender to the IMF.

Rousseff's first challenge in 2011 was to make crucial short-term decisions about how to deal with an overheating economy. Her administration has not had much policy space to bring the economy under control through budgetary spending cuts. One of Rousseff's greatest fiscal challenges is dealing with the commitments of social policies put in place under da Silva at a time when the government was running a budgetary surplus. There has been a pending minimum wage increase. All pensions are geared to minimum wage, so that when minimum wage rises, pensions rise accordingly. Also, da Silva passed marginal reform for social security in 2003 that once implemented, will still not apply to existing civil servants whose numbers grew tremendously under Lula. On top of social commitments, including the successful *bolsa familia*, is the fact that public spending on infrastructure has increased significantly as the federal government has worked to accomplish improvements leading up to this year's World Cup and the Olympics in 2016. These commitments remain, as Brazil has gone from having a marginal fiscal surplus in 2010 to patterns of deficit under Rousseff. In her characteristic no-nonsense style, Dilma has already made some cuts in primary fiscal expenditure, such as cuts made to the national development bank (BNDES). But, even as she advocates fiscal responsibility, in keeping with the PT platform, Rousseff favors new spending for social programs that benefit the poor. To keep her promise to do both, Rousseff has been challenged to address larger fiscal issues that Lula left unresolved, such as public sector pension reform. But, the political danger is that such reforms run the risk of alienating the PT base.

The Rousseff administration has not had to contend with major debt overhang during the recent global crisis. During the present global recession, Brazil has carried relatively low levels of international debt, and to some extent, financing needs have been met in the domestic market. However, one short-term risk has been an increasing current account deficit, which has exposed the broader economy to financial risk. Also, private and public consumption has continued to grow at a faster rate than GDP, putting pressure on domestic industry and on prices. Brazil experienced rapid growth in recent years, with record growth of 7.5% in 2010, which marked a peak. Subsequent decreases in growth have led to larger macroeconomic problems.

During the recent period of growth, the main driver of the Brazilian economy has been the rise in consumption, leveraged indirectly by capital inflows. The majority of capital inflows were liquid, and the product of steadily improving terms of trade. Increasing commodity prices plus high domestic interest rates translated into a large increase in foreign reserves. The government was obliged to purchase much of the currency entering the country to maintain monetary stability. Excessive capital inflows and foreign exchange intervention caused foreign currency reserves to reach US$ 300 billion by the end of 2011. As a result, the already overvalued Real continued to creep upward as did interest rates and inflation. By 2011, the consequences of an overvalued Real showed up in decreased manufactures compared to increased consumption. Domestic inflationary trends kicked in by mid-2011, and inflation began to run well about the 4.5 +/- percent target. Technically, the Central Bank's policy of inflation targeting places limits on the government's ability to adjust interest rates to manage capital inflows. This policy making constraint in part allowed the value of the Real to increase substantially over 2011. Real interest rates were increase to as high as 11.25% to compensate for inflation. From there forward, the Central Bank began a pattern of cutting interest rates from their high of 12.5% in August 2011 down to 7.25% in October 2012, as a way to stem deflationary pressures and reignite growth.

By 2012, Rousseff had taken a series of macroeconomic steps to bolster the economy. However, there being no such thing as a "free lunch," rather than restoring growth, Rousseff's monetary stimulus and resultant inflation brought about additional macroeconomic challenges that were subsequently given temporary fixes. First, generous increases in the minimum wage were granted, which caused real earnings to rise substantially, bringing inflation in the non-tradable sector to the 8% range. Second, to control inflation, the government began to directly manage key prices in the economy, namely for electricity, gasoline and public transport fares. These price controls have not been without consequences, as they have placed burdens on producers, including in the electricity, hydrocarbons and ethanol sectors. For example, profits at Petrobras have fallen sharply as the government requires it to subsidize prices for downstream petroleum product prices amidst rising consumer demand. Furthermore, these government subsidies intended to control inflation have, like it or not, dampened Brazil's investment climate.

The recent commodities boom of 2003–2007 was a socioeconomic transformation, if not in terms of actual restructuring, then in terms of societal expectations. The boom's export-led prosperity caused many more Brazilians to take an optimistic view of their future prospects for raising their standard of living. Many poor Brazilians began to live like they had entered the middle class, even if just barely. A case in point is that, given certain parameters of income, between 2002 and 2008 at least 10 million Brazilians joined the ranks of the middle class almost entirely due to upward social mobility. Over the same time period, improving terms of trade made Brazilians feel wealthier and has caused them to spend more money. An expanding middle class has driven demand and consumption. Middle class "C" growth has been caused by higher levels of employment and greater access to consumer credit, which led to high consumer confidence. As a result, domestic industrial output has been negatively impacted by increased demand for imports. As a consequence, to the Rousseff government's list of concerns has also been added the threat of de-industrialization. In the short-run, the threat of de-industrialization comes from an overvalued exchange rate, especially against China, and from fact that there are a higher percentage of commodities in Brazil's export profile. Brazil currently generates more foreign exchange by exporting commodities and primary products than it does by exporting industrial goods. As in past commodities cycles, Brazil is now experiencing a terms-of-trade shock, and inflation rates will likely only get stronger.

Aside from interest rates, the Rousseff government has employed other policy alternatives to tighten monetary policy and control the real exchange rate. Bank reserve requirements were increased, consumer credit standards were tightened and the Brazilian sovereign wealth fund was authorized to trade currencies and derivatives. However, neither increased reserve requirements nor capitalization ratios have proven to be consistently effective stopgaps against inflation in the past or present. Inflation is not showing signs of abatement in Brazil nor is it stabilizing. Brazilians might again have to accept inflation rates above 6%. Brazil's economy has an inverse relationship between credit and inflation growth. On the one hand, interest rates will need to rise to bring inflation down. On the other hand, interest rates cannot be raised too much because to do so would make it difficult to sustain a favorable exchange rate for exports.

Brazil is facing a complicated macroeconomic tradeoff. The public sector must meet spending commitments at a time when public saving is needed. Moreover, based on Brazil's current expenditure, it is not easy to see where the government could cut spending because the bulk of current expenditures are in public payroll, social security, pensions and conditional cash transfer programs. This leaves fewer funds left for investment and infrastructure or to produce a primary surplus. The government cannot slash the budget. This set of expenditure policies carried over from the da Silva administration are at the heart of the current macroeconomic equation, which needs to boost investment. Rousseff will likely need to reexamine the federal budget and attempt to create some balance between social and economic equality and economic growth. Essentially, the social contract is being undermined by inflation.

By 2013, it became apparent that Brazil was experiencing a recession in economic productivity due to lessening aggregate demand. The indications were seen in the balance of payments as a result of capital inflow and outflow adjustment. Brazil has been exporting less and importing less, and the economy has gone from a surplus to a deficit. The adjustments have affected trade and investment, through both the current and the capital accounts. Unlike previous crises, Brazil was not a contributing cause of this crisis, but neither can it export its way out of the problem. As a result, tremendous belt tightening will be required in the area of spending, as external obligations must continue to be repaid. While Brazil was not involved with highly risky investments in this cycle, it is not without financing commitments.

Balance of payments problems gave way to other macroeconomic challenges in Brazil in 2013. The shift in overall productivity has translated to a stagnant GDP. In 2014, the Rousseff government was witness to increasing levels of unemployment and decreasing levels of private sector consumption. The unemployment rate has risen by 2.5 to 3 percentage points. In what are now decreasing terms of trade, society is also being made to feel poorer. A sustained decrease of capital inflows has the potential to lead to currency devaluation. Moreover, the government will experience diminishing growth of the fiscal account. An added complication is that the government did not set aside fiscal revenues as contingency funds. Should the fiscal situation deteriorate, Brazil could be exposed to risk of losing its investment grade rating.

Brazil has enjoyed a very favorable period in the global spotlight ever since its major discovery of billions of barrels of ultra-deep water hydrocarbon reserves in 2006. However, in 2014, Brazil's favorability is diminishing in the eyes of global investors because macroeconomic instability has returned, and the *pre-sal* hydrocarbon reserves still remain undeveloped as the international price of petroleum has plummeted to below $40 per barrel. Economic policy is currently Brazil's main challenge, as annual GDP growth has recently decelerated from an impressive 7.5% in 2010 to a paltry 0.9% in 2012. This downward trend is the result of both external and domestic factors. Despite the continuation of PT rule at the executive level, whether deserved or not, Brazil's recent economic decline is also being attributed to the 2010 transition in political leadership from da Silva to his protégé, Rousseff. Overall, in 2014, Brazil faced a unique set of economic and trade-related challenges.

It has been argued that Rousseff's heightened monetary interventionism is part of a broader and more transformational economic strategy being put in place under her watch. Those with this view expect Rousseff to impose stricter regulation, lower rates of return for financial and physical capital, and greater protection of domestic industry, ranging from higher import tariffs to larger credit subsidies. In Rousseff's defense, rates of return and interest rates in Brazil had clearly become too high in an international climate where currency devaluations have become commonplace, including in the US. Nevertheless, interest rate adjustments in the absence of reforms are a risky proposition for Brazil as they are likely to spook investors who hold mobile capital. New regulatory steps lend additional support to the alarmist theory that Rousseff is carrying out a broader interventionist agenda. New regulation has already been imposed at the sectoral level, including in electricity, railways, ports and airports, including reductions in the rate of return that private investors could obtain from their investments. In the banking sector, the government imposed standards to bring down interest rate spreads on bank loans.

So far, the results of Rousseff's efforts to right Brazil's economy have been mixed. Investment and output have been on a downward trend since the end of 2011. Advocates of government initiatives argue that the market that has not yet recognized the benefits of new policies. Critics argue that no such recognition has occurred. Even with the increased liquidity brought about by looser monetary policy, fiscal stimulus, and freer availability of credit via public banks, Brazil's economy is expected to grow minimally, while inflation will remain above target. Therefore, government subsidies and price controls, including fuel subsidies, until the plunge in petroleum prices, remained in place to control inflation. Critics of Rousseff contend that the government has a greater chance of turning around Brazil's economy by adhering more closely to market efficiency fundamentals. This, these same detractors argue, would involve a commitment to private sector investment, which could be encouraged by reducing inflation and offering realistic rates of return in concession contracts at the sectoral level.

In any case, in 2015, Brazil cannot be viewed as a benefactor of the global economy, and the government is showing indications that it cannot manage the development of the ultra-deep water pre-sal oil reserves on its own. Despite being its largest trading partner, China is now seen as a threat as much as it is an opportunity, due to its erosion of the Brazilian manufacturing export sector and deindustrialization. Despite its best efforts, Brazil is now susceptible to Dutch disease, because it has resorted to exporting soy, iron ore, and oil. Brazil's trade deficit sits at $8 billion, and its current account deficit is estimated at 4% for 2014. The average Selic (short-term interest) rate was at 9% in 2014. These conditions tend to pose a threat of capital flight.

Brazil's languishing economy combined with a failure to anticipate societal discontent reached a crescendo in June 2013, when massive protests unfolded nationwide in Brazil. Increases in public transportation fares were just the tip of the iceberg. Brazilians had also been long fed-up with the inadequacy of public schools and hospitals as they watched the government spend billions of Reais in preparation to host the World Cup and the upcoming Olympics in 2016. *Bolsa familia* brought 40 million Brazilians into the middle class and this process raised societal expectations, resulting in current demands for different policies. Insulated politicians were ignorant to rising expectations of the populace, and to the emergent demand for greater public transparency and accountability in government.

Faced with mounting political pressure in the run-up to the presidential election in 2014, the Rousseff government also found itself under growing fiscal pressure. Fiscal space was temporarily created by the lowering of interest rates and the decline in the net public debt to GDP ratio, which combined to yield lower interest payments. Flexibility also stemmed from the large pool of liquidity that still exists in the global economy. However, over the medium-term, the government's policy will likely need to change once again. Despite a flood of uncertainty surrounding her reelection prospects, Rousseff came out the victor in the October 2014 runoff election.

The protests reveal an electorate demanding redistribution and greater attention to education and health care. Rousseff's coalition is made up of sixteen different parties, and this has required lots of pork distribution to hold that coalition together. PT has some important decisions to make now that Rousseff has been reelected. In many ways, Rousseff is caught between a rock and a hard place, due to the lack of growth, and Brazil's staggering lack of global competitiveness that cannot seem to be turned around. Ironically for Rousseff, Brazil is experiencing the results of da Silva's second term, and her first term was indicative of this fact.

CONCLUSION

An analysis of Brazil's hydrocarbon and sugarcane sectors, makes it is possible to highlight three important elements of domestic energy policy for a given country: (1) which energy resources will be relied upon to fuel the economy and society; (2) provisions for infrastructure needed to deliver the resources chosen to be developed; and (3) recognition of the impact of price on consumer decisions to make use of energy resources.

First, to achieve energy security, there must be an ability to produce domestically in a manner that does not conflict with other economic sectors. With its vast territory, half of which is arable, and its tropical climate, Brazil is in the enviable position of being able to make sugarcane the feedstock for a thriving renewable bio-fuel sector. Twenty percent of Brazil's arable land is being used for farming today. Remarkably, over the last thirty years, Brazil has displaced half of its gasoline consumption by farming sugarcane on 1% of its arable land.

A second lesson to be learned from Brazilian energy policy is that government has a key role to play in terms of building infrastructure. In the area of sugarcane, supply infrastructure has actually been installed twice in Brazil because of competing gasoline and ethanol prices. With flex-fuel, a workable model has now been established. Since 2003, cars with flex-fuel systems have grown to represent 90% of automobile ownership. This allows the consumer to decide at the individual level which fuel to choose. As a result, fueling stations in Brazil are like restaurants with a menu of fuel options. Overall, industry and the government have cooperated to put the infrastructure in place to realize the benefits of a diversified energy matrix.

A third lesson to be learned from Brazil's energy policy is that prices for alternative fuels must be competitive. Consumers respond to prices, and Brazil has been no exception. Gasoline is currently priced two times that of the price of a liter of ethanol. Petrobrás has actually kept prices down to compete with ethanol. Petrobrás would like to sell more gasoline at a higher price but subsidies won't allow for it. Yet, while gasoline prices in Brazil have been flat over the last few years, factors in the international oil market have the ability to bring about unexpected price shifts, which further justifies the need for a flex-fuel system. Under the current system and price structure, it is gasoline that has become Brazil's alternative fuel. This is partly due to the lack of an external market for ethanol. If there were more demand abroad, the domestic price for ethanol would rise. Ethanol in the Brazilian domestic market is selling at below market price due to subsidized gasoline.

In the case of Brazil, it can be argued that protectionist policies early on, followed by gradual market reforms, appear to have paid off with success in both sugarcane and hydrocarbons. With sugarcane, this is apparent in the growth of ethanol as a renewable fuel for the domestic market. With hydrocarbons, this can be seen in the discovery of a new offshore development frontier, and in the growing influence of Petrobrás as both a producer and a target of investment in international markets.

Brazil has been touted as a "peaceful" alternative to the Middle East and Africa, for foreign investment in oil; but development of the new offshore reserves is not without uncertainties. While demand for oil remains strong in spite of global financial uncertainty, international price remains an important variable. Less than a decade ago, a price in the range of $50 per barrel was considered an ideal target to attract foreign direct investment in the sector. Today, that price point has nearly doubled to $100 per barrel. However, even at this price, investors have been few and far between. The most likely contenders for development of Brazil's ultra-deep offshore oil resources has been China, through its state-owned oil companies.

In domestic terms, the prospect of increased oil development contributing to a higher GDP and greater fiscal revenues presents Brazil with an opportunity to promote broader based national development, as has been the case in Norway. Yet, as suggested by Venezuela and Nigeria, enhanced oil production also presents a new set of potential risks. Most immediate among these risks is "Dutch disease," a significant increase in value of the domestic currency in international markets caused by accumulation of oil rents, and a "crowding out" of other economic sectors. It is a fact that the Real's value will appreciate with an export structure made up more significantly of oil. As the price of oil increases, it will generate further pressure to devalue the Real. The risk of Dutch disease is especially high for Brazil in 2015, due to a commodity-heavy export profile. Moreover, rents from Petrobrás are being steadily directed to the federal government and eventually to consumption, rather than into a sovereign wealth fund. Another criticism leveled at da Silva and at Rousseff has been the PT's promotion of refined petroleum exports as a way to guarantee more jobs in the sector for their political base. It can also be seen in terms of control over the Petrobrás budget and debates over how to manage greater anticipated oil revenues, with a rising sentiment of resource nationalism becoming more than apparent. This is understandable, however, because oil has already financed many of the government's social policies.

Still, in comparative perspective, Petrobrás is the only successful and efficiently run NOC in Latin America. This is attributed to the unique structure of internal governance within Petrobrás, which has enabled it to distance itself from political interference and private interests more so than other NOCs. This unique structure has been explained by a legacy of the period of mid-twentieth century military leadership and bureaucratic authoritarianism dating back to 1965. This same legacy, in conjunction with Brazil's unique form of coalition-oriented democracy also explains the success of Petrobrás in the wake of liberalization and privatization reforms of the late-1990s. This makes it unlikely that major changes will be seen in the ownership and investment structure of Petrobrás.

NOTES

1. Brazil's economic performance in the pre-reform era of the 1980s was characterized by low growth, high inflation, and substantial fiscal imbalances. Economic growth averaged 3% and inflation averaged near 300%. Fiscal policy was expansionary with perpetual budget deficits that averaged 5 to 7% of GDP leading up to 1989. Weak fiscal performance led to an increase in net public debt from 24% of GDP in 1981 to almost 40% of GDP in 1989. These deteriorating conditions put pressure on authorities to alter Brazil's import-substitution industrialization (ISI) development strategy, and to liberalize the economy, including the energy sector.
2. MercoPress South Atlantic News Agency. October 31, 2006.

3. Marcelo Giugale, World Bank, 2009: http://web.worldbank.org/WBSITE/EXTERNAL/COUNTRIES/LACEXT/0,,contentMDK:22057190~pagePK:146736~piPK:146830~theSitePK:258554,00.html.

4. Since the introduction of the Brady Plan in 1989, international bonds and other highly liquid securities issued by Latin American governments have replaced unsecuritized commercial bank loans as a major source of finance for the region. The market for international bonds is more liquid than it was for the types of commercial bank loans that led to the 1980s debt crisis. This puts considerable pressure on governments to maintain domestic political and economic stability, or risk an investor sell-off of their countries' bonds, which leads to an elevated external interest rate. There is a direct connection between international perception and local interest rates that can lead to capital flight. The increased external interest rate makes it more difficult for the country to finance new debt, while the increase in country risk will cause a spike in domestic interest rates, hindering domestic investment and growth. Yet, because Latin American countries are relatively closed in terms of trade flows, dependent upon commodity exports, rank very low in terms of national savings rates, and also have underdeveloped domestic financial resources, they are also highly dependent upon international bondholders for investment, which leaves them no real financing alternatives. Confirming Latin America's lack of trade openness relative to other countries is the fact that not one country in the region is listed among the top 25 in the 2008 World Economic Forum Index of Competitiveness. Chile ranks the highest, at 28th. This scenario becomes particularly risky in times of global economic recession because sources of external credit are scarcer (Liliana Rojas-Suarez, 2009).

BIBLIOGRAPHY

Birdsall, Nancy. "Income Distribution Effects on Growth and Development," Working Paper 118: Center for Global Development, Washington, D.C. (April, 2007).

Bresser-Pereira, Luiz Carlos. "The *value* of the exchange rate and the Dutch disease" *Brazilian Journal of Political Economy*. 33, 3 (132), Jul-Sep 2013, pp. 371–387.

Brollo, Fernanda, Tommaso Nannicini, Roberto Perotti and Guido Tabellini. "Brazil: The Political Resource Curse," http://www.voxeu.org/article/political-resource-curse?quicktabs_tabbed_recent_articles_block=1.

Hagopian, Frances, Carlos Gervasoni and Juan Andres Moraes. "From Patronage to Program: The Emergence of Party-Oriented Legislators in Brazil." *Comparative Political Studies*. 42, 360, 2009.

Hunter, Wendy. *The transformation of the Workers' Party in Brazil, 1989–2009*. Cambridge, UK: Cambridge UP, 2010.

International Monetary Fund. "Case Studies on Energy Subsidy Reform: Lessons and Implications," January 28, 2013.

Monteiro, Joana, and Claudio Ferraz. "Does Oil Make Leaders Unaccountable? Evidence from Brazil's offshore oil boom." Unpublished manuscript, Department of Economics, PUC-Rio (2012).

Montero, Alfred P. *Brazil: Reversal of Fortune*. Cambridge, UK: Polity, 2014.

Power, Timothy J. "Brazilian Democracy as a Late Bloomer: Reevaluating the Regime in the Cardoso—Lula Era." *Latin American Research Review*. 2010.

Rojas-Suarez, Liliana (editor). *Growing Pains in Latin America: An Economic Growth Framework as Applied to Brazil, Colombia, Costa Rica, Mexico, and Peru*. Washington, D.C.: Center for Global Development, 2009.

Conclusions

Bruce M. Bagley, Dina Moulioukova, and Hanna Samir Kassab

On September 26th, 2014, the Department of International Studies at the University of Miami hosted a conference for analysts and scholars of energy policy. Inspired by the theme, "The Impact of Emerging Economies on Global Energy and the Environment: Challenges Ahead," the conference explored the changing energy landscape within the context of rapid and dramatic transformation in the international system.

This volume focuses specifically on the impacts of the BRICS nations on world energy markets. In the analyses presented here, a variety of factors, such as national patterns of production and consumption, international energy trade, and the extent to which alternative and competing policies maintain or bring about change in the international order, are examined.

The conference participants drew a number of compelling conclusions that serve as the underpinnings of this volume. These conclusions range across international security and foreign policy issues through individual country energy security dilemmas to problems of environmental and human security. Broadly, this volume examines seven distinct but related energy themes and issues proximate to the study of international relations in the twenty-first century.

First, it dispels the conventional wisdom that the BRICS alliance coordinates Global South policy in a unified way by highlighting a number of inconsistencies among the BRICS that have been glossed over or ignored in many journalistic, academic, and policy-analyses. Our contributors revealed that there are key differences and unresolved conflicts among the members of the BRICS alliance that prevent them from acting in unified fashion on most energy issues, beginning with the fact that some—for example, China and India—are major energy importers while others—for example, Russia and Brazil—are sizeable energy exporters. Moreover, in the context of energy security, there are significant power asymmetries that differentiate the principal actors or core powers, of this alliance—China and Russia—from the other members—Brazil, India and South Africa—that remain on the periphery of the alliance. The volume underscores the reality that the BRICS member states are more disunited than united and that their specific roles in global energy markets ultimately depend on their relative economic size, distinct power resources, resilience, and military capabilities. In short, the BRICS cannot accurately be portrayed as a new counter-hegemonic block that has arisen to challenge the hegemony of the United States and its Western European allies. Perceptions that define the BRICS must change on a fundamental level. Each BRICS member occupies a different point on the spectrum of global power. As a result, the BRICS alliance must be re-conceptualized as a loose alignment among a diverse set of Global South state actors that is characterized by significant differences and disagreements that have existed for decades and that seriously inhibit unified collective action on energy and a host of other issues.

Second, participants analyzed the implications for energy security of the growing conflict between the United States and Russia. Will Russia move further away from the West to embrace China as a partner in systemic transformation? Scholars point to the sanctions placed on Russia over the Ukraine conflict as well as the oil and gas agreements between Russia and China. This relationship may spill over into other areas of collaboration: China and Russia share a common border and, within the context of the BRICS discourse, further cooperation may generate spillover culminating in the creation of a counter-hegemonic bloc determined to take the title of global hegemon from the United States and its allies. Energy policy, therefore, remains essential to any interpretation of international politics in the twenty-first century.

Third, the North American energy revolution has been profoundly disruptive of global energy markets. This development was largely unexpected but has proven to be highly significant. A major transformation is taking place in the energy production in the Western Hemisphere that carries with it global implications. The world is only now beginning to grasp the consequences of this "Revolution" for oil producing countries around the globe, especially for the members of OPEC. OPEC is seemingly falling apart as key member countries—especially Saudi Arabia—have proven unable to agree on energy production levels. The cartel may well be approaching irrelevance in the second decade of the twenty-first century. OPEC members and other states dependent on oil and gas exports for economic growth and geostrategic power and influence, specifically Russia, will likely bear the brunt of the consequences of the North American energy revolution. In the USA and Canada, this fact is celebrated while it simultaneously serves as a warning to the other states that lack economic diversification and, hence, unable to respond effectively to systemic level transformations.

Fourth, it is clear that the North American energy revolution presents formidable challenges to North America as well. The infrastructure for energy development on the North American continent will have to be developed as considerable financial costs and will pose major environmental risks. If North American energy is to be exported to the world, pipelines or other means of transport will have to be developed. Unquestionably, North American energy exports across the Atlantic could open up new avenues for profit as well as for geostrategic power projection toward Russia. Such a development would be a cause of resentments and added frictions in the international system, on the part of the Russians particularly given new competition and on the part of many Americans and Western Europeans given inevitable environmental costs.

Fifth, the volume emphasizes that there are inherent security linkages tied to changing energy policy and practice—at the levels of national security, economic security, environmental security and human/individual security. For instance, as Venezuela continues to suffer deepening economic instability as a result of the precipitous drop in world oil prices, US and other world leaders must begin to plan for the possibility that Venezuela will become a failed or collapsed state. Other states around the globe exhibiting similar levels of economic dependence on their energy exports could suffer similar fates: the possibilities of increased criminal networks of drug and human trafficking and so forth will create increased instability for the world. The energy revolution is causing major disequilibrium for states dependent on oil and the unintended consequences could be profoundly disruptive in many regions of the world system.

Sixth, expressing cautious optimism, many contributors anticipated the establishment of alternative renewable energy sources among hydrocarbon-poor states. These will have to be developed against the perpetual seduction of hydrocarbons such as coal and petroleum, in addition to atomic energy. To what extent do cleaner energy and the development of technologies present possibilities to move away from fossil fuels? Several authors speculated that the search for alternatives to hydrocarbons will be a slow butt continual process over the medium term. However, the only way for these alternatives to replace their carbon- or nuclear-based counterparts is if the price of renewable energy technology becomes cheaper in relative terms. Changes in the price structure must be facilitated by changes in governmental incentive structure to help facilitate the viability of alternative energies. The declining price of oil and natural gas may continue to displace other forms of cleaner energy sources.

Connected to these considerations is the seventh conclusion. Environmental consequences of continued hydrocarbon use are very real and threatening to the survival of the human race. May of the volumes' contributors argued that if proper, renewable techniques and technologies are not implemented effectively, there may not be much of a future to look forward to, particularly for the planet's heavily populated coastal areas. Any new regulatory environment must be based on collective action, and therefore uniform and require significant attention by all actors throughout the world. It is imperative for different kinds of regulatory practices, or best practices, to be imagined. We can easily see a growing, significant reaction to the new world of alternative energy. Yet what the world currently has is an inefficient and dangerous dependence on crude oil. Therefore, there must be a commitment by states to policies that make alternative energy affordable. Lastly, the volume puts forward a list of important considerations that analysts, policymakers, and scholars will inevitably confront in the next decade. Geopolitically, developing countries invite their own demise as they enter into or uphold exiting neo-dependency relationships with the producers of oil and gas. As a result, and given the energy revolution in the Western Hemisphere, are we going to see more failing and collapsing states due to the drop in the price of oil? States dependent on oil may see serious economic difficulties if the oil revolution continues. The world may see the creation of a 4th world, the world of the perpetually poor. These states, most likely most of the former OPEC nations, may fail due to their crippling dependence on the price of oil being high.

Going forward, we must keep in mind equity and social inclusion for the creation of a better, cleaner world. In our next volume, we intend to cover additional areas of the world such as energy starved nations of Latin America, the Caribbean, the Middle East and Central Asia. These very important geographic areas are equally relevant in terms of their own energy policies and role in the changing structure of the international energy system. The contributors to this volume would again like to thank all conference participants. Without your input and thought, it would not have been possible to produce this volume.

About the Contributors

Bruce Bagley: Professor at the Department of International Studies at the University of Miami, where he is served as former Chair of the Department of International Studies. He has also served as Director of the Center for Latin American Studies. Prior to this, Dr. Bagley was a professor at the School of Advanced International Studies (SAIS) at Johns Hopkins University, in Washington, D.C., and Universidad de los Andes, in Bogotá. He has served as a consultant for the United Nations, the US government and several governments in Latin America, on issues of drug trafficking, money laundering and public security. His current research focuses on US-Latin America relations, with an emphasis on drug trafficking and security issues. Dr. Bagley is the author and editor of several books, the latest being La desmovilizacion de los paramilitares en Colombia: Entre la esperanza y el escepticismo (co-edited, 2011).

Nashira Chavez: PhD candidate at the University of Miami. Her research interests center on International Relations, particularly foreign policy, security, and Latin American politics. She has researched on the US War on Drugs in the Andes and its impact in Ecuador security policies (published by Flacso-Abya-Yala), Drugs and the prison system in Ecuador, and Latin American countries at the Trans-Pacific Partnership and Pacific Alliance. Currently, Nashira is interested in studying Latin America national responses to China's increase presence in the region. Before coming to UM, Nashira worked for the Ecuadorian government.

Eric Farnsworth: Vice President Americas Society/Council of the Americas. Eric Farnsworth is Vice President of the Council of the Americas and the Americas Society, heading the Washington office since 2003. He is a recognized expert on hemispheric affairs and US foreign and trade policy, having given congressional testimony numerous times, and he is a frequent commentator in print, visual, and new media. From 1995–1998, he oversaw policy and message development for the White House Office of the Special Envoy for the Americas. He served at the US Department of State and the office of the US Trade Representative beginning in 1990, and was awarded the Superior Honor Award three times and the Meritorious Honor Award once.

Edward Glab: Director Global Energy Security Forum at School of International and Public Affairs, Florida International University

Christian Gómez, Jr.: Director of energy at the Washington office of the Council of the Americas. In this capacity, he coordinates the activities of the Council's Energy Action Group, a private-public dialogue on hemispheric energy issues that develops strategic energy policies and publishes working papers on key hemispheric energy topics. Prior to joining COA in 2013, Gómez was a consultant at the International Finance Corporation (IFC), focusing on business models that reach the base of the pyramid worldwide. He was previously program assistant for democratic governance at the Inter-American Dialogue and served as executive director of Princeton in Latin America (PiLA). Gómez received an MBA from the Wharton School of the University of Pennsylvania, an MA from the Paul H. Nitze School of Advanced International Studies (SAIS) of Johns Hopkins University, and an AB with honors in Political Economy and Latin American Studies from Princeton University.

Christ Kraul: Kraul is a former foreign correspondent for the Los Angeles Times, Chris Kraul is a freelance writer based in Bogota who specializes in economic, cultural and environmental topics.

Roger Kanet: Professor in the Department of Political Science of the University of Miami, where he served as Dean of the School of International Studies 1997–2000 and a member of the Department of International Studies 1997–2014. Prior to 1997, he taught at the University of Illinois at Urbana-Champaign, where he was a member of the Department of Political Science and served as Head of that Department, 1984–1987, and as Associate Vice Chancellor for Academic Affairs and Director of International Programs and Studies (1989–1997). He has authored more than two hundred scholarly articles and book chapters and edited or coedited thirty books. He is a member of the Council on Foreign Relations, New York.

Hanna Samir Kassab, PhD: Visiting Assistant Professor at Northern Michigan University. He is the author of "Weak States in International Relations Theory: The Cases of Armenia, St. Kitts and Nevis, Lebanon, and Cambodia" and "The Power of Emotion in Politics, Philosophy and Ideology" both published by Palgrave.

Maxime Larive, PhD: is a Research Associate at the EU Center of Excellence at the University of Miami. He has published several EU related articles in peer-reviewed journals such as European Security, Perceptions, and Whitehead Journal of Diplomacy and International Relations. His book "Debating European Security and Defense Policy" is scheduled for publication in October 2014 with Ashgate. Dr. Larivé also writes weekly commentaries on the EU and transatlantic relations for Politipond

Louise Léger: Consul General of Canada in Miami. Louise Léger (BSc [Economics], Université de Moncton, 1980) joined the Department of Industry, Trade and Commerce as a trade commissioner in 1981. Her career at Foreign Affairs and International Trade Canada has included postings to Colombia, Switzerland and Israel. In 1995, she was appointed ambassador to Panama, and in 2001 ambassador to Costa Rica, with concurrent accreditation to Nicaragua and Honduras. In Ottawa, Ms. Léger has held the positions of deputy director of the Eastern Europe Commercial Relations Division as well as executive director and deputy director general at the Summit of the Americas Bureau. Ms. Léger has also twice served as director general of Trade Commissioner Service, leading the Operations Bureau as well as the Client Services Bureau. Ms. Léger began her tenure as Consul General of Canada in Miami in September 2009.

Dina Moulioukova: Dina Moulioukova is an advanced PhD student at the Department of International Studies at the University of Miami. For a number of years she conducted research and led Eastern Observatory at the European Union Center at the University of Miami. She comes from legal background and holds law Degree in Russian Civil Law and International Law from Kazan State University in Russia. As a recipient of Cambridge Oversees Trust Scholarship for Russian Federation she completed her Master of Law (LL.M.) Degree at Cambridge University in England with specialization in international Law. In the United States, Ms. Moulioukova continued her training in International Law with focus on international trade regulations and regional trade agreements in Latin America. She has extensive experience working with various projects at US Agency for International Development and the Library of Congress. Prior to that, Ms. Moulioukova oversaw operations of multinational corporation for Eurasian Region. Ms Moulioukova research interests include Issues of Global Security, Energy and Role of the Emerging Economic in Global Order.

Olga Nazario: Senior Researcher at Florida International University in Miami, Florida. She is an independent consultant on matters of governance, transparency and accountability. She also served as a foreign policy analyst for US Information Agency.

Vladimir Rouvinski: Director of the CIES Interdisciplinary Research Center and Associate Professor of the Department of Political Studies at Icesi University in Cali, Colombia. His research interests include relations between Latin America, Russia, and East Asia; international cooperation for development in Latin America; and international migration studies. Recent publications include "Studies of Latin American Foreign Policy in the Soviet Union and Russia: Old and New Challenges" in Perspectivas Internacionales, and "Russia and ALBA: A Marriage of Convenience?" in Decline of US Hegemony and Latin American Integration in the 21st Century: New Patterns of Regional Cooperation and Conflict? ALBA and UNASUR. Rouvinski received a PhD and MA in Development and Cooperation Studies from Hiroshima University, Japan; a post-doctoral degree from the Institute for Peace Studies in Hiroshima, Japan; and an MA in History and International Relations from Irkutsk State University, Russia.

Diana Soller: Researcher at the European Union Center and PhD Student at the Department of International Studies University of Miami.

John Twichell*:* PhD student at the Department of International Studies at the University of Miami.

Gonzalo Vazquez: PhD student at the Department of International Studies at the University of Miami. He holds a BA in Economics (Minor Political Science) from the Advance Corporate Finance Program (Programa Avansado de Gerencia Financiera) at the Instituto de Estudios Superiores de Administracion (IESA) and MA in International Political Economy (International Studies) from the University of Miami. He has extensive experience in corporate finance and in the areas of international and domestic capital markets, particularly with the design and implementation of debt financing arrangements, the restructuring domestic and foreign liabilities, the syndication of loans, the instrumentation of import/export financing and managing equity issues (IPOs).

Harrie Vredenburg: Professor of Strategy at the University of Calgary's Haskayne School of Business and holds the Suncor Energy Chair in Strategy and Sustainability, a Haskayne research chair affiliated with the University's Institute for Sustainable Energy, Environment & Economy. He holds an appointment as an International Research Fellow at Oxford University's Saïd Business School. He is Academic Director of the Global Energy Executive MBA for mid-career energy industry managers, a program with modules in Calgary, Houston, London, Doha and Beijing/Shanghai and online, offered jointly by Haskayne and IHS, the Cambridge Massachusetts-based global energy consultancy. Harrie has authored or co-authored more than 50 papers on strategy and sustainability in journals such as Strategic Management Journal, MIT Sloan Management Review, Organization Science, and Harvard Business Review.

Wenyan Wu: PhD student at the Department of International Studies at the University of Miami.

Lilian Yaffe: Lilian Yaffe is Director of Graduate Studies in the Program of International Studies at the University of Miami, where she teaches courses in international studies and Latin American political economy.

Marcelo Zorovich: PhD student at the Deparment of International Studies at the University of Miami. His main focus is on issues related to Brazil and China. He has been working as professor of international business at ESPM (Escola Superior de Propaganda e Marketing), in São Paulo, Brazil, since 2007. In the 15 years prior, he worked for The Nielsen Company and Kantar Worldpanel, and led various international consulting projects for some government agencies as well as for Cadbury Adams, Hewlett Packard, Nestle, P&G, Unilever, Philip Morris, British American Tobacco, Johnson & Johnson, Melitta, TATA Consultancy Services, Kraft and Hugo Boss throughout Latin America, United States, Canada, China and India.

www.ingramcontent.com/pod-product-compliance
Ingram Content Group UK Ltd.
Pitfield, Milton Keynes, MK11 3LW, UK
UKHW012035150926
475560UK00001B/22